T0319390

INTEGRATED VEHICLE DYNAMICS AND CONTROL

INTEGRATED VEHICLE DYNAMICS AND CONTROL

Professor Wuwei Chen
Hefei University of Technology, China

Dr. Hansong Xiao
Hanergy Product Development Group, China

Professor Qidong Wang
Anhui University of Science and Technology, China

Dr. Linfeng Zhao
Hefei University of Technology, China

Dr. Maofei Zhu
Hefei Institutes of Physical Science, Chinese Academy of Sciences, China

Library of Congress Cataloging-in-Publication data applied for

ISBN: 9781118379998

A catalogue record for this book is available from the British Library.

Cover Image: © gettyimages.com

Set in 10/12.5pt Times by SPi Global, Pondicherry, India
Printed and bound in Singapore by Markono Print Media Pte Ltd

1   2016

Contents

Preface

As "the machine that changed the world", the vehicle has been developed for more than one hundred years. When examining the development history of vehicle technologies, people will find that the development of vehicle technologies had mainly relied on the improvement of mechanical structures and mechanisms during the incipient stage of vehicle. However, by taking advantage of the rapid development of energy, microelectronic, computer, and communication technologies in recent years, it is believed that the vehicle is now experiencing a significant transformation to have attributes of being electric, intelligent, and networked.

Vehicle dynamics studies the basic theory of the motions of various vehicle systems and the performance of the system integration of the vehicle. The design of the dynamic performance of the modern vehicle must meet the multiple requirements of ride comfort, handling stability, safety, and environment-friendliness. These requirements present great challenges to automotive researchers and engineers since the vehicle itself is a complex system consisting of up to several thousands of parts, and often driven under various unpredictable working conditions. Moreover, the traditional vehicle design has focused on improving primarily the mechanical parts and systems through finding a compromised solution amongst the conflicting performance requirements. Thus it is difficult to optimize simultaneously all the performance requirements of the vehicle. To overcome these difficulties, the electronics technology and control theory have been applied to improving the dynamic performance of vehicle systems, especially the vehicle chassis subsystems. As a result, the research topic on vehicle dynamics and control has attracted great attention in recent years. A number of chassis control subsystems have been developed, for example the active suspension, ABS (antilock brake system), and EPS (electrical power steering system), etc.

The dynamic chassis subsystems including tyres, brakes, steering, and suspension, etc., fulfill complex and specific motions, and transmit and exchange energy and information by means of dynamic coupling, and hence realize the basic functions of the vehicle chassis. The fundamental study of the chassis system dynamics focuses on the development of the nonlinear tyre model and nonlinear coupling dynamic model of the full vehicle that describe the combined effects of the longitudinal, lateral, and vertical motions of the tyre and vehicle through analyzing the tyre–road interactions and the coupling mechanisms amongst the brake, steering, and suspension subsystems under different driving conditions.

To date, there are quite a few textbooks and monographs addressing vehicle system dynamics and control. However, most of them explore mainly the stand-alone control of the individual chassis subsystem to improve solely the performance of the subsystem, i.e. brake control, steering control, suspension control, etc. In addition, there have been numerous achievements in the theoretical research and engineering applications on the multi-objective, multivariate integrated control of multiple vehicle subsystems over the past two decades. It is also demonstrated by a large volume of research articles in this field, covering the modeling of full vehicle dynamics, the architecture of integrated control system, the integrated control strategies, and the decoupling control methods for the integrated system. However, there are few monographs investigating the dynamic model and simulation, design, and experimental verification of the vehicle integrated control systems.

This book provides an extensive discussion on the integrated vehicle dynamics control, exploring the fundamentals and emerging developments in the field of automotive engineering. It was supported by the following research projects: the general programs of the National Natural Science Foundation of China (NSFC), including "Research on integrated control of vehicle electrical power steering (EPS) and active suspension system" (No. 50275045), "Research on methods and key technologies of integrated control of vehicle chassis based on generalized integration" (No. 50575064), "Research on methods and key technologies of integrated control of vehicle chassis integration" (No. 51075112), "A study on control methods of human-machine sharing and key technology for vehicle lateral movement" (No. 51375131), and "Research on integrated control and key technologies of ESP and EPS based on security border control" (No. 51175135). It was also supported by the project of International Cooperation and Exchanges NSFC "Research on methods and key technologies of active integrated control of vehicle chassis system" (No. 50411130486, 50511130365, 50611130371).

The topics presented in this book have been organized in 9 chapters, covering the background of vehicle system dynamics modeling, tyre dynamics, longitudinal vehicle dynamics and control, vertical vehicle dynamics and control, lateral vehicle dynamics and control, analysis of system coupling mechanism, model of full vehicle dynamics, centralized integration control, and multi-layer coordinating control. The major contents of the book are based on the research practice of the authors over the last ten years. Chapters 1, 2 and 3 are written by Professor Qidong Wang, Anhui University of Science and Technology; Chapter 4, Sections 7.1, 7.3, 7.4, 7.5, 7.6, 7.9, Sections 8.1, 8.2, 8.3, 8.7, 8.8, 8.9, and Chapter 9 are written by Professor Wuwei Chen, Hefei University of Technology; Chapter 5 is written by Dr. Linfeng Zhao, associate professor with Hefei University of Technology; Chapter 6 is written by Dr. Maofei Zhu, associate professor with Hefei Institutes of Physical Science, Chinese Academy of Sciences; Sections 7.2, 7.7, 7.8 and Sections 8.4, 8.5, 8.6 are written by Dr. Hansong Xiao, Hanergy Product Development Group. This book is finally compiled and edited by Professor Wuwei Chen.

This book cites a number of research articles published in domestic and international journals, and the scientific publications, all of which enrich the contents of this book. We would like to express gratitude to all the authors of the related references.

Wuwei Chen
Hansong Xiao
Qidong Wang
Linfeng Zhao
Maofei Zhu
January 2016, Hefei

1

Basic Knowledge of Vehicle System Dynamics

1.1 Traditional Methods of Formulating Vehicle Dynamics Equations

Traditional methods of formulating vehicle dynamics equations are based on the theories of Newtonian mechanics and analytical mechanics. Some of the definitions used in dynamics are presented first.

1. *Generalized coordinates*

   Any set of parameters that uniquely define the configuration (position and orientation) of the system relative to the reference configuration is called a set of generalized coordinates. Generalized coordinates may be dependent or independent. To a system in motion, the generalized coordinates that specify the system may vary with time. In this text, column vector $\mathbf{q} \equiv [q_1, q_2, \ldots, q_n]^T$ is used to designate generalized coordinates, where n is the total number of generalized coordinates.

   In Cartesian coordinates, to describe a planar system which consists of b bodies, $n = 3 \times b$ coordinates are needed. For a spatial system with b bodies, $n = 6 \times b$ (or $n = 7 \times b$) coordinates are needed.

   The overall vector of coordinates of the system is denoted by $\mathbf{q} \equiv \left[\mathbf{q}_1^T, \mathbf{q}_2^T, \ldots, \mathbf{q}_b^T \right]^T$, where vector $\mathbf{q}_i$ is the vector of coordinates for the ith body in the system.

2. *Constraints and constraint equations*

   Normally, a mechanical system that is in motion can be subjected to some geometry or movement restrictions. These restrictions are called constraints. When these restrictions

Integrated Vehicle Dynamics and Control, First Edition. Wuwei Chen, Hansong Xiao, Qidong Wang, Linfeng Zhao and Maofei Zhu.

are expressed as mathematical equations, they are referred to as constraint equations. Usually these constraint equations are denoted as follows:

$$\Phi \equiv \Phi(\mathbf{q}) = 0 \qquad (1.1)$$

If the time variable appears explicitly in the constraint equations, they are expressed as:

$$\Phi \equiv \Phi(\mathbf{q}, t) = 0 \qquad (1.2)$$

3. *Holonomic constraints and nonholonomic constraints*
   Holonomic and nonholonomic constraints are classical mechanics concepts that are used to classify constraints and systems. If constraint equations do not contain derivative terms, or the derivative terms are integrable, these constraints are said to be called holonomic. They are geometric constraints. However, if the constraint equations contain derivative terms that are not integrable in closed form, these constraints are said to be nonholonomic. They are movement constraints, such as the velocity or acceleration conditions imposed on the system.
4. *Degrees of freedom*
   The generalized coordinates that satisfy the constraint equations in a system may not be independent. Thus, the minimum number of coordinates required to describe the system is called the number of degrees of freedom (DOF).
5. *Virtual displacement*
   Virtual displacement is an assumed infinitesimal displacement of a system at a certain position with constraints satisfied while time is held constant. Conditions imposed on the virtual displacement by the constraint equations are called virtual displacement equations. A virtual displacement may be a linear or an angular displacement, and it is normally denoted by the variational symbol δ. Virtual displacement is a different concept from actual displacement. Actual displacement can only take place with the passage of time; however, virtual displacement has nothing to do with any other conditions but the constraint conditions.

1.1.1 Newtonian Mechanics

The train of thought used to establish the vehicle dynamics equations using Newton's law can be summarized in a few steps. According to the characteristics of the problem at hand, first, we need to simplify the system and come up with a suitable mathematical model by representing the practical system with rigid bodies and lumped masses which are connected to each other by springs and dampers. Then, we isolate the masses and bodies and draw the free-body diagrams. Finally, we apply the following formulas to the masses and bodies shown by free-body diagrams.

The dynamic equations of a planar rigid body are:

$$m \frac{d^2 \mathbf{r}}{dt^2} = \sum \mathbf{F}_i \qquad (1.3)$$

$$J\dot{\omega} = \sum M_i \qquad (1.4)$$

where m is the mass of the body, r is the displacement of the center of gravity, F_i is the ith force acting on the body, J is the mass moment of inertia of the body about the axis through the center of gravity, ω is the angular velocity of the body, and M_i is the moment of the ith force acting on the center of gravity of the body.

1.1.2 Analytical Mechanics

In solving the dynamics problems of simple rigid body systems, Newtonian mechanics theories have some obvious advantages; however, the efficiency will be low if dealing with constrained systems and deformable bodies. Analytical mechanics theories have been proven to be a useful method in solving these problems. This theory contains mainly the methods of general equations of dynamics, the Lagrange equation of the first kind, and the Lagrange equation of the second kind; the latter being the most widely used.

For a system with b particles (or bodies), and n DOF, $q_1, q_2, \ldots, q_n$ is a set of generalized coordinates. Then, the Lagrange equation of the second kind can be expressed as

$$\frac{d}{dt}\left(\frac{\partial T}{\partial \dot{q}_k}\right) - \frac{\partial T}{\partial q_k} + \frac{\partial V}{\partial q_k} = 0 \quad (k = 1, 2, \cdots n) \qquad (1.5)$$

where T is the kinetic energy, and V the potential energy of the system.

1.2 Dynamics of Rigid Multibody Systems

1.2.1 Birth and Development

The history of the development of classical mechanics goes back more than 200 years. In the past two centuries, classical mechanics has been successfully used in the theoretical study and engineering practice of relatively simple systems. However, most modern practical engineering problems are quite complicated systems consisting of many parts. Since the middle of the 20th century, the rapid development of aerospace, robotics, automotive and other industries has brought new challenges to classical mechanics. The kinematics and dynamics analysis of complicated systems becomes difficult. Thus, there was an urgent need to develop new theories to accomplish this task.

In the late 1960s and early 1970s, Roberson[1], Kane[2], Haug[3], Witternburg[4], Popov[5] and other scholars put forward methods of their own to solve the dynamic problems of complex systems. Although there were some differences between these methods in describing the position and orientation of the systems, and formulating and solving the equations, one characteristic was common among them: recurring formularization was adopted in all these methods. Computers, which help engineers to model, form, and solve differential equations of motion, were used analyze and synthesize complex systems. Thus, a new branch of mechanics called multibody dynamics was born. This developing and crossing discipline arises from the combination of rigid mechanics, analytical mechanics, elastic mechanics, matrix theory, graph theory, computational mathematics, and automatic control. It is one of the most active fields in applied mechanics, machinery, and vehicle engineering.

Multibody systems are composed of rigid and/or flexible bodies interconnected by joints and force elements such as springs and dampers. In the last few decades, remarkable advances have been made in the theory of multibody system dynamics with wide applications. An enormous number of results have been reported in the fields of vehicle dynamics, spacecraft control, robotics, and biomechanics. With the development and perfection of the multibody formalisms, multibody dynamics has received growing attention and a considerable amount of commercial software is now available. The first International Symposium on multibody system dynamics was held in Munich in 1977 by IUTAM. The second was held in Udine in 1985 by IUTAM/IFTOMM. After the middle of the 1980s, multibody dynamics entered a period of fast development. A wealth of literature has been published[6,7].

The first book about multibody system dynamics was titled *Dynamics of System of Rigid Bodies*[4] written by Wittenburg, was published in 1977. *Dynamics: Theory and applications* by Kaneand Levinson came out in 1985. In *Dynamics of Multibody System*[8], printed in 1989, Shabanacomprhensively discusses many aspects of multibody system dynamics, with a second edition of this book appearing in 1998. In *Computer-aided Analysis of Mechanical Systems*[9], Nikravesh introduces theories and numerical methods for use in computational mechanics. These theories and methods can be used to develop computer programs for analyzing the response of simple and complex mechanical systems. Using the Cartesian coordinate approach, Haug presented basic methods for the analysis of the kinematics and dynamics of planar and spatial mechanical systems in *Computer Aided Kinematics and Dynamics of Mechanical Systems*[3].

The work of three scholars will also be reviewed in the following section.

1. Schiehlen, from the University of Stuttgart, published his two books in 1977 and 1993 respectively. *Multibody System Handbook*[10] was an international collection of programs and software which included theory research results and programs from 17 research groups. *Advanced Multibody Dynamics*[11] collected research achievements of the project supported by The German Research Council from 1987 to 1992, and the latest developments in the field of multibody system dynamics worldwide at that time. The content of this book was of an interdisciplinary nature.

2. In *Computational Methods in Multibody Dynamics*[12], Amirouche Farid offered an in-depth analysis of multibody system dynamics with rigid and flexible interconnected bodies, and provided several methods for deriving the equations of motion. Computer methods of tree-like systems and systems with closed loops and prescribed motion were fully discussed.

3. In *Multi-body Dynamics: Vehicles, machines and mechanisms*[13], Rahnejat guided readers through different topics from dynamics principles to the detailed multibody formulation and solution approach. Model analytic solutions were provided for a variety of practical machines and mechanisms such as the suspension of a vehicle and the rotor of helicopter. State-of-the-art modeling and solution methods were presented to investigate complex

dynamics phenomena in the behavior of automobiles and aircraft. Optimal control of multibody systems were also discussed.

Multibody dynamics research in China started late but developed quickly. The inaugural meeting of the Multibody System Dynamics group, part of the General Mechanics Committee of Chinese Society of Mechanics, was held in Beijing in August 1986. Since then, many books on multibody system dynamics have come out. Many researchers have published high-quality papers on modeling theory, computational methods, and other subjects of multibody system dynamics[14,15].

1.2.2 Theories and Methods of Multi-Rigid Body System Dynamics

Formulism methods and numerical algorithms are the two most important aspects in multibody system dynamics research. Over the past few decades, many methods have appeared. For example, the New-Euler method by Schiehlen[10], the Kane method by Kane and Huston[2,16], the graph theory method by Roberson and Wittenburg[1,17], and the Lagrangian method by Haug[3] are representative. According to the difference in coordinates adopted, formulism methods can be divided into two categories: minimum number of coordinates method and maximum number of coordinates method. The minimum number of coordinate method uses joint coordinates, taking the relative angular or displacement of the adjacent bodies as generalized coordinates. The main advantage of this method is that fewer variables are used, and higher calculation efficiency can be obtained. However, the construction process of coefficient matrix of differential equations is very complex, including a large amount of nonlinear operations. The maximum number of coordinates method uses the Cartesian coordinates of the center of mass of bodies and the Euler angles or Euler parameters as generalized coordinates, combining Lagrangian multipliers to constraint equations to formulate the equations of motion. This method can be easier implemented for coding, but with the features of more variables and lower calculation efficiency.

1. *Graph theory (R-W)*
   Roberson and Witternburg introduced graph theory into the research of multi-rigid body system dynamics. This method applies some basic concepts and mathematical tools to describe the topological structure of a multibody system. The relative displacements of the adjacent bodies are taken as generalized coordinates, and the unified mathematical formula of the complex tree-like structure is derived and the general forms of the dynamic equations are formulated for multibody systems. Code MESA VERDE based on this method has been developed.
2. *Lagrangian method*
   This method uses the Cartesian coordinates of the center of mass of bodies and the Euler angles or Euler parameters that describe the orientation of the system as generalized coordinates, combining Lagrangian multipliers to constraint equations to formulate the equations of motion. This method has the characteristic of being easier for programming purposes. Orlandea, Chace, Haug, and Nikravesh developed their general purpose codes ADAMS, DADS, DAP. There are still some differences

between them in the detailed formulism and algorithm which is mainly reflected in the different coordinates used. In ADAMS, Cartesian coordinates of the center of mass of bodies and Euler angles that describe the orientation of the system are used as generalized coordinates. In DADS, Cartesian coordinates of the center of mass of bodies and Euler parameters that describe orientation of the system are used as generalized coordinates.

3. *Multibody dynamics method in ADAMS*

   For a spatial system with b bodies, the Cartesian coordinates of the center of mass of body i are x_i, y_i, z_i, the Euler angles that describe orientation of the body are ψ_i, θ_i, ϕ_i, the generalized coordinates of the body can be expressed with a vector q_i, such as $q_i = [x, y, z, \psi, \theta, \phi]_i^T$.

   If vector q is used to denote all of the coordinates of the system, then

   $$q = [q_1, q_2, \ldots\ldots q_b]^T$$

   If the system contains holonomic and non-holonomic constraints, based on the Lagrangian equations with multipliers, the equations of motion, which are a set of differential–algebraic equations (DAE), can be obtained.

   $$\frac{d}{dt}\left(\frac{\partial \mathbf{T}}{\partial \dot{q}}\right)^T - \left(\frac{\partial \mathbf{T}}{\partial \mathbf{q}}\right)^T + \varphi_q^T \rho + \theta_{\dot{q}}^T \mu = \mathbf{Q} \qquad (1.6)$$

   with holonomic constraints equations

   $$\varphi(q, t) = 0$$

   and with non-holonomic constraints equations

   $$\theta(\mathbf{q}, \dot{q}, t) = 0$$

   where, T is the kinetic energy of the system, Q is the vector of generalized forces, ρ is the Lagrange multiplier vector corresponding to holonomic constraints, μ is the Lagrangian multiplier vector corresponding to the non-holonomic constraints.

   If the kinetic energy is expressed with velocity and mass, the equations can be written in matrix form.

4. *Multibody dynamics methods in DADS*

   For a spatial system with b bodies, the Cartesian coordinates of the center of mass of body i are x_i, y_i, z_i, the Euler parameters that describe the orientation of the body are $p_i = [e_{0i}, e_{1i}, e_{2i}, e_{3i}]^T$, and the generalized coordinates of the body can be expressed with a vector q_i, and $q_i = [x, y, z, e_0, e_1, e_2, e_3]_i^T = [r, p]_i^T$.

   If a vector q is used to denote all of the coordinates of the system, then

   $$q = [q_1, q_2, \ldots\ldots q_b]^T$$

   For body i, the mass is m_i, the inertia matrix in the local coordinate system J_i' is composed of moments of inertia and products of inertia, the mass characteristics $N_i = diag(m, m, m)_i$, the generalized forces consist of forces and torques acting on body I, such as $Q_i = \begin{bmatrix} f & n' \end{bmatrix}_i^T$, and the angular velocity matrix in local coordinate system is made up by the diagonal matrix ω_i' with the entries of angular velocities around the axis.

The constraints equations in compact form $\Phi(q,t) = 0$, with the Jacobian matrix Φ_q are defined here:

$$M = \begin{bmatrix} N_1 & & & \\ & J_1' & & 0 \\ & & \ddots & \\ & 0 & & N_b \\ & & & & J_b' \end{bmatrix}, Q = [Q_1, Q_2 \cdots Q_b]^T,$$

$$L_i = \begin{bmatrix} -e_1 & e_0 & e_3 & -e_2 \\ -e_2 & -e_3 & e_0 & e_1 \\ -e_3 & e_2 & -e_1 & e_0 \end{bmatrix}_i,$$

$$\dot{h} = \begin{bmatrix} \ddot{r}_1 & \dot{\omega}_1' & \cdots & \ddot{r}_b & \dot{\omega}_b' \end{bmatrix}^T, b = \begin{bmatrix} 0 & \tilde{\omega}_1' J_1' \omega_1' & \cdots & 0 & \tilde{\omega}_b' J_b' \omega_b' \end{bmatrix}^T$$

$$B = \begin{bmatrix} \Phi_{r1} & \dfrac{1}{2}\Phi_{p1} L_1^T & \cdots & \Phi_{r1} & \dfrac{1}{2}\Phi_{pb} L_b^T \end{bmatrix}^T$$

The equations of motion can be expressed as:

$$\begin{bmatrix} M & B^T \\ B & 0 \end{bmatrix} \begin{bmatrix} \dot{h} \\ -\lambda \end{bmatrix} + \begin{bmatrix} b \\ 0 \end{bmatrix} = \begin{bmatrix} Q \\ \gamma \end{bmatrix} \tag{1.7}$$

where λ is the Lagrangian multiplier, and γ is the right side of the acceleration equations.

$$\gamma = -\left(\Phi_q \dot{q}\right)_q \dot{q} - 2\Phi_{qt}\dot{q} - \Phi_{tt} \tag{1.8}$$

5. *Algorithms for solving equations*

Normally, the equations of motion of multibody systems are a set of mixed differential-algebraic equations, with the characteristics that their coefficient matrices are quite sparse. There are three main algorithms to solve them: direct integration, coordinate partitioning, and stiff differential equations. The main steps of direct integration are explained below.

To a set of first order differential equations with initial values

$$\begin{cases} \dot{y} = f(y,t) \\ \quad y = y_0 \end{cases}$$

Many algorithms can be used to solve them. For example, the Runge-Kutta algorithm, the explicit and implicit multistep algorithm, and the predictor-corrector algorithm are commonly chosen. According to theories of differential equations, n number of ordinary second order differential equations can be expressed as 2n number of first order differential equations:

$$y = \begin{bmatrix} location\,coordinates \\ velocity \end{bmatrix} = \begin{bmatrix} q \\ \dot{q} \end{bmatrix} \quad \dot{y} = \begin{bmatrix} velocity \\ acceleration \end{bmatrix} = \begin{bmatrix} \dot{q} \\ \ddot{q} \end{bmatrix}$$

The matrix form of the equations of motion can also be written as

$$\begin{bmatrix} \mathbf{M} & \varphi_q^T \\ \varphi_q & 0 \end{bmatrix}\begin{bmatrix} \ddot{q} \\ \lambda \end{bmatrix} = \begin{bmatrix} Q \\ \gamma \end{bmatrix} \tag{1.9}$$

The following algorithm is adopted to solve the equations:

1. Specify the initial conditions of q and $\dot{q}$.
2. Transfer the contents of q and $\dot{q}$ into vector $y = \left[q^T, \dot{q}^T \right]^T$.
3. Call the numerical integration subroutine to solve the differential equation $\dot{y} = f(y,t)$.

In the process of numerical integration, $f(y,t)$ must also be calculated. The procedure is outlines as follows:

1. Transfer y to q and $\dot{q}$. Assemble the configuration in time.
2. At a typical time, calculate M, $\dot{\phi}_q$, Q, and γ.
3. Calculate $\dot{\phi} = \phi_q \dot{q}$ at the typical time.
4. Obtain $\ddot{q}$ and λ by solving equation (1.9).
5. Transfer $\dot{q}$ and $\ddot{q}$ to y.
6. Return.

1.2.3   An Example of the Application of Multi-Rigid Body Dynamics Method in Vehicle System Modeling

There are two options for formulating the equations of motion. One is by using existing commercial software; another is using relevant theories whereby researchers do the modeling themselves. The second way will be demonstrated in this chapter. The steps for modeling multi-rigid body systems are basically the same no matter which method is used. These are:

1. Simplify practical problems to a system composed of bodies, joints, and force elements.
2. Establish the global and local coordinate systems, and determine the generalized coordinates describing the system.
3. Calculate the number of degrees of freedom of the system.
4. Set up the constraint equations of the system.
5. Build the Jacobian matrix, and calculate the right side of the acceleration equation.
6. Construct the equations of motion.

The method mentioned above is applied to an automotive electric power steering system.

The diagram of a typical EPS steering system, in this case the type of force assistance on steering column using a rack and pinion mechanical steering mechanism, is shown in Figure 1.1.

First, the system is simplified. The flexibility of the steering column is neglected, the motor is simplified by reducing the mechanism to a rigid body revolving around a fixed axis. The wheel and the pitman arm are taken as one body. So, as shown in Figure 1.1, the practical system can be treated as a planner multibody system consisting of 7 bodies, 6 revolute joints, 1 translational joint, 1 rack and pinion joint, 1 gear joint, and 2 cylindrical joints. Body 1 and ground, body 2 and body 1, body 2 and body 3 are connected by revolute joints respectively. Body 3 and ground are connected by a translational joint. Body 4 and body 3 are connected by a rack and pinion joint. Body 5 and body 4 are connected by a gear

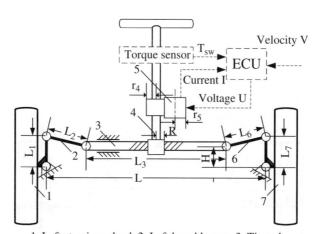

1. Left steering wheel 2. Left knuckle arm 3. Tie rod
4. Steering column 5. Motor and reduction mechanism
6. Right knuckle arm 7. Right steering wheel

Figure 1.1   Structure of EPS system.

joint. Body 5 and ground are connected by a cylindrical joint. Body 6 and body 3, body 6 and body 7, body 7 and ground are connected by revolute joints respectively. External forces include the steering resistance moment, the input force from the steering wheel, and the assist torque.

The global and local coordinate systems are now established as shown in Figure 1.2. In Cartesian coordinate system, three coordinates are needed to describe the position and orientation of the system. So, the total number of generalized coordinates of the system is 21, which can be expressed as $\mathbf{q} = \begin{bmatrix} q_1 & q_2 & \cdots & q_{21} \end{bmatrix} = \begin{bmatrix} x_1 & y_1 & \delta_1 & \cdots & x_7 & y_7 & \delta_7 \end{bmatrix}^T$. The joints provide 20 constraint equations; therefore, the number of degree of freedomn of the system is reduced to 1.

The constraint equations of the revolute joints between body 1 and ground, body 2 and body 1, body 2 and body 3 are:

$$\Phi_1 : x_1 = 0 \tag{1.10}$$

$$\Phi_2 : y_1 = 0 \tag{1.11}$$

$$\Phi_3 : x_1 + \cos(\delta_1) \times L_1 - x_2 - \cos(\delta_2) \times L_2 / 2 = 0 \tag{1.12}$$

$$\Phi_4 : y_1 + \sin(\delta_1) \times L_1 - y_2 - \sin(\delta_2) \times L_2 / 2 = 0 \tag{1.13}$$

$$\Phi_5 : x_2 + \cos(\delta_2) \times L_2 / 2 - x_3 - \cos(\delta_3) \times L_3 / 2 = 0 \tag{1.14}$$

$$\Phi_6 : y_2 + \sin(\delta_2) \times L_2 / 2 - y_3 - \sin(\delta_3) \times L_3 / 2 = 0 \tag{1.15}$$

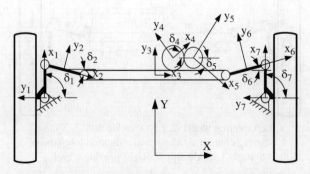

Figure 1.2   Plane coordinate system of steering.

The constraint equations of the translational joints between body 3 and ground are:

$$\Phi_7 : y_3 - H = 0 \tag{1.16}$$

$$\Phi_8 : \delta_3 = 0 \tag{1.17}$$

The constraint equation of the rack and pinion joint between body 4 and body 3 is:

$$\Phi_9 : x_4 - x_3 - R \times \delta_4 = 0 \tag{1.18}$$

The constraint equation of the gear joint between body 5 and body 4 is:

$$\Phi_{10} : \delta_4 \times r_4 - \delta_5 \times r_5 = 0 \tag{1.19}$$

The constraint equations of the cylindrical joints between body 4 and ground are:

$$\Phi_{11} : x_4 - L / 2 = 0 \tag{1.20}$$

$$\Phi_{12} : y_4 - H - R = 0 \tag{1.21}$$

The constraint equations of the cylindrical joints between body 5 and ground are:

$$\Phi_{13} : x_5 - x_4 - (r_4 + r_5) = 0 \tag{1.22}$$

$$\Phi_{14} : y_5 - y_4 = 0 \tag{1.23}$$

The constraint equations of the revolute joints between body 6 and body 3 are:

$$\Phi_{15} : x_3 + \cos(\delta_3) \times L_3 / 2 - x_6 - \cos(\delta_6) \times L_6 / 2 = 0 \tag{1.24}$$

$$\Phi_{16} : y_3 + \sin(\delta_3) \times L_3 / 2 - y_6 - \sin(\delta_6) \times L_6 / 2 = 0 \tag{1.25}$$

The constraint equations of the revolute joints between body 6 and body 7 are:

$$\Phi_{17} : x_6 + \cos(\delta_6) \times L_6 / 2 - x_7 - \cos(\delta_7) \times L_7 / 2 = 0 \tag{1.26}$$

$$\Phi_{18} : y_6 + \sin(\delta_6) \times L_6 / 2 - y_7 - \sin(\delta_7) \times L_7 / 2 = 0 \tag{1.27}$$

The constraint equations of the revolute joint between body 7 and ground are:

$$\Phi_{19} : x_7 - L = 0 \tag{1.28}$$

$$\Phi_{20} : y_7 = 0 \tag{1.29}$$

The Jacobian matrix can be obtained by performing the partial derivative of the items in the 20 equations.

$$\Phi_q = \begin{bmatrix} \dfrac{\partial \Phi_1}{\partial q_1} & \cdots & \dfrac{\partial \Phi_1}{\partial q_{21}} \\ \vdots & & \vdots \\ \dfrac{\partial \Phi_{20}}{\partial q_1} & \cdots & \dfrac{\partial \Phi_{20}}{\partial q_{21}} \end{bmatrix} \in \mathbf{R}^{20 \times 21}$$

Using the formula $\gamma = -\left(\Phi_q \dot{q} \right)_q \dot{q} - 2\Phi_{qt} \dot{q} - \Phi_{tt}$, the right side of the acceleration equation can be calculated. After the moment of inertia, mass matrix, and generalized force vector are determined, the equations of motion can be constructed by applying the formulas (1.7) or (1.9).

For more details about equations establishment, literature such as[3],[9],[14], on multibody system dynamics can be consulted.

1.3   Flexible Multibody Dynamics

Flexible multibody dynamics studies the dynamic behavior of systems composed of flexible and rigid multibodies undergoing large-scale spatial motion. The scope of rigid multibody dynamics covers the interaction between the motion of rigid bodies and the influence on dynamic characteristics. The subject of flexible multibody dynamics is the interaction and coupling of flexible bodies' deformation and overall rigid motion, and the influence on the dynamic behavior of systems. The core feature of flexible multibody system dynamics is that deformation and rigid motion of the object of study occur simultaneously. To flexible bodies, parameters like inertia tensors are functions of time.

Modi[18], Frisch[19] and others have carried out plenty of research in this field. In early 1970s, Bodley et al. established equations using the method of Lagrange multipliers. This theory is quite effective when dealing with flexible multibody systems with constraints, but the derivation process is complicated.

In 1983, NATO-NSF-ARD conducted a seminar on computer-aided analysis and optimization of mechanical systems in which flexible multibody dynamics was one of the important themes. Since the late 1980s, a series of conferences on flexible multibody dynamics have been held in China which have helped to promote the development of the discipline. Most of the studies focused on the dynamics of mechanical systems.

Since 1980s, flexible multibody dynamics have been gradually applied to high-speed flexible mechanisms. Shabana[8], Haug[20] and others have made plenty of contributions in these aspects. Some Chinese researchers applied dynamics analysis to solar energy panels, flexible mechanical arms, and space-borne deployable antennas[21]. The earlier method dealing with the problem of flexible multibody system dynamics was kinematic-elasto dynamics method (KED method). Zhang Ce[22] made the detailed introduction to this method in his book. Using the vector variational method and the principle of virtual work, adopting relative coordinates pulsing modal coordinates of flexible bodies, Hang et al. put forward the modeling method of constrained open-loop, closed-loop mechanical systems and open-loop, closed-loop flexible multibody systems. Chang and Shabana brought forward nonlinear finite element methods for modeling elastic plates undergoing large-scale motion.

Because elastic bodies have an infinite number of degrees of freedom, an accurate solution to these dynamic problems cannot be obtained. The most common approach to handling these problems is to discretize the body to a model with finite degrees of freedom. The main discretization algorithms include: Rayleigh-Litz method, finite element method, modal analysis, and synthesis method.

Mainly, there are three methods for flexible multi-body system dynamics analysis: the Newton Euler method, the Lagrange equation method, and the modification of the two methods, such as the Kane method.

References

[1] Roberson R E, Schwertassek R. Dynamics of Multibody Systems. Springer-Verlag, Berlin, 1988.
[2] Kane T R, Likins P W, Levinson D A. Spacecraft Dynamics. McGraw-Hill Book Company, New York, 1983.
[3] Haug E J. Computer Aided Kinematics and Dynamics of Mechanical Systems. Allyn and Bacon, Boston, 1989.
[4] Wittenburg J. Dynamics of Systems of Rigid Bodies. B. G. Teubner, Stuttgart, 1977.
[5] Popov, E.P. Theory of Nonlinear Systems of Automatic Regulation and Control, Nauka, Moscow, 1979.
[6] Yu Q, Hong J Z. Some topics on flexible multibody system dynamics. Chinese Journal of Advances in Mechanics. 1999, 29(2): 145–154.
[7] Huang Wen-hu, Chen Bin. The Recent Development of General Mechanics (Dynamics, Vibration and Control). Science Press. Beijing, 1994.
[8] Shabana A A. Dynamics of Multibody Systems. John Wiley & Sons, New York, 1989.
[9] Nikravesh P E. Computer-aided Analysis of Mechanical Systems. Prentice Hall, Engelwood Cliffs, NJ, 1988.
[10] Schiehlen W. Multibody Systems Handbook. Springer-Verlag, Berlin, 1990.
[11] Schiehlen, W. Advanced Multibody System Dynamics. Kluwer Academic Publishers, Amsterdam, 1992.
[12] Amirouche F.M.L. Computational Methods in Multibody Dynamics. Prentice Hall, Englewood Cliffs, New Jersey, 1992.
[13] Rahnejat H. Multi-body Dynamics: Vehicles, Machines and Mechanisms, Professional Engineering Publications (IMechE), 1998.
[14] Hong Jia-zhen, Jia Shu-Hui. Multibody Dynamics and Control. Beijing Institute of Technology Press, Beijing, 1996.
[15] Liu Yan-zhu. On dynamics of multibody system described by fully Cartesian coordinates. Chinese Journal of Theoretical and Applied Mechanics, 1997, 29(1): 84–94.
[16] Huston R L. Multibody Dynamics. Butterworth-Heinemann, Boston, 1990.
[17] Wittenburg J. Dynamics of Multibody Systems. Springer, Berlin, 2008.
[18] Modi V J. Attitude dynamics of satellites with flexible appendages: A brief review. J Spacecraft and Rockets, 1974, 11(11): 743–751.
[19] Frisch H P. A Vector-Dyadic Development of the Equations of Motion for N-Coupled Flexible Bodies and Point Masses. NASA TND-8047, 1975.
[20] Yoo W S, Haug E J. Dynamics of flexible mechanical systems using vibration and static correction modes. J Mech, Trans and Automation, 1986, 108(3): 315–322.
[21] Liu Zhu-yong, Hong Jia-zhen. Research and prospect of flexible multi-body systems dynamics. Chinese Journal of Computational Mechanics, 2008, 25(4): 411–416.
[22] Zhang C. Analysis and Synthesis of Elastic Linkage Mechanism. China Machine Press, Beijing, 1997.

2

Tyre Dynamics

2.1 Tyre Models

Tyres are important components of automobiles; they support the weight of the car and reduce the impact from ground, amongst other things. At the same time, through the interaction between the tyre and the ground, forces and torques which change the car's motion can be produced. The forces and torques include tractive force, braking force, and aligning moment. The dynamic behavior of tyres is critical to the car's handling, ride comfort, traction, and braking performance. The accuracy of dynamic performance analysis and the success of the chassis control system design rely heavily on the accuracy of the vehicle dynamics model and the tyre dynamic model.

Modern tyres are complex viscoelastic structures. The relationships between forces, deformation, and motion are highly nonlinear; thus, it is very difficult to establish a precise mathematical model. Since the 1930s many scholars have carried out plenty of research and achieved fruitful results on tyre modeling. In summary, tyre models fall into three categories: theoretical models, empirical models, and semi-empirical models. Before explaining these models in detail, some terminology and concepts are introduced in advance.

2.1.1 Terminology and Concepts

1. *Tyre axis system and six-component wheel force*

   The standard tyre axis system defined by the U.S. Society of Automotive Engineers (SAE) is the most commonly used in tyre modeling. Three forces and three moments

Integrated Vehicle Dynamics and Control, First Edition. Wuwei Chen, Hansong Xiao, Qidong Wang, Linfeng Zhao and Maofei Zhu.
© 2016 John Wiley & Sons Singapore Pte. Ltd. Published 2016 by John Wiley & Sons, Ltd.

acting on the tyre by the road can also be shown in this axis system. They are also known as the tyre 6-component wheel forces, shown in Figure 2.1.

The origin of the coordinate system is the center of the tyre road contact patch. The X axis is defined as the line of intersection of the wheel plane and the road plane with its positive direction being forward. The Y axis is perpendicular to the X axis and parallel to the pivot of the wheel, with its positive direction being to the right when viewed from the positive X direction. The Z axis is perpendicular to the road plane, with its positive direction pointing downwards.

The forces found on a tyre are: the longitudinal force F_x, the lateral force F_y, and the normal force F_z. The longitudinal, lateral, and normal forces are in the X, Y, Z direction respectively. The positive directions of these forces are shown in Figure 2.1. The moments found on a tyre are: the overturning moment M_x, the rolling resistance moment M_y, and the aligning moment M_z. The overturning moment, the rolling resistance moment, and the aligning moment are applied on the X, Y, and Z axes respectively; their positive directions are shown in Figure 2.1.

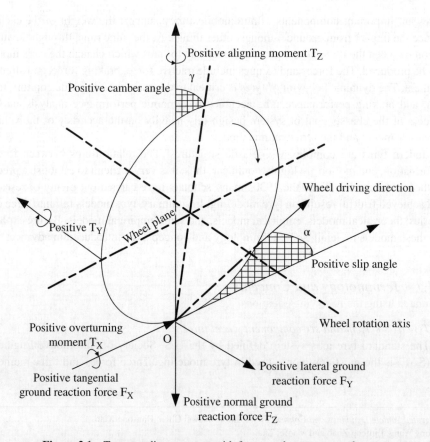

Figure 2.1 Tyre coordinate system with forces and moments on the tyre.

2. *Slip ratio*

Normally, a rolling wheel on the ground will slide (slip or skid). In order to quantify the proportion of the sliding to the wheel rolling movement, the concept of slip ratio is introduced. If the rolling radius of the wheel is r_d, the travelling speed of the wheel is u_w, the angular velocity of the wheel is ω, and then the slip ratio is defined as:

$$s = \frac{|u_w - r_d\omega|}{u_w} \times 100\% \tag{2.1}$$

3. *Tyre slip angle*

The tyre slip angle is the angle between the actual travelling direction of a rolling wheel and the direction towards which it is pointing. If the wheel travel speed is u_w, and the lateral velocity is v_w, then the tyre slip angle can be expressed as:

$$\alpha = \arctan\left(\frac{v_w}{u_w}\right) \tag{2.2}$$

2.1.2   Tyre Model

Tyres are modelled as a set of mathematical expressions which describe the relationship between the tyre's 6-component forces and wheel parameters. The output of these models are 6-component tyre forces, and the input parameters are the slip ratio, slip angle, radial deformation, camber angle, wheel speed, and yaw angle. The relationship between the inputs and the outputs is highly nonlinear. Here are some commonly-used models.

2.1.2.1   Unified Semi-empirical Tyre Model

This model was proposed by Professor Konghui Guo (Fellow of Chinese Academy of Engineering)[1]. It is suitable for the conditions of longitudinal slip, side slip, and combined slip when the tyre is rotating in a steady-state condition. It can also be extended to non-steady cases through the use of "effective slip ratio" and "quasi-steady" concepts.

In the steady-state condition, under pure longitudinal slip, pure side slip, and combined slip conditions, the tyre's longitudinal force, lateral force, and aligning moment can be calculated using the following equations.

1. Longitudinal force under pure longitudinal slip condition

$$F_x = -\frac{\phi_x}{|\phi_x|}\mu_x F_z \overline{F}_x \quad (\text{if } \phi_x = 0, \ F_x = 0) \tag{2.3}$$

where ϕ_x is the relative longitudinal slip ratio, $\phi_x = K_x s_x / \mu_x F_z$, μ_x is the longitudinal friction coefficient, $\mu_x = b_1 + b_2 F_z + b_3 F_z^2$, $\overline{F}_x$ is the non-dimension longitudinal force,

$\overline{F}_x = 1 - \exp\left(-|\phi_x| - E_1|\phi_x^2| - \left(E_1^2 + \dfrac{1}{12}\right)|\phi_x^3|\right)$, K_x is the longitudinal slip stiffness, s_x is the longitudinal slip ratio, and E_1 is the curvature factor $E_1 = 0.5/[1 + \exp(-(F_z - a_1)/a_2]$.

2. *Lateral force under pure lateral slip condition*

$$F_y = -\frac{\phi_y}{|\phi_y|}\mu_y F_z \overline{F}_y \quad (\text{if } \phi_y = 0, F_y = 0) \tag{2.4}$$

where ϕ_y is the relative lateral slip ratio, $\phi_y = K_y \tan\alpha/\mu_y F_z$, μ_y is the lateral friction coefficient, $\mu_y = a_1 + a_2 F_z + a_3 F_z^2$, $\overline{F}_y$ is the non-dimension lateral force, k_y is the cornering stiffness, $\overline{F}_y = 1 - \exp\left(-|\phi_y| - E_1|\phi_y^2| - \left(E_1^2 + \dfrac{1}{12}\right)|\phi_y^3|\right)$, and α is the side slip angle.

3. *Aligning moment under pure lateral slip condition*

$$M_z = -F_y D_x \tag{2.5}$$

where $D_x = (D_{x0} + D_e)\exp\left(-D_1|\phi_y| - D_2|\phi_y^2|\right) - D_e$, D_{x0}, D_e, D_1, D_2 are the parameters related to the vertical loads, $D_{x0} = c_1 + c_2 F_z + c_3 F_z^2$, $D_e = c_4 + c_5 F_z + c_6 F_z^2$, $D_1 = c_7 \exp(-F_z/c_8)$, $D_2 = c_9 \exp(-F_z/c_{10})$.

4. *Longitudinal force, lateral force, and aligning moment under combined slip condition*

$$\begin{cases} F_x = -\mu_x F_z \overline{F}\varphi_x/\varphi \\ F_y = -\mu_y F_z \overline{F}\varphi_y/\varphi \\ M_z = -F_y D_x - F_x D_y \end{cases} \tag{2.6}$$

where $\overline{F}$ is the non-dimensional shear force, $\overline{F} = 1 - \exp\left[-\varphi - E_1\varphi^2 - \left(E_1^2 + \dfrac{1}{12}\right)\varphi^3\right]$, ϕ is the relative slip ratio, $\phi = \sqrt{\phi_x^2 + \phi_y^2}$, D_y is the tyre lateral offset, $D_y = F_y/K_{cy}$, K_{cy} is the lateral stiffness, and $a_1, a_2, \ldots b_1, b_2, \ldots d_1, d_2$ can be obtained from tyre test data.

2.1.2.2 "Magic formula" Tyre Model

In recent years, the "magic formula" tyre model has been widely used in the study of vehicle dynamics and control systems. The "Magic formula" was proposed by Professor Pacejka[2] from Delft University of Technology. A combination of trigonometric functions was used to fit tyre experimental data. The formula can calculate longitudinal force, lateral force, and the aligning moment. Its general form is shown below.

$$y = D\sin\left\{C\arctan\left[Bx - E(Bx - \arctan Bx)\right]\right\} \tag{2.7}$$

where y is the longitudinal force, lateral force, or aligning moment, x is the tyre slip angle or longitudinal slip ratio, D is the peak value which represents the maximum value of the curve, C is the shape factor which indicates which force the curve represents (lateral force, longitudinal force, or aligning moment), B is the stiffness factor, and E is the curvature factor which represents the shape of the curve near the peak value.

Under pure driving or braking conditions, the longitudinal forces, lateral forces, and aligning moments on the tyre can be expressed as follows.

1. *Lateral force*

$$F_y = \left(D \sin \left(C \arctan \left(BX_1 - E \left(BX_1 - \arctan \left(BX_1 \right) \right) \right) \right) \right) + S_v \qquad (2.8)$$

where $X_1 = \alpha + S_h$, $D = a_1 F_z^2 + a_2 F_z$, $C = a_0$, $BCD = a_3 \sin (a_4 \arctan (a_5 F_z))(1 - a_6 |\gamma|)$, $B = BCD/CD$, $E = a_7 F_z^2 + a_8 F_z + a_9$, S_v is the curve shift in the vertical direction, $S_v = \left(a_{10} F_z^2 + a_{11} F_z \right) \gamma$, S_h is the curve shift in the horizontal direction, $S_h = a_{12} \gamma$, F_z is the vertical load, γ is the camber angle, and a_i ($i=0,1,2 \ldots 12$) the parameters fitted by the 12 data points.

2. *Longitudinal force*

$$F_x = \left(D \sin \left(C \arctan \left(BX_1 - E \left(BX_1 - \arctan \left(BX_1 \right) \right) \right) \right) \right) + S_v \qquad (2.9)$$

where $X_1 = s + S_h$, $D = b_1 F_z^2 + b_2 F_z$, $C = b_0$, $BCD = \left(b_3 F_z^2 + b_4 F_z \right) e^{-b_5 F_z}$, $B = BCD/CD$, $E = b_6 F_z^2 + b_7 F_z + b_8$, $S_h = b_9 F_z + b_{10}$, $S_v = 0$, and S is the longitudinal slip ratio.

3. *Aligning moment*

$$M_z = \left(D \sin \left(C \arctan \left(BX_1 - E \left(BX_1 - \arctan \left(BX_1 \right) \right) \right) \right) \right) + S_v \qquad (2.10)$$

where $X_1 = \alpha + S_h$, $D = c_1 F_z^2 + c_2 F_z$, $C = c_0$,

$$BCD = \left(c_3 F_z^2 + c_4 F_z \right) \left(1 - c_5 |\gamma| \right) e^{-c_5 F_z}, B = BCD/CD,$$

$$E = \left(c_7 F_z^2 + c_8 F_z + c_9 \right) \left(1 - c_{10} |\gamma| \right),$$

$$S_v = \left(c_{11} F_z^2 + c_{12} F_z \right) \gamma + c_{13} F_z + c_{14}, S_h = c_{15} \gamma + c_{16} F_z + c_{17}.$$

2.2 Tyre Longitudinal Mechanical Properties

When a vehicle is in motion under different conditions, the properties of the longitudinal forces of different tyres are different. Under driving conditions, the longitudinal forces are driving forces and rolling resistances on the driving wheels, and rolling resistances on the driven wheels. Under braking conditions, the longitudinal forces on the wheels are braking force and rolling resistance.

The method for calculating tyre driving and braking forces in different conditions has been presented in the previous section. Rolling resistance will be mainly discussed in the following section.

The force-resisting automobile motion which is generated between the tyre and the road is called rolling resistance. This force consists of three components: tyre rolling resistance, road resistance, and side slip resistance. When a pneumatic tyre rolls on a straight, dry, and hard surface, the resistance to the automobile is caused by the hysteresis of the tyres and the friction between the tyres and the road; this phenomenon is called tyre rolling resistance. The resistance caused by an uneven pavement, pavement deformation, wet roads, and other road conditions is called road resistance. The tyre longitudinal resistance due to tyre cornering is called side slip resistance.

2.2.1   Tyre Rolling Resistance

As can be seen from the definition, tyre rolling resistance is composed of tyre elastic hysteresis resistance, which is caused by the elastic hysteresis, friction resistance between the tyre and the road caused by relative sliding, and fan effect resistance generated by air loss due to the rotational movement of the tyre[3]. For convenience, the detailed causes of rolling resistance of a moving tyre on a flat, dry, and hard surface will be discussed.

1. *Elastic hysteresis resistance*
   The rolling process of a tyre on a road can be described by using the equivalent model of a wheel system. Suppose the wheel is composed of a number of radial springs and dampers between the outer surface and the rim, and a series of tangential springs and dampers which represent the tyre tread (see Figure 2.2).

   When a wheel rolls on a road, every unit of the wheel experiences repeated cycles of deformation and recovery. The unit entering the contact area is undergoing a deformation process, while the unit leaving the contact area is going through a recovery process.

   Experiments show that the work done to the tyre in the deformation process is greater than the work released during the recovery process (see Figure 2.3). The energy lost is converted into heat and dissipated into the air. Thus, the energy of an automobile is wasted and the vehicle is subject to resistance. Because the relative movement between adjacent layers will produce damping work, for tyres with same ply material, the more layers there are, the greater the damping of the tyre. Elastic hysteresis resistance is denoted by $F_{R,\text{Elastic hysteresis}}$.

2. *Frictional resistance*
   When a tyre unit is continuously entering into a tyre–road contact area, relative sliding in both the longitudinal and lateral directions will take place between the road and the tyre. Thereby, it causes additional friction resistance between the tyre and the road, which is denoted by $F_{R,\text{friction}}$.

3. *Fan effect resistance*
   Just like the action of a rotating fan, the rotational motion of a tyre will cause air loss, resulting in "fan effect resistance", which is denoted by $F_{R,\text{fan}}$.

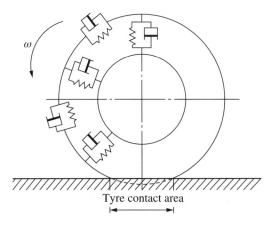

Tyre contact area

Figure 2.2   Wheel system equivalent model.

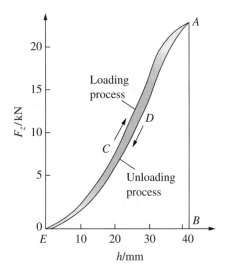

Figure 2.3   Tyre 9.00-2.0 radial deformation curve (tyre hysteresis loss).

Thus, when a tyre rolls straight on a dry, hard, flat road, the rolling resistance can be expressed as

$$F_R = F_{R,\text{Elastic hysteresis}} + F_{R,\text{friction}} + F_{R,fan} \tag{2.11}$$

2.2.2   Road Resistance

When an automobile moves on an uneven, plastic and wet road, the tyre rolling resistance will increase. The increased resistance due to the road conditions is called "road resistance".

1. *Uneven road*

   When a vehicle is travelling on an uneven road, the road roughness in the vertical direction acts as an excitation. The tyre is undergoing continuous deformation and recovery processes whilst moving up and down relative to the body of the vehicle. The energy released during the process of recovery is less than the work done to the deformation, thus the part of the vehicle that drives the power of the car is consumed by the road resistance due to the uneven road.

2. *Plastic road*

   When tyres roll on soil, sand, grass, or snowy roads which possess greater plasticity, the additional resistance caused by the road plastic deformation is defined as plastic road resistance, denoted by $F_{R,\text{plastic}}$. It includes three parts: compression resistance, bulldozing resistance, and shear resistance. Detailed explanations of these resistant forces of vehicles driving on a plastic soil road are given below.

3. *Compression resistance*

   When tyres are rolling on plastic road surfaces, soil deformation occurs in the vertical direction as the soil is compressed and only a small portion can recover. The car power will be consumed for the compression of the soil, which causes additional tyre rolling resistance, which is called compression resistance.

4. *Bulldozing resistance*

   When tyres roll on a soft soil surface, some soil will be pushed to the front of the tyre, which is then pushed to the side or is finally compacted. The longitudinal and lateral movement of the soil will cause additional tyre rolling resistance; this is called bulldozing resistance.

5. *Shear resistance*

   Under vertical loads, tyre tread grooves insert into the soil, thus the rotation of the wheels tend to cut the soil. Friction generated between the tyre and the road results in additional tyre rolling resistance, which is called shear resistance.

6. *Wet pavement*

   In the longitudinal plane of a car, the rolling tyre tread wading area can be divided into three parts: water film district, transition zone, and direct contact region, as shown in Figure 2.4. In the water film district, the tread does not actually touch the road and most of the water is

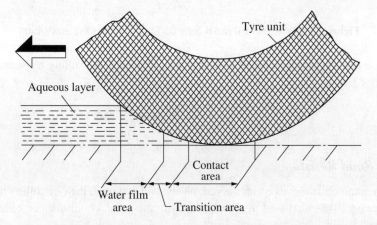

Figure 2.4   The wading area of tyres on wet pavement.

drained. In the transition zone, the tyre tread makes contact with the road partially, and the tyre is deformed. In the direct contact region, the tread and the road surface directly contact each other, and a small amount of water between the tyre treads is pressed out.

In the wading area, additional resistance will be produced to the rolling tyre during the process of drainage and contacting the road surface directly, as the so-called "turbulence resistance". Tests show that the turbulence resistance depends basically on the volume of the water drained away per unit time which is determined by the drain water depth h, the wheel tyre width W_t, and the wheel rolling speed. The additional tyre resistance on a wet road is expressed as $F_{R,thrash}$.

2.2.3 Tyre Slip Resistance

The preceding discussion on the causes of rolling resistance is based on the assumption that the wheel rolls parallel to the vehicle longitudinal plane. However, when a car is in motion, additional travel resistance is generated because of its structural features.

1. *Influence of tyre slip*
   The tyre side slip phenomenon appears when pneumatic tyres roll. The component of the lateral tyre force in the vehicle longitudinal plane is called the additional rolling resistance, which is denoted by $F_{R,sideslip}$.
2. *Influence of wheel alignment*
   In order to ensure good vehicle handling and stability, vehicles are usually calibrated with certain wheel orientation angles, such as the toe and camber angle. Similar to the phenomenon of tyre slip, the tyre rolling resistance increases due to the wheel toe angle and camber, which are denoted as $F_{R,toe}$ and $F_{R,camber}$ respectively.

2.2.4 Overall Rolling Resistance of the Tyres

As can be seen from the analysis above, the overall rolling resistance applied to a rolling tyre consists of a variety of resistances with complex causes. Some of them inevitably accompany the wheel rolling motion, such as tyre rolling resistance, and tyre slip resistance. Other additional resistances are caused by special road conditions such as road resistance. Thus, the elements of the total rolling resistance of a wheel are: $F_{R,elastic\ hysteresis}$, $F_{R,friction}$, $F_{R,fan}$, $F_{R,plasticity}$, $F_{R,uneven}$, $F_{R,thrash}$, $F_{R,sideslip}$, $F_{R,toe}$ and $F_{R,camber}$.

Under actual vehicle driving conditions, the overall resistance applied to the rolling tyres of a vehicle is composed partially or fully of the resistances introduced above. For convenience, the following formula is used to calculate the rolling resistance for general purposes:

$$F_R = fF_z \qquad (2.12)$$

where f is the rolling resistance coefficient, and F_z is the vertical wheel load. As can be seen from the equation above, when the vertical load is certain, the rolling resistance coefficient is the factor that affects the rolling resistance.

2.2.5   Rolling Resistance Coefficient

1. *The influential factors*

   The tyre rolling resistance coefficient is affected by the tyre construction, materials, pressure, vehicle speed, and road conditions. The rolling resistance coefficient of a vehicle at low speed on a given road is shown in Table 2.1.

   Figure 2.5 shows the influence of the tyre pressure on the rolling resistance coefficient. As the tyre pressure increases, the rolling resistance coefficient decreases. Since the tyre stiffness increases with the tyre pressure, under a constant wheel load condition, the tyre deformation decreases accordingly. As a result, the elastic deformation energy is reduced. At the same time, the length of the tyre–ground contact area decreases, and the component of the tyre frictional resistance is also reduced.

   The impact of vehicle speed on the rolling resistance coefficient is illustrated in Figure 2.6. As the speed increases, the rolling resistance coefficient increases slightly at lower vehicle speeds. Gradually, the graph shows a more significant gradient as the speed increases. The reason for this is that when the vehicle reaches a critical speed, the deformation of the tread which just leaves the contact area cannot recover immediately. The tyre is no longer circular, and a standing wave phenomenon occurs at its rim.

   The tyre rolling resistance also depends on the structure of the tyre, materials, and rubber used. The rolling resistance of the radial tyre is generally smaller compared with a bias tyre. In addition, the structural design of the tyre tread also has an impact on the rolling resistance; shallow tread patterns and a well-designed tread profile can reduce the rolling resistance. However, as the speed increases, the impact of the tread patterns becomes smaller.

Table 2.1   The rolling resistance coefficient of a vehicle at low speed on a given road.

Pavement type		Rolling resistance coefficient
Good asphalt or concrete pavement		0.010–0.018
General asphalt or concrete pavement		0.018–0.020
Gravel road		0.020–0.025
Good gravel road		0.025–0.030
Pebble potholes pavement		0.035–0.050
Pressed dirt road	Dry	0.025–0.035
	Rainy	0.050–0.150
Muddy dirt road (the rainy season or thaw)		0.100–0.250
Dry sand		0.100–0.300
Wet sand		0.060–0.150
Icy roads		0.015–0.030
Compacted ski track		0.030–0.050

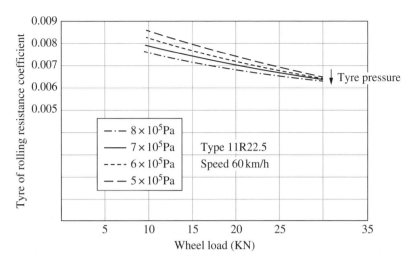

Figure 2.5   Influence of the tyre pressure on the rolling resistance coefficient.

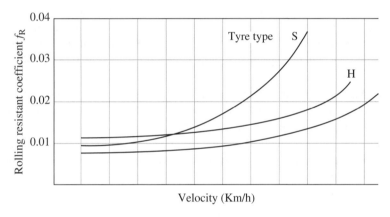

Figure 2.6   The impact of speed on the rolling resistance coefficient.

2. *Measurement of rolling resistance coefficient*

There are two different methods to measure the tyre rolling resistance coefficient: the vehicle road test and the indoor bench test. The typical method for the indoor bench test is to place the wheel with the test tyre on a movable rolling surface. Test data can be captured by a force sensor on the wheel fixtures (such as the connecting rod system and rims). According to the rolling surface condition, a tyre test bench can be categorized into the following three types[4] (see Figure 2.7): external support bench, internal support bench, and flat bench. The most commonly used is the external support bench, the advantages of which are its relatively low cost, high load carrying capacity, and compact structure, leaving more space around the wheel. Thus, it not only can accommodate a variety of linkages to ensure the wheel alignment, but also it facilitates the installation of the wheel.

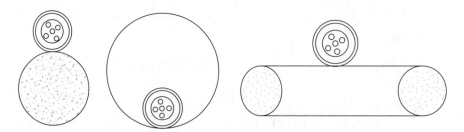

Figure 2.7   Structure and characteristics of the tyre test rig.

2.3   Vertical Mechanical Properties of Tyres

When a normal load acts on a pneumatic tyre, a vertical deformation occurs. The relationship between the tyre vertical deformation and the normal load is the key when studying the vertical mechanical properties of a tyre which have a great influence on the ride comfort of the vehicle.

The "equivalent model of a wheel system" has been mentioned when the tyre rolling resistance was discussed. This model assumes that the wheel is composed of a number of radial springs and dampers between the outer surface and the rim, and a series of tangential springs and dampers which represent the treads. If a study is focused on the vertical dynamic behaviors of the tyre rolling on an uneven road surface, the model can be simplified to a point-contact linear spring–damping model as shown in Figure 2.8. In this model, the tyre's vertical stiffness and damping are the main physical parameters that affect the vertical mechanical properties of the tyre. Compared with the damping of the suspension system, the tyre damping is much smaller, therefore it can be ignored in some models where only the vertical stiffness is then considered.

According to the test conditions, three different kinds of tyre vertical stiffness are defined: static stiffness, non-rolling dynamic stiffness and rolling dynamic stiffness. A brief introduction of static stiffness and rolling dynamic stiffness is given below.

1. *Static stiffness*

   Tyre static vertical stiffness is determined by the slope of the relationship curve between the static vertical load and deformation. Figure 2.9 shows the relationship curve between the vertical load and deformation of a passenger radial tyre. When the inflation pressure is constant, the relationship between the load and deformation is essentially linear, except under very low loads. So, it can be considered that in the actual load range the tyre vertical stiffness does not vary with the load. Figure 2.10 shows the relationship between the stiffness of a 165 × 13 type radial passenger tyre and inflation pressure[5].

2. *Rolling dynamic stiffness*

   Among the tyre design parameters, the tyre structure parameters (such as the crown cord angle, tread width, tread depth, and number of plies) and materials have a more significant influence on the stiffness of the tyre. Among the tyre operating condition parameters, the tyre inflation pressure, speed, normal load, and wear have a great influence on the stiffness of the tyre. There has been no general applicable conclusion about the quantitative description of the relationship between the tyre static stiffness and dynamic stiffness. Some studies have reported that, for a passenger car tyre, the roll

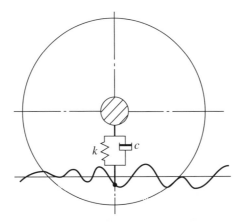

Figure 2.8 Point-contact linear spring–viscous damping model.

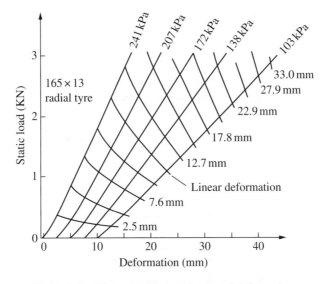

Figure 2.9 Diagram of the tyre loads and deformation.

dynamic stiffness is usually 10–15% less than the stiffness obtained by the static load deformation curve; for a heavy-duty truck tyre, the dynamic stiffness is approximately 10% less than the static stiffness[6].

Figure 2.11 shows the curve that describes the influence of the car's speed on the rolling dynamic stiffness. When the car's speed is slow, the tyre dynamic stiffness decreases significantly with the increase of car speed; but when the speed exceeds about 20 km/h, car speed has almost no effect on the dynamic stiffness.

The damping of a pneumatic tyre is mainly caused by the material hysteresis, and the magnitude of the damping force depends on the design and structure of the tyres, as well as on the tyre working conditions. Generally, it is not simple Coulomb or viscous damping, but the combination of both.

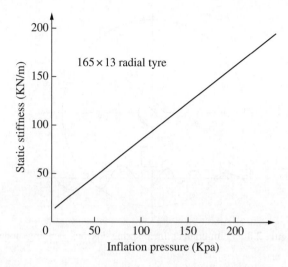

Figure 2.10   Diagram of the static stiffness and tyre inflation pressure.

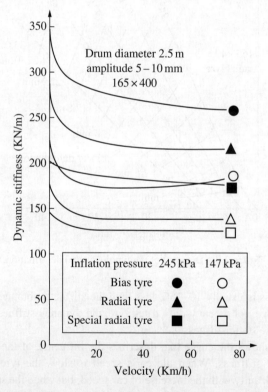

Figure 2.11   The influence of car speed on the rolling dynamic stiffness of a tyre.

If a tyre is simplified as a single degree of freedom system that consists of a spring and damping, as shown in Figure 2.8, the theory of vibration analysis can be applied and the equivalent viscous damping coefficient C_{eq} and dynamic stiffness k_z of the tyre can be obtained by dynamic test methods.

2.4   Lateral Mechanical Properties of Tyres

The lateral mechanical properties of a tyre are mainly about the relationship between the lateral force, aligning moment of the tyre, and parameters such as the tyre slip angle, vertical load, and camber[7].Tyre models reflect the analytical relationships between the parameters just mentioned. A more detailed analysis of the properties and influencing factors is given below.

To a specific tyre, the main factors which influence the tyre lateral force are the side slip angle, vertical load, and camber. The side slip angle is determined by the motion parameters of the tyre, such as the vehicle longitudinal velocity, lateral velocity, yaw rate, steering angle, and other factors.

The vertical load of a tyre is determined by the vehicle axle load distribution when the vehicle is designed. The weight is transferred when the vehicle is under running conditions. The initial value of the camber is determined during vehicle design. However, the rolling motion of a moving vehicle has a certain impact on the wheel camber value.

Figure 2.12 (a), (b), (c) illustrates the relationship between lateral force and sideslip angle, vertical load and camber under the condition of pure steering, respectively. As can be seen in the figures, when the slip angle, vertical load, and camber angle are relatively small, the lateral force can be considered as having a linear relationship with them. When they exceed certain values, this linear relationship will no longer exist.

Figure 2.13(a), (b), (c) illustrates the relationships between aligning moment and sideslip angle, vertical load and camber under the condition of pure steering, respectively. The aligning moment increases gradually with the increase of the slip angle when the sideslip

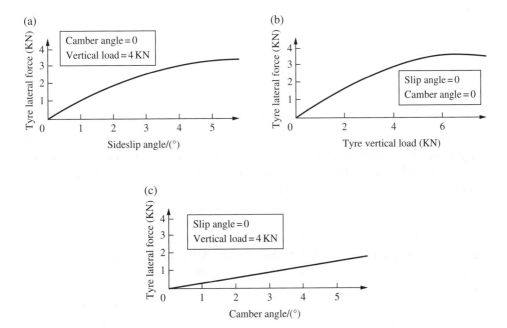

Figure 2.12   The tyre lateral force characteristics of a car. (a) The relationship between tyre lateral force and slip angle. (b) The relationship between tyre lateral force and vertical load. (c) The relationship between tyre lateral force and camber angle.

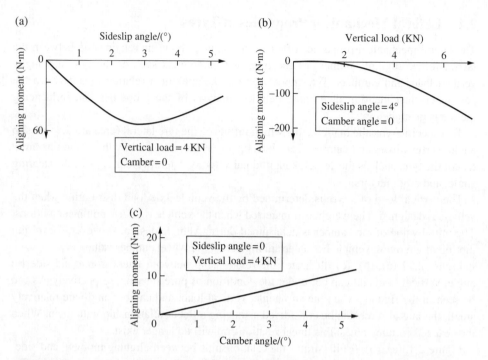

Figure 2.13   The tyre aligning torque characteristics of a car. (a) The relationship of aligning moment and sideslip angle. (b) The relationship of aligning moment and vertical load. (c) The relationship between aligning moment and camber angle.

angle is relatively small. The aligning moment decreases with the slip angle after the aligning torque reaches its maximum value. The aligning moment increases with the increase of the vertical load.

2.5   Mechanical Properties of Tyres in Combined Conditions

Tyre models describe the mathematical relationship between the tyre's 6-component forces (output) and the wheel parameters (input). When a car is moving, the outputs of the tyre model, such as the longitudinal force, lateral force, vertical force, rolling moment, rolling resistance moment, and aligning moment are determined by the longitudinal slip ratio of the wheel, side slip angle, radial deformation, camber angle, wheel speed, and yaw angle. Because the relationships between the inputs and outputs of the tyre models are strongly nonlinear, and the vertical tyre force, lateral force, and longitudinal force affect each other, the 6-component forces are coupled. In the case of vehicle acceleration, or braking when cornering on an uneven road, all thetyre characteristics described above must be taken into consideration simultaneously.

   If the vertical force is constant, whenever the slip ratio is small, the longitudinal force is a linear function of the longitudinal slip ratio. But when the longitudinal slip ratio reaches a certain value, with the increase of the longitudinal slip ratio, the tyre longitudinal force decreases slightly. However, when the longitudinal slip ratio remains unchanged, the tyre longitudinal force increases linearly with the increase of the vertical force.

If the vertical force is fixed, whenever the slip angle is small, the lateral force is approximately linear with the tyre slip angle. The lateral force reaches a maximum value at a certain tyre slip angle. Then, with the increase of tyre slip angle, the lateral force remains almost unchanged. At this point, the tyre lateral force reaches saturation. When the tyre slip angle remains constant, the tyre lateral force increases approximately linearly with the increase of the vertical force. But when the vertical force becomes larger, the lateral force shows a strongly nonlinear relationship with the increase of the vertical force.

When the sideslip angle and the vertical force are fixed, the tyre longitudinal force and the lateral force interact with each other. The tyre longitudinal force increases with the increase of the longitudinal slip ratio at the beginning. When the slip ratio reaches a certain value, the longitudinal force decreases gradually and tends to be stable. However, the tyre lateral force gradually decreases with the increase of the longitudinal slip ratio. When the longitudinal slip ratio approaches 100%, the tyre lateral force is almost zero.

The interaction between the lateral force and the longitudinal force can also be interpreted by the "friction ellipse". Studies have shown that the resultant force in the tyre–road contact print is constant. The tyre resultant force is usually indicated by a family of curves, as shown in Figure 2.14. The figure shows relationship curves of the tyre lateral force and the longitudinal force with some given series of slip ratio or side slip angle. The envelope line of these curves is similar to an ellipse, which is why it is called "friction ellipse" or "adhesion ellipse". Figure 2.14 also shows that the tyre cannot obtain the maximum tyre

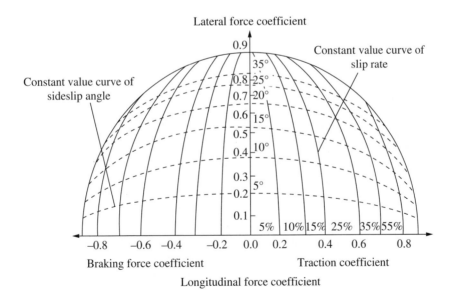

Figure 2.14 Distribution of the tyre longitudinal force and lateral force with combined conditions (friction ellipse).

lateral force and the maximum longitudinal force simultaneously as a result of the limitation of the maximum friction. No lateral force is available when the tyre driving force or the braking force is at its maximum, the lateral force achieves its maximum value only when the longitudinal force is zero.

References

[1] Guo K H, Ren L. A Unified Semi-empirical Tyre Model with Higher Accuracy and Less Parameters. SAE Technical Paper Series, 1999-01-0785.
[2] Pacejka H B. Tyre and Vehicle Dynamics. Butterworth-Heinemann, London, 2002.
[3] Wallentowitz H. Longitudinal Dynamics of Vehicles. Insitut Fur Kraftfahrwesen, Aachen, 1997.
[4] Clark S K. Mechanics of Pneumatic Tyres. DOT HS 805 952. U.S. Dept of Transportation, NHTSA, Washington DC, 20590, 1981.
[5] Overton J A, Mills B, Ashley C. The vertical response characteristics of the non-rolling tyre. Proceedings of IMechE, 1969, 184(2):25–40.
[6] Fancher P S, etc. A Fact Book of the Mechanical Properties of the Components for Single Unit and Articulated Heavy Trucks. Report No. HS 807 125, National Highway Traffic Safety Administration, U.S. Department of Transportation, 1986.
[7] Crolla D, Yu F. Vehicle Dynamics and Control. People's Communications Press, Beijing, 2004.

3

Longitudinal Vehicle Dynamics and Control

3.1 Longitudinal Vehicle Dynamics Equations

3.1.1 Longitudinal Force Analysis

The dynamics problems along the x-axis in a vehicle coordinate system are called the longitudinal dynamics problems. The characteristics of vehicle acceleration and deceleration are the main research topics, and these two states of motion correspond to, respectively, the traction and braking of the vehicle.

3.1.1.1 Traction

In the case of traction, the force on a car pointing to the positive direction of the x-axis is the longitudinal force produced by the interaction between the driving wheels and the road, which is represented by F_x.

When the tyre is on a high adhesion road, the longitudinal force Fx is equal to the force of traction F_t.

$$F_t = \frac{T_t}{r_0} = \frac{T_e i_g i_0 \eta_t}{r_0} \tag{3.1}$$

where T_t is the torque transmitted to the driving wheels, T_e is the engine torque, i_g is the transmission ratio, i_0 is the main gear ratio, η_t is the driveline efficiency, and r_0 is the driving wheel radius.

Integrated Vehicle Dynamics and Control, First Edition. Wuwei Chen, Hansong Xiao, Qidong Wang, Linfeng Zhao and Maofei Zhu.
© 2016 John Wiley & Sons Singapore Pte. Ltd. Published 2016 by John Wiley & Sons, Ltd.

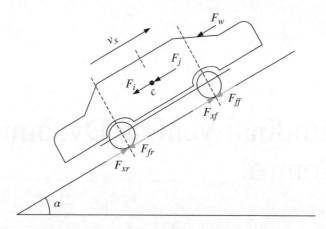

Figure 3.1   Forces of the uphill accelerating conditions.

When a tyre is running on a poor adhesion road, the longitudinal force F_x is equal to the tyre longitudinal force, which is actually provided by the ground. It can be calculated through the tyre longitudinal force and slip ratio curve, or through tyre models when the tyre motion status is given. The longitudinal slip ratio has a linear relationship with the tyre longitudinal force when the ratio is small; the scale factor is called the longitudinal tyre stiffness[1].

The forces acting on a car driving uphill are shown in Figure 3.1, The longitudinal forces along the negative x-axis direction, the driving resistance, includes the rolling resistance F_f (the sum of the front rolling resistance F_{ff} and the rear rolling resistance F_{fr}), aerodynamic drag force F_w, ramp resistance F_i, and the inertial resistance. It should be noted that when the car is accelerating the inertial moment of some rotating masses which include the rotating elements of the engine, driveline, and tyre contribute to the driving resistance. For convenience's sake, the rotating masses are converted to an equivalent translation mass, and only one expression used to denote the inertial resistance. The symbol F_j normally donates inertial resistance.

3.1.1.2   Braking

When a car brakes while driving uphill, the longitudinal forces along the positive x-axis direction are the inertial forces which are generated by the vehicle deceleration. The forces along the negative x-axis direction include the tyre longitudinal forces, rolling resistance, aerodynamic drag force, and ramp resistance. The tyre longitudinal force is less than the brake force; it is calculated in the same way as in the traction force.

3.1.2   *Longitudinal Vehicle Dynamics Equation*

To establish the longitudinal vehicle dynamics equation, the status of two movements should be considered: traction and braking. This section focuses on establishing the longitudinal vehicle dynamics equation when the vehicle is in traction.

3.1.2.1 Dynamics Equation

Moving uphill, if the driving force is greater than the sum of the rolling resistance, aerodynamic drag force, and ramp resistance, the excess part of the driving force can be used to accelerate the vehicle.

The dynamics equation is

$$F_t - F_f - F_w - F_i = F_j \tag{3.2}$$

Equation (3.2) has some advantages in vehicle dynamics analysis but, when studying longitudinal dynamics control problems, a more accurate model is needed which includes the model of the engine, transmission, and tyre.

Taking the Acceleration Slip Regulation control (ASR) as an example, a longitudinal dynamics model can be established in the following way. The static torque of the engine is converted to the dynamic torque to the transmission after a first-order delay, and the transmission model calculates the half axle torque and rotational speed of the driving wheels. The normal force model calculates the normal load on each tyre. The tyre model takes the transmission model and the output from other modelst as its input, calculating the longitudinal velocity, longitudinal acceleration, and other parameters, which are fed back to the transmission model, the normal force model, and the tyre model. Each model will be described in subsequent chapters, and the longitudinal vehicle dynamics model is given here.

3.1.2.2 General Vehicle Longitudinal Dynamics Equation

When a car drives uphill, during both traction and braking, the vehicle longitudinal dynamics equation can be expressed by one equation:

$$m\dot{v}_x = F_{xf} + F_{xr} - F_f - F_w - F_i \tag{3.3}$$

where F_{xf} is the sum of the longitudinal forces of the front wheels, F_{xr} is the sum of the longitudinal forces of the rear wheels. The method of calculating these two forces has already been discussed.

The directions of the longitudinal forces are different in traction and braking, the signs of which are indicated in the tyre models.

When braking, all the forces affecting vehicle motion are in the same direction. Compared with the longitudinal tyre force, rolling resistance, aerodynamic drag force, and inertia moment of a rotating mass generated by the car's deceleration which are much smaller, and can be ignored when establishing the equation.

3.2 Driving Resistance

A vehicle accelerating uphill is in a typical longitudinal running condition. The resistances acting on the vehicle include the tyre rolling resistance, the aerodynamic drag, the ramp resistance, and the acceleration resistance. The tyre rolling resistance has been described in Section 2.2 2. Three other resistances are discussed here.

3.2.1   Aerodynamic Drag

The aerodynamic drag F_w is the component of the air drag force acting on the vehicle along the running direction when the vehicle moves in a straight line. It is made up of pressure resistance and frictional resistance. The pressure resistance is the component of the resultant normal air force acting on the vehicle surface along the running direction. The air, which has a certain viscosity as other fluids, produces friction between air micelles and the vehicle surface as it flows through it; the friction forms the resistance, and its component force along the running direction is called the frictional resistance.

Pressure drag is composed of the shape drag, disturbance drag, internal circulation drag, and induced drag. When a vehicle is moving, air flows through the vehicle surface generating eddy flows in the area where the air velocity changes sharply, building negative pressure in the back and positive pressure in the front. This part of pressure drag caused by eddy flows is related to the shape of the body, so it is called the shape drag, and it accounts for about 58% of the total drag. Disturbance drag is caused by projecting parts of the vehicle body, such as the rearview mirrors, door handles, and projecting parts below the chassis, and it accounts for about 14% of the total drag. Internal circulation drag refers to the resistance caused by air flowing through the inside of the vehicle body, for the purposes of engine cooling, passenger compartment ventilation, and airconditioning. It accounts for 12% of the total drag. When air flows through the asymmetric upper and lower surfaces, it creates a lift force which is perpendicular to the direction of the air velocity, whose component force along the running direction is called the induced drag, which comprises about 7% of the total drag[2]. The aerodynamic drag force is expressed as follows:

$$F_w = \frac{1}{2} C_D A \rho v_r^2 \tag{3.4}$$

where C_D is the aerodynamic drag coefficient, ρ is the mass density of air (normally $\rho = 1.2258\,Ns^2/m^4$), A is the frontal area of the vehicle, v_r is the relative velocity of the vehicle where there is no wind condition.

Considering the car moving in windless conditions, if v_a (km/h) is used to denote the running speed and A (m²) for the frontal area of the car, the air drag F_w (N) can be expressed as

$$F_w = \frac{C_D A v_a^2}{21.15} \tag{3.5}$$

3.2.2   Ramp Resistance

When moving uphill, the component of a vehicle gravity force along the ramp is called the ramp resistance F_i,

$$F_i = G \sin \alpha = mg \sin \alpha \tag{3.6}$$

where G is the gravity of vehicle, m is the vehicle mass, g is the acceleration of gravity, and α is the slope angle.

When going up the slope, the component of the vehicle gravity force perpendicular to the ramp is $mg\cos\alpha$. So, the rolling resistance F_f can be expressed as

$$F_f = fmg\cos\alpha$$

Since the ramp resistance F_i and rolling resistance F_f are related to the road conditions, the sum of these two resistances is called the road resistance, represented by F_ψ:

$$F_\psi = F_f + F_i = Gf\cos\alpha + G\sin\alpha \qquad (3.7)$$

If i is used to represent the slope of the road, by definition, it can be expressed as

$$i = \tan\alpha$$

When the slope of the road is relatively small, then $\cos\alpha \approx 1$, $\sin\alpha \approx i$, then

$$F_\psi = Gf\cos\alpha + G\sin\alpha = G(f+i)$$

where $f+i$ is called the road resistance coefficient, which is represented by Ψ,

$$F_\psi = G\Psi \qquad (3.8)$$

3.2.3 Inertial Resistance

As described in Section 3.1.1, both the translational mass and the rotating mass are accelerating when the vehicle is speeding up, and in order to use one expression to express the dynamics equation, the rotating mass is converted to the equivalent translational mass. So, the inertial resistance F_j is

$$F_j = \delta m \frac{dv_x}{dt} \qquad (3.9)$$

where δ is the rotating mass conversion coefficient[2], and $\dfrac{dv_x}{dt}$ is the acceleration.

$$\delta = 1 + \frac{1}{m}\frac{\Sigma I_w}{r_0^2} + \frac{1}{m}\frac{I_f i_g^2 i_0^2 \eta}{r_0^2} \qquad (3.10)$$

where I_w is the inertia moment of the wheel, and I_f is the inertia moment of the flywheel.

Thus, the dynamics equation can also be expressed as:

$$\frac{T_t i_g i_0 \eta_T}{r_0} - Gf\cos\alpha - \frac{C_D A}{21.15}v_a^2 - Gi\sin\alpha = \delta m\frac{dv_x}{dt} \qquad (3.11)$$

3.3   Anti-lock Braking System

3.3.1   Introduction

Anti-lock braking system (ABS), one of the most important automotive active safety technologies, is able to improve vehicle direction stability and steering performance, and shorten the vehicle braking distance through preventing the wheels from being locked.

Usually, wheels slip on a road's surface during braking. The relationship between the longitudinal adhesion coefficient, lateral adhesion coefficient, and slip ratio is shown in Figure 3.2. Experiments show that the lower the slip ratio, the larger the lateral adhesion coefficient with the same sideslip angle. A large longitudinal and lateral adhesion coefficient can be obtained at the same time when the slip ratio is kept at an appropriate value (generally 15–25%), and at this time, both good braking and lateral stability performance can be achieved.

Research on ABS started in the early 20th century. In China, ABS research began in the early 1980s. Today, ABS is widely used and has become standard equipment in many vehicles.

3.3.2   Basic Structure and Working Principle

ABS mainly consists of wheel speed sensors, an electronic control unit (ECU), and a hydraulic pressure control unit (HCU)[3]. It also includes the brake warning light, anti-lock brake warning light, and some other components.

According to its braking pressure control mode, ABS can be classified into mechanical ABS and electronic ABS. Nowadays, most ABS are electronically controlled. According to the arrangement of the brake pressure regulating device, ABS can be divided into the integral type and split type. In the integral type ABS, the HCU and the master cylinder are integrated in one unit, while in the split type ABS, the HCU and the master cylinder are

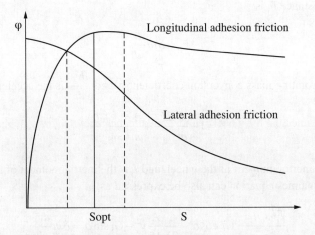

Figure 3.2   Adhesion coefficients and slip ratio curves.

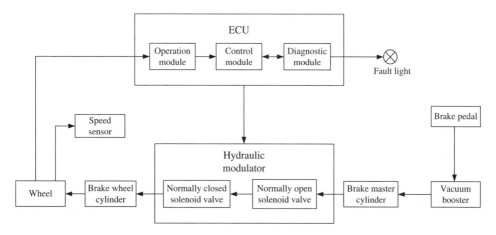

Figure 3.3   ABS system structure diagram.

separated. In addition, according to the arrangement of brake pipe lines, ABS can also be classified into single, dual, three- and four-channel type.

Take as an example a typical ABS system[4] where speed sensors are installed on every wheel. The speed information of the wheels is sent to the ECU. The ECU processes this data and analyses the movement of the wheels, issuing control commands to the brake pressure regulating device. The brake pressure regulating device, consisting of a regulating solenoid control valve assembly, electric pump assembly, and liquid reservoir, connects to the brake master cylinder and the wheel cylinders through the brake pipe lines. Under the commands of the ECU, the braking pressure of each wheel cylinder is adjusted. Taking a single wheel as an example, the principle of each stage of the braking process is shown in Figure 3.3.

The ABS working process can be divided into normal braking stage, brake pressure holding stage, brake pressure reduction stage, and brake pressure increase stage.

1. *Normal braking stage*

   In the normal braking stage, the ABS is not involved in the brake pressure control. Each inlet solenoid valve in the pressure regulating solenoid valve assembly is not powered and is normally in an open state, while each outlet solenoid valve is not powered and is normally in a closed state. The electric pump is not running for electricity. The brake pipe lines connecting the wheel cylinder and the master cylinder are unobstructed, while the brake pipe lines connecting the wheel cylinder and the liquid reservoir are impeded. The pressure of each wheel cylinder changes with the output pressure of the master cylinder.

2. *Pressure holding stage*

   When the ECU judges that a wheel tends to get locked when braking, the normally open solenoid valve is electrified and closed, and the brake fluid from the master cylinder can no longer flow into the wheel cylinder. The normally closed valve is still not energized and stays closed, while the brake fluid in the wheel cylinder does not flow out. Thus, the pressure in the wheel cylinder stays stable.

3. *Pressure reduction stage*

   If the ECU determines that a wheel still has a tendency of being locked even if the pressure in the wheel cylinder stays unchanged, the normally closed solenoid valve is opened. Then part of the brake fluid in the wheel cylinder flows back into the reservoir through the opened solenoid valve, which makes the brake pressure in the wheel cylinder decrease rapidly. Thus, the locking trend of the wheel is eliminated.

4. *Pressure increase stage*

   As the liquid pressure in the wheel cylinder decreases, the wheel accelerates gradually under the inertia force. When the ECU judges that the locking trend of the wheel has been completely eliminated, the normally open valve opens and the normally closed valve closes. At the same time, the electric pump begins working and pumps the brake fluid back to each wheel cylinder. Brake fluid from the master cylinder is pumped into the wheel cylinder together through the normally open valve. Thus, the pressure of the wheel cylinder increases rapidly, while the wheel begins to decelerate.

   By adjusting the pressure in the wheel cylinder to experience the cycle of pressure holding decreasing and increasing, the ABS keeps the slip ratio in the ideal range until the vehicle speed is very much reduced, or the output pressure of the master cylinder is not big enough to make the wheel trend to lock. Generally, the range of the frequency of the brake pressure adjustment cycle is 3–20Hz. The braking pressure of the wheel cylinder can be adjusted independently in a 4-channel ABS, so that the four wheels do not trend to be locked.

3.3.3   Design of an Anti-lock Braking System

ABS control is a complex, nonlinear control problem and many control methods have been applied in order to control of ABS successfully. Logic threshold control, PID control, sliding mode variable structure control and fuzzy control methods have all been used for this purpose.

Logic threshold control method is currently the most widely used and the most mature control algorithm of ABS. Firstly, it does not involve a specific control mathematical model, which eliminates a lot of mathematical calculations, and improves the system's real-time response. Thus, this complicated nonlinear problem can be simplified. Secondly, because this algorithm uses fewer control parameters, the vehicle speed sensors can be eliminated, which makes the ABS simple and leads to a great reduction in the cost. In addition, its actuator is also relatively easy to implement. The disadvantage is that the control logic is complex, the control system is not stable and, due to its lack of an adequate theoretical basis, the threshold values are obtained through mass data tests. Furthermore, the ABS using a logic threshold algorithm has poor interchangeability between different models of vehicles, which means that it takes lots of time and many tests to determine and adjust the control logic and parameters to achieve the best anti-lock braking effect for newly-developing vehicles.

3.3.3.1 Mathematical Model of Brake System

1. *Brake model*

Since brakes produce torques on wheels when braking, the brake model is used to calculate the braking torque of each wheel under a certain hydraulic pressure. Taking an X-type diagonal dual-circuit brake system as an example, the modeling method is described. The system construction is shown in Figure 3.4.

The relationship between the pedal force F_b and the wheel cylinder pressure P_0 can be obtained by modeling the pipe lines, and then the relationship between the wheel cylinder pressure P_0 and time t can be calculated. Finally, the relationship between the braking torque T_b and time t can be obtained. Ignoring the flexibility of the brake fluid, and the hydraulic transmission lag, the spring return force of the drums and discs, the relationship between the pedal force F_b and the output hydraulic pressure P_0 is:

$$P_0 = 4F_b i_b \eta B / \pi D_m^2 \qquad (3.12)$$

where i_b is the rake lever ratio, η is the efficiency of the control mechanism, B is the boosting ratio, and D_m is the diameter of the master cylinder.

The maximum braking torque of a single front brake disc is

$$T_{b1} = n\pi P_0 C_f D_f D_{wf}^2 / 4 \qquad (3.13)$$

where n is the number of clamp brake cylinders, C_f is the braking efficiency factor of the disc, D_f is the working radius of the disc, and D_{wf} is the front wheel cylinder piston diameter.

The maximum braking torque of a single rear brake drum is

$$T_{b2} = \pi P_0 C_r D_r D_{wr}^2 / 4 \qquad (3.14)$$

where C_r is the braking efficiency factor of the drum, D_r is the working radius of the brake drum, D_{wr} is the rear wheel cylinder piston diameter.

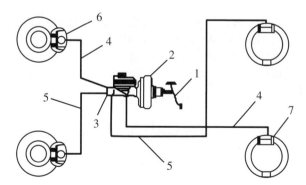

Figure 3.4 Diagonal dual-circuit brake system. (1) brake pedal; (2) vacuum booster; (3) tandem master cylinder; (4, 5) brake pipe lines; (6) disc brakes; (7) drum brakes.

(a)

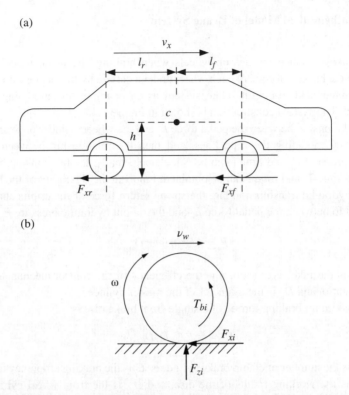

(b)

Figure 3.5 Longitudinal motion of a vehicle and rotary motion model of a wheel. (a) Longitudinal motion of a vehicle; (b) Rotary motion.

2. *Tyre model*

   When modeling an ABS control system, tyre models are needed to get the relationship between the tyre adhesion and other parameters related, the relationship is usually represented by the function between the adhesion coefficient and the parameters.

3. *Half vehicle model*

   Ignoring the roll motion of the vehicle body, the body itself is simplified to a 3 degrees of freedom model which includes a longitudinal movement along the x-axis, where the wheels rotate about their axis, as shown in Figure 3.5.

   According to Figure 3.5, the equations of the vehicle longitudinal motion and wheel's rotational motion are:

$$m\dot{v}_x = -\left(F_{xf} + F_{xr}\right) \tag{3.15}$$

$$I_1\dot{\omega}_1 = F_{xf}r_0 - T_{b1} \tag{3.16}$$

$$I_2\dot{\omega}_2 = F_{xr}r_0 - T_{b2} \tag{3.17}$$

where m is the vehicle mass, v_w is the vehicle longitudinal velocity, F_{xf} and F_{xr} are the longitudinal forces of the front and rear wheels respectively, I_1 and I_2 are the movements of inertia of the front and rear unsprung masses, ω_1 and ω_2 are the angular velocities of the front and rear wheels, and r_0 is the front and rear rolling radius.

3.3.3.2 Anti-lock Braking System Controller Design

1. *Principle of logic threshold control*[5]

   Take the wheel angular acceleration (or deceleration) as the main control threshold value and the wheel slip ratio as an auxiliary control threshold value. Using any of them alone has some limitations. For example, if the wheel angular acceleration (or deceleration) is taken as the control threshold alone, when the car strongly brakes at high speeds on a wet road, the wheel's deceleration may reach the control threshold, even if the wheel slip ratio is far away from the unstable region. In addition to the driving wheels, if the clutch is not disengaged during braking, the large inertia of the wheel system will cause the wheel slip ratio to reach the unstable region while the wheel's angular deceleration seldom reaches the control threshold value, which will affect the control effect seriously. If the wheel slip ratio is taken as the control threshold alone, it is difficult to ensure the best control effect in a variety of road conditions because the slip ratios corresponding to the peak adhesion coefficients vary widely (from 8% to 30%) in different road conditions. Thus, taking the wheel's angular acceleration (or deceleration) and the wheel slip ratio as the control thresholds together will help the road identification and improve the adaptive capacity of the system. Anti-lock logic enables the wheel slip ratio to fluctuate around the peak adhesion coefficient, to obtain larger wheel longitudinal and lateral forces, and guarantees a shorter braking distance with good stability. The whole control process is shown in Figure 3.6.

   In the initial stage of braking, the brake pressure rises as the driver depresses the brake pedal, and the wheel's angular deceleration gradually increases until it reaches a set threshold value $-a$. This is the first stage of the whole loop. In order to avoid the pressure being decreased while the slip ratio is still within the range corresponding to the wheel's stable region, the reference slip ratio should be compared with the threshold value S_{opt} (see Figure 3.2). If the reference wheel slip ratio is less than S_{opt}, the wheel slip ratio is still small, and the control process goes to the second stage (the pressure holding stage). The inertia effect on the vehicle makes the wheel slip ratio greater than S_{opt}, which indicates that the wheel has entered the unstable region. Thus comes the third stage of the control process (the pressure decrease stage). As the wheel pressure decreases, the wheel begins to accelerate due to the vehicle inertia until the deceleration is smaller than $-a$ (absolute values are compared). Then the fourth stage (the pressure holding stage) starts. Hereafter, because of inertia, the wheel is still accelerating until the acceleration reaches the threshold value $+A$. The pressure increases until the acceleration drops to $+A$, and the wheel is in the stable region again. The fifth stage begins. In order to keep the wheel in the stable region for longer, the shifting between the pressure increase and hold should be quick, and a low increasing rate of the pressure is maintained. This is the sixth stage. This stage lasts until the wheel's angular deceleration

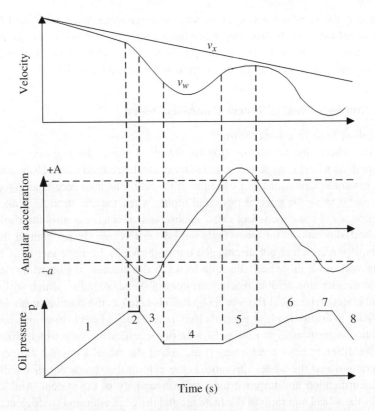

Figure 3.6   ABS control system's process on a high adhesion surface.

reaches the threshold value again. Then, the pressure decrease stage begins and the next pressure regulation loop starts.

2. *Logic threshold control flow*

   According to the logic threshold control theory described above, the control flow diagram is shown in Figure 3.7.

3. *Simulation and analysis*

   In order to verify the effect of the anti-lock braking system and improve the control algorithm, simulation or experimentation should be carried out. As an example, a simulation was done to a vehicle (parameters are shown in the appendix) in a MATLAB/Simulink environment to compare the braking performance with an uncontrolled braking system and with a logic threshold control anti-lock braking system. Assuming that the car moves on a road with a high adhesion coefficient with the initial speed of 20 m/s, and an expected slip ratio value of 0.16. The ODE45 algorithm was used and an automatic step size was chosen. The simulation results are shown in Figures 3.8–3.13.

   Figure 3.8 shows the comparison of the longitudinal acceleration in two cases. When the car is equipped with a logic threshold control anti-lock braking system, it can take full advantage of the road's adhesion coefficient, and have larger longitudinal acceleration and a much shorter braking distance (see Figure 3.9).

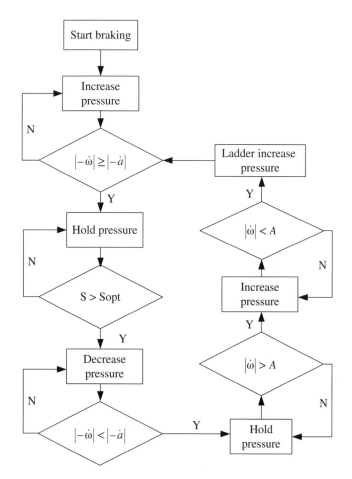

Figure 3.7   The ABS logic threshold control flow.

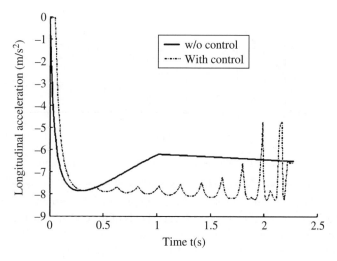

Figure 3.8   Comparison of the longitudinal acceleration.

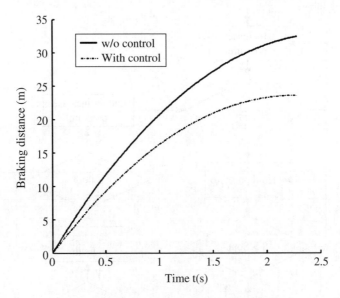

Figure 3.9   Comparison of the braking distance.

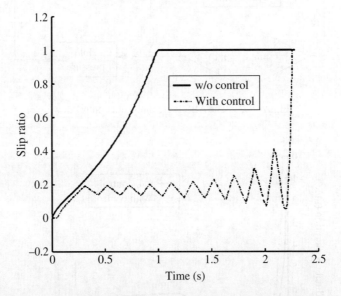

Figure 3.10   Comparison of the front wheels slip ratio.

Figure 3.10 and Figure 3.11 show the comparison of the slip ratios of the front and rear wheels in two cases. The front and rear wheels of the car with an uncontrolled brake system are almost locked in about 1 second, and then they start slipping; the wheels of the car with the control system would not be locked during braking. The slip ratio slightly changes around the optimum value.

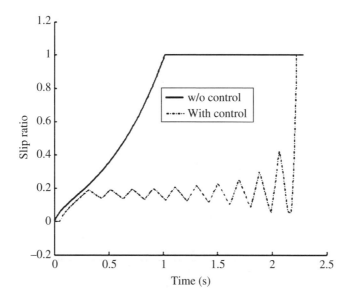

Figure 3.11   Comparison of the rear wheel slip ratio.

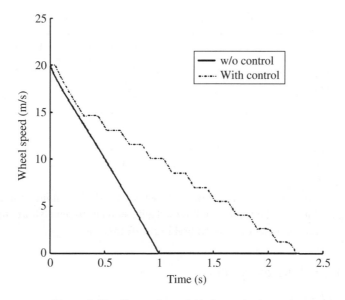

Figure 3.12   Comparison of the front wheel speed.

Figure 3.12 and Figure 3.13 show the comparison of the front and rear wheel speed. The results are same as those shown in Figure 3.10 and Figure 3.11.

Table 3.1 shows the comparison of the parts of the braking performance. The RMS value of the longitudinal acceleration increases from 6.6740 m/s^2 to 7.5570 m/s^2, and the front and rear wheel slip ratio's RMS reduces from 0.8101 and 0.8054 to 0.2237 and

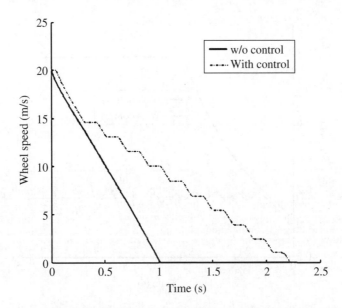

Figure 3.13   Comparison of the rear wheel speed.

Table 3.1   Partial simulation results (rms).

Performance	Logic threshold control	Uncontrolled
Longitudinal acceleration (m/s^2)	7.5570	6.6740
Front wheel slip ratio	0.2237	0.8101
Rear wheel slip ratio	0.2574	0.8054

0.2574, which means that the performance is increased or decreased by 11.68%, 72.39% and 68.04% respectively. In addition, the braking distance was reduced from 27.418 m to 23.554 m.

As can be seen, the rational design of the wheel anti-lock braking system controller makes it possible to make full use of the road adhesion coefficient, shortening the braking distance and guaranteeing a good braking performance.

3.4   Traction Control System

3.4.1   Introduction

When a vehicle accelerates on a low adhesion coefficient road or on a bisectional road, the driving wheels may slip resulting in a sideslip for rear-wheel driving cars and a difficulty in controlling the direction for front-wheel driving cars. The car's traction, handling stability, safety, and comfort will all influence this phenomenon.

The traction control system (TCS) is also known as the Acceleration Slip Regulation system (ASR). Its function is to: prevent the driving wheels from slipping excessively;

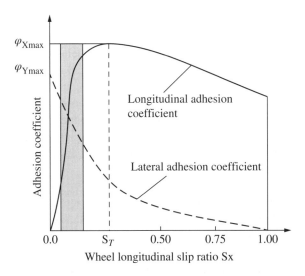

Figure 3.14   Slip ratio and adhesion coefficient curve.

keep the driving wheel slip ratio in the optimal range; guarantee direction stability; ensure the vehicle handling and dynamic performance; and to improve the vehicle driving safety.

When accelerating, the longitudinal adhesion coefficient increases along with the driving wheel slip ratio, where it reaches to a peak when the slip ratio increases to S_T. It begins to decrease when the slip ratio increases further from this point, as shown in Figure 3.14. Here, the relationship between the lateral adhesion coefficient and the longitudinal slip ratio is also indicated. The lateral adhesion coefficient decreases rapidly when the longitudinal slip ratio increases. So, the car can get a greater longitudinal and lateral adhesion coefficient if the slip ratio is between 0 and S_T. When the slip ratio is greater than S_T, the longitudinal and lateral adhesion coefficient decreases rapidly, which will cause the loss of steering ability or even sideslip. Therefore, in order to guarantee good traction, handling, safety, and comfort in a starting or accelerating vehicle, the driving wheel slip ratio should be controlled by S_T.

3.4.2   Control Techniques of TCS[6]

Since the slip ratio of the wheel cannot be directly regulated, the most common technique is to adjust the torque on the wheel in order to control the motion of the wheel. The torque adjustment of the wheel can be realized by manipulating the engine output torque, transmission output torque, and braking torque on the wheel.

1. *Adjusting the engine output torque*
   For internal combustion engine cars, the torque transmitted to the driving wheels can be varied by adjusting the engine output torque, and then control of the driving wheel slip ratio can be achieved. There are three main ways to regulate the engine output torque.

One is to control the angle of the throttle to change the engine output torque. This method has the advantage of being smooth and continuous, but it has to combine with other control methods to overcome the shortcoming of being slow to respond. The second is to adjust the ignition parameter. The engine output torque can be controlled by reducing the ignition timing or even temporarily stopping the ignition. This method has the advantage of having a rapid response, but it may cause an incomplete combustion and worsen the emissions. Another option is to alter the fuel supply. The engine output torque can be reduced by reducing or suspending the fuel supply; this is a relatively easy approach to implement but it may cause an abnormally working engine and influence its life and emissions.

2. *Differential lock control*

A feature of the common symmetric differential is that it can differently distribute the rotational speeds but equally distribute the torques. So, when a car is driving on a bisectional road, the driving force on a high adhesion coefficient road can't be fully used if the driving wheel slip ratio on the low adhesion coefficient road is too large. The traction performance of the car may be affected. The differential locking method can control the locking valve to lock the differential moderately according to the left and right driving wheel slip conditions and the road condition to keep the slip ratios of the driving wheels in a reasonable range. The driving forces of the wheels on a high adhesion coefficient road can be fully used, but the cost of this method is relatively high.

3. *Clutch or transmission control*

When the driving wheel slip ratios are too large, the torque transmitted to the drive wheels can be reduced by controlling the engagement of the clutch, thus preventing excessive slip of the wheels. In addition, reducing the torque transmitted to the drive wheels can also be realized by reducing the transmission ratio. This control method should not be used alone because of the shortcomings of the slow response and the unsmooth changes of the torques.

4. *Braking torque adjustment*

For a large slip ratio driving wheel, the slip ratio can be kept in the optimum range by controlling the braking torque. This is a commonly-chosen technique, which is normally used in conjunction with the engine torque adjustment method. The engine output torque must be regulated when the brake is applied to avoid excessive consumption of engine power. This control method has quite a good effect when the speed is not too high (less than 30 km/h), and the two driving wheels are on the bisectional road. However, it cannot be used for a long time and should not be used when the vehicle speed is high.

Figure 3.15 shows the simplified schematic of a TCS with integrated control methods of the engine throttle opening control and the wheel braking torque control. The system consists of sensors, hydraulic components, ECU, brake adjusters, and other components. When the car starts, the signals collected by the wheel speed sensors are sent to the ECU. After these signals are processed, the ECU gives commands to control the brake adjuster and the throttle angle according to the designed control strategy. The slip ratios are then regulated.

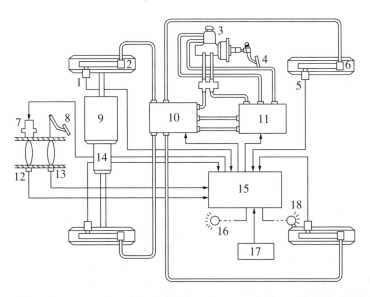

Figure 3.15   Structure of a TCS. 1. Front wheel speed sensor. 2. Front wheel brake. 3. Hydraulic components. 4. Brake pedal. 5. Rear wheel speed sensor. 6. Rear wheel brakes. 7. Sub-throttle actuator. 8. Throttle pedal. 9. Transmission. 10. ABS brake adjuster. 11. TCS brake adjuster. 12. Sub-throttle position sensor. 13. Main throttle position sensor. 14. Engine. 15. ECU. 16. Warning light. 17. Cut-off switch. 18. Working light.

3.4.3   TCS Control Strategy

Just as with the ABS, different methodologies have been used for the controlling of the TCS. At present, the main control strategies are: logic threshold control; PID control; optimal control; sliding mode variable structure control; fuzzy control; and neural network control[7].

1. *Logic threshold control*

   Logic threshold control is now the most widely-used method in vehicle control. The main idea is to first set a target slip ratio value, and then compare the driving wheel slip ratio with the reference one. If the driving wheel slip ratio exceeds the threshold or the target value, the control system sends the output regulating instructions to keep the driving wheel slip ratio or wheel acceleration close to the target value. When the driving wheel slips excessively once again, the control system begins to work until the slip ratio comes down to the target value. This is a conventional control method. It has the benefits of a simple structure, the small amount of computational effort that is needed, and it has a short time lag. However, it has the shortcomings of having fluctuations, poorer stability, and lacking adaptive capability.

2. *PID control*

   The PID control method can control the slip ratio by regulating the engine output torque or the braking torque. The system's input is the difference between the target slip ratio

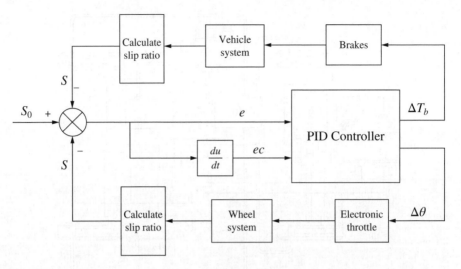

Figure 3.16 Block diagram of the PID controller of TCS. S_0 Target slip ratio; S Actual slip rate; e Error; ec Error rate of change; $\Delta\theta$ Throttle opening degree increments; ΔT_b Braking torque increment.

and the actual slip ratio. To keep the actual slip ratio close to the target slip ratio, this method feedbacks the throttle angle, which is calculated according to the input, to the engine to adjust the output torque of the engine, or it controls the braking system to change the braking torque. The conventional PID controller is simple and easy to implement. However, the control effect is not ideal in solving complex and nonlinear problems such as traction control, because of the difficulty in the mathematical modeling and the uncertainty of the vehicle parameters and environmental factors. Therefore, in order to get a better control effect, a conventional PID algorithm is usually used together with other control methods in the traction control problem. The block diagram of a PID controller is shown in Figure 3.16[8].

3. *Fuzzy logic control*
   It has been mentioned before that a car is a complex dynamic system and its working environment is complicated, so the traction control system is a complex nonlinear system. It is difficult to get the desired control results if conventional control methods are used. Fuzzy logic control does not require the precise mathematical model of the system. It is robust enough to accept changes in the system parameters and it allows time-varying, nonlinear, and complex systems. Therefore, it is suitable for the traction control system. In addition, fuzzy logic control theory has a better effect in solving the control problem of complex systems if used in conjunction with other control methods. The fuzzy PID control method has been currently applied in traction control systems. When using the PID method, the parameters K_p, K_i, and K_d need to be adjusted online constantly, and the adjustment of these parameters is a difficult task. However, after a reasonable fuzzy control rule table is designed, the fuzzy controller is a feasible and practical method for adjusting the three parameters of a PID controller.

4. *Sliding mode variable structure control*

The characteristics of sliding mode variable structure control determine that the change of parameters and the external environment disturbance will not affect the control results when the controlled system is in the sliding mode motion. This method is therefore very robust, and it is suitable for use on the traction control system.

To design a sliding mode controller, the driving wheel slip ratio should be taken as the control target, and the target ratio S_0 as the input threshold; the braking torque, the engine output torque, or the throttle angle should be taken as the control parameters. The controller regulates the control parameters to adjust the driving torque and thus leads to the change of the drive wheel rotational speed. The actual slip ratio can be kept close to the target slip ratio.

3.4.4   Traction Control System Modeling and Simulation

1. *Dynamic model*

The dynamic model of a traction control system includes the sub-models such as the engine model, driveline model, tyre model, brake system model, and vehicle model. The tyre model, brake system model, and vehicle model can be established in the same way as in the ABS control system.

The engine model expresses the relationship between the output torque and the rotation speed. The modeling approach is to use a high-order polynomial to fit the mathematical relationship between the output torque and the speed on the basis of the experimental values, or the engine load characteristic curves corresponding to different throttle angles. A third order polynomial is commonly used, as follows:

$$T_e = a_0 + a_1 n_e + a_2 n_e^2 + a_3 n_e^3 \tag{3.18}$$

where T_e is the engine torque (Nm), a_0, a_1, a_2, a_3 are the fitting coefficients, and n_e is the engine speed (r/min).

The throttle angle changes rapidly when the car is starting or changing speed, so it takes time for the engine to reach steady-state from a transient state. Therefore, the engine lags in response to the step inputs of the throttle angles. To describe the dynamics of its output characteristics, the engine is generally treated as a first order model with pure delay[8], and the transfer function can be expressed as:

$$T_{eq} = \frac{T_e e^{-sT_1}}{1 + sT_2} \tag{3.19}$$

where T_e is the steady-state torque, T_{eq} is the dynamic torque, T_1 is the system lag time constant, T_2 is the system time constant, and s is the Laplace variable.

2. *Simulation and analysis*

The literature[8] analyzes a bus TCS system with different control strategies. Figure 3.17 shows the slip ratio simulation results from the application of fuzzy PID control, and

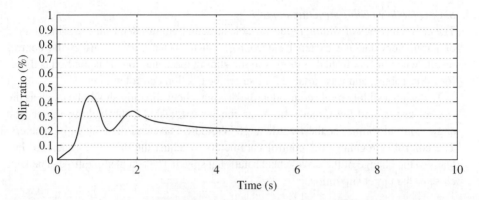

Figure 3.17   Slip ratio curve.

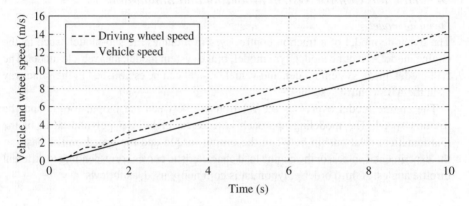

Figure 3.18   Vehicle speed and wheel speed curves.

Figure 3.18 shows the wheel speed and the vehicle speed curve varying with time. As can be seen, the slip ratio stays stable near the desired value rapidly for the bus with TCS, which gives it better acceleration.

3.5   Vehicle Stability Control

With the extensive application of ABS and TCS in the 1980s, it was noticed that proper modification of their control algorithms could give the systems the ability to actively maintain the vehicle body stability. Then the idea of vehicle stability control (Vehicle Stability Control, VSC, also known as the electronic stability program, ESP) was put forward. By adding active yaw control, longitudinal and lateral dynamics control could be achieved over the vehicle and accidents avoided when steering. In 1992, based on the ABS and TCS system, McLellan et al.[9] proposed the idea of the motion control of a vehicle in GM. The differences between the car's motion parameters and the control targets were used as the feedbacks to control the wheel slip ratio, and thus the motion of the car could be controlled[9].

In the mid-1990s, Bosch launched the first VSC product, which marked the technology of the yaw stability control and its advances towards maturity[10]. Based on the ABS and TCS systems, the Bosch VSC system required the installation of a steering wheel angle sensor, brake master cylinder pressure sensor and throttle angle sensor in order to obtain the driver's intention, and then the expected tracking curve could be calculated using a pre-established driver model. The information of the yaw rate, lateral, and longitudinal acceleration could be obtained through the gyro in the VSC system. Based on all this information, if the VSC system judged that the car's current trajectory deviated from the driver's expectations, it initiated the direct yaw control (DYC) to maintain the vehicle stability and to avoid accidents.

Toyota and BMW also introduced their own VSC products. The lateral acceleration sensors and yaw rate sensors were used to measure the vehicle position directly in their systems, and application of the stability control system has been extended greatly.

After the statistical analysis of a large number of traffic accident data, the National Highway Traffic Safety Administration of the United States reached the conclusion that vehicle stability control systems are the most effective active safety control technology for vehicles at present. The collision accident ratio of passenger cars installed with VSC systems have declined by 34%, and rollover accidents dropped 71%[11].

3.5.1 Basic Principle of VSC

One of the basic principles of any VSC system is to identify the driver's intention and represent the expected motion of the car by sensors and arithmetic logic. The actual motion of the car is then measured and estimated. If the difference between the actual and the desired motion is greater than a given threshold value, the VSC begins to regulate the longitudinal forces on the wheels according to the set control logic. The yaw acting on the car is changed, which may lead to the car's actual motion becoming closer to the expected motion. A cornering car is used as the example to explain the VSC control principle, shown in Figure 3.19. When the vehicle has a tendency to oversteer, the VSC imposes the brakes on the external

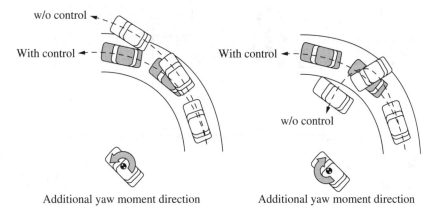

Figure 3.19 VSC system control principle.

front wheel; this generates a yaw moment in the opposite direction to the vehicle as it corners which will take the car back to the desired path.

When the car tends to understeer there are two ways in which the VSC can interfere with its motion. One is to apply the brakes on the inner rear wheel to generate a yaw moment in the same direction as the vehicle cornering to increase the vehicle yaw motion. The other is to reduce the engine output torque to change the lateral forces of the front and rear axles, which will produce a yaw moment in the same direction as the vehicle cornering.

3.5.2   Structure of a VSC System

As shown in Figure 3.20, the VSC is generally comprised of three parts: sensors, an electronic control unit (ECU) and a hydraulic control unit (HCU)[12]. The sensors are steering angle sensors, wheel speed sensors, gyroscopes, throttle opening sensors and master cylinder and wheel cylinder pressure sensors. The ECU generally uses embedded systems such as ARM or DSP, and also includes the sensor signal conditioning circuit, the solenoid valve control circuit, the pump motor control circuit, CAN bus communication circuit, detection and protection circuit, and other parts. The HCU is mainly composed of solenoid valves, a pump motor, and a special hydraulic circuit module. Figure 3.21 shows the internal VSC structure when the controller and the actuator are separated.

The sensors used on a VSC system are of two types. One type is used to perceive the driver's actions, such as the steering wheel angle sensor, master cylinder pressure sensor, and throttle angle sensor, which can measure the action and the extent of steering, braking,

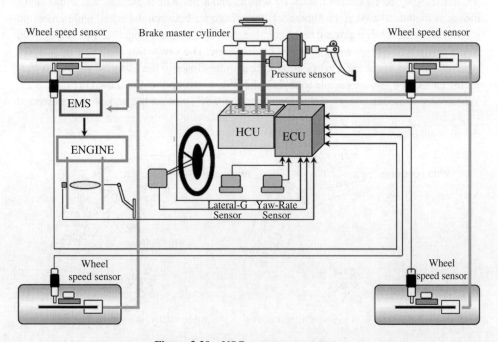

Figure 3.20   VSC system components.

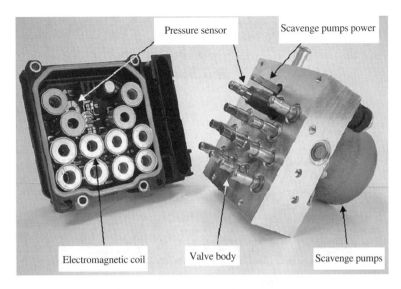

Figure 3.21   Internal structure of a VSC system.

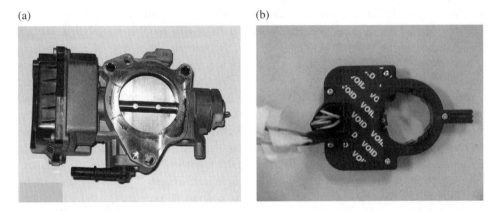

Figure 3.22   Throttle angle sensor and steering angle sensor. (a) The throttle angle sensor, (b) The steering wheel angle sensor.

and accelerating that the driver applies to the car. The throttle angle sensor and steering angle sensor are shown in Figure 3.22. The other type is used to gather information about the vehicle running state, such as the wheel speed sensors, the wheel cylinder pressure sensors, and the gyroscope. The gyroscope shown in Figure 3.23 can measure the tri-axial angular velocities and accelerations.

An ECU generally has two microprocessors, one for dealing with the control logic, and the other for fault diagnosis and treatment. The two microprocessors exchange information with each other via an internal bus. In addition to the microprocessors, the ECU also includes a power management module, the sensor signals input module, a hydraulic

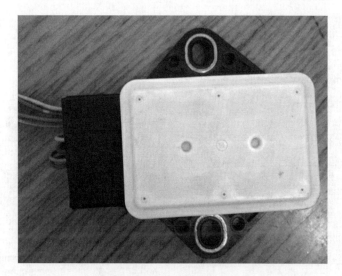

Figure 3.23   Vehicle gyroscope.

regulator driving module, a variety of indicators interface, and a CAN bus communication interface. Most ECUs are mounted together with the hydraulic regulator through the electromagnetic coupling between the solenoid coils and the spool of the solenoid valve.

According to the information obtained from the steering wheel angle sensor and the master cylinder pressure sensor, the ECU determines the driving intention and calculates the ideal motion state (such as the desired yaw rate, etc.). After comparing the measured actual motion with the ideal car motion, the amount of yaw moment that should be applied to keep the car stable can be determined by the control logic. Finally, the ECU adjusts the brake cylinders of the braking system through the hydraulic regulator to generate the needed yaw moment. If necessary, the ECU communicates with the engine management system (EMS) to change the driving force of the driving wheels to change the car's motion. The sensors measure the updated car motion parameters and send the information to the ECU to conduct the next control cycle in order to continue to keep the car stable.

Figure 3.24 shows the working principle of the VSC–hydraulic actuator unit. The VSC control logic requires that each wheel can be braked at any necessary time, so the hydraulic brake pipeline in the car equipped with VSC must guarantee that the brakes can be applied to every single wheel without affecting other wheels. Normally, the hydraulic pipeline of the VSC is constructed by adding four solenoid valves on that of the ABS. Two normally open valves between the main pipeline and the master cylinder will be energized to close the main oil pipeline immediately the HCU starts working, and then the driver's braking operation will not function. A further two valves, which are normally closed, are located in two additional pipelines connecting the master cylinder and the oil return pump, and these lines are normally closed. Only when the driver steps on the brake pedal will the brake fluid flow into the wheel cylinders. If the driver applies the brakes when the VSC works, the normally closed valve will be electrified to an opened state to keep the brake fluid flowing into the low pressure accumulator.

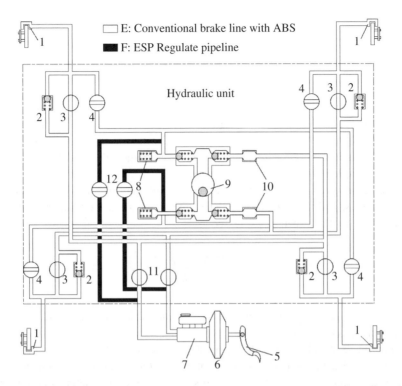

Figure 3.24 The hydraulic brake pipeline layout in a vehicle with VSC. 1 Brake caliper. 2 Brake valve. 3 Entry for solenoid valve. 4 Exhaust gas solenoid valve. 5 Brake pedal. 6 Brake booster. 7 Brake master cylinder. 8 Accumulator. 9 Jet pump. 10 Buffer. 11 Transform solenoid valve. 12 Master solenoid valve.

The active increase of pressure of the VSC is carried out by the oil return pump. When the brakes need to be applied on one wheel, the two normally open solenoid valves on the main pipeline will be closed, and the normally open valves on the pipeline with the other wheel cylinders will be closed too; the oil from the return pump can only flow to the wheel cylinder which needs higher pressure. Thus, the pressure in one single wheel cylinder should be increased.

3.5.3 Control Methods to Improve Vehicle Stability

When a vehicle is moving with the lateral acceleration of less than 0.4g (linear region), the driver can effectively control the vehicle without entering the unstable region. When the motion of the car is in the unstable region, the vehicle stability control system helps to decrease the sideslip angle rapidly, and to return the car back into the stable region quickly[13].

There are many ways to improve vehicle stability, amongst which active steering control, front and rear axle roll stiffness distribution control, and direct yaw moment control are those most commonly used.

1. *Active steering control*

   There are many possible methods of applying active steering control, such as 4-wheel steering (4WS), Steer by Wire (SBW), and active front wheel steering (AFS). By introducing motion parameters such as yaw rate or lateral acceleration as the feedback, the active steering control method can improve the handling and stability of the car in the linear region and, to some extent, prevent the sideslip angle from appearing. However, adjusting the car's attitude by the steering would be very difficult if the lateral forces of the tyres tend to reach saturation. Therefore, the effect of this method is not obvious when steering in the nonlinear region and with very large sideslip angles.

2. *Cornering stiffness distribution control*

   Vehicle stability has a considerable relationship with the cornering stiffness of the front and rear tyres, and this relationship determines the car's steering characteristics. The vehicle stability factor can be expressed as:

   $$K = \frac{m\left(l_r k_r - l_f k_f\right)}{l k_f k_r} \tag{3.20}$$

   where k_f is the front tyre cornering stiffness, k_r is the rear tyre cornering stiffness, and m is the vehicle mass.

   Equation (3.20) shows that the sign of K is determined by $(l_r k_r - l_f k_f)$, which is closely related to the cornering stiffness of the front and rear tyres. Practically, the tyre cornering stiffness changes with the working load. By introducing the motion state parameters as the feedback to a vehicle with an active suspension system, the system can adjust the axle load distribution which will change the cornering stiffness of the front and rear tyres. However, this control method has serious limitations. First, the vehicle must have an active suspension system. Second, this method is only effective when the lateral acceleration is large. And finally, the vehicle longitudinal acceleration generated by the longitudinal force has a great impact on the axle load transfer.

3. *Direct yaw moment control*

   Direct yaw moment control, also known as differential braking control, is an active control method which changes the vehicle state by applying different braking forces on different wheels. This is a very effective way of changing the yaw moment and adjusting the motion attitude of the car. This method functions when the car is braking, driving, steering, and even in a combination of conditions; it also works effectively when the sideslip angle is large. So differential braking is the most suitable method for vehicle stability control, especially when the tyre's adhesion reaches the limit[14].

3.5.4   Selection of the Control Variables

The VSC system needs to solve two main problems, track keeping and vehicle stability, which can be described by the sideslip angle and the yaw rate respectively. The sideslip angle and yaw rate reflect the essential characteristics of the vehicle steering, and the relationship between them determines the stable state of the car and characterizes the stability from different sides. The sideslip angle represents the lateral velocity and track deviation of the car when steering, which emphasizes the description of the track keeping problem. The yaw rate is the reflex of the changing speed of the heading angle when steering, which determines the steering characteristics of the car: understeering or oversteering. The sideslip angle can be used to describe stability.

1. *Relationship between yaw rate and vehicle stability*
   The two degrees of freedom linear vehicle model shows that the motion of a car is mainly described by the longitudinal velocity, lateral velocity, and yaw rate. Longitudinal velocity and lateral velocity determine the vehicle sideslip angle, and the integral of the yaw rate is the yaw angle. Take θ as the car heading angle, it is equal to the sum of the sideslip angle and yaw angle[1].

$$\theta = \beta + \psi = \beta + \int r dt \qquad (3.21)$$

   where ψ is the vehicle yaw angle.
   When the sideslip angle is small, the car's heading angle is determined by the yaw angle or yaw rate mainly. The larger the heading angle, the smaller the turning radius will be, and vice versa. Understeering or oversteering, which are the two key steering characteristics of a car, can be estimated from the turning radius. Therefore, when the sideslip angle is small, the yaw rate can represent the car's state of stability. Also, because the yaw rate signal can be directly obtained from the sensors on the car, the yaw rate is an important control variable in most vehicle stability control strategies.

2. *The relationship between sideslip angle and vehicle stability*
   The analysis above is based on the assumption that the sideslip angle is small, but the sideslip angle is generally large when severe sideslip occurs. In this case, the vehicle stability cannot be accurately described by using the yaw rate, but the sideslip angle can be used instead. Therefore, the sideslip angle is normally limited to relatively small values in vehicle stability control systems.
   Some scholars have analyzed the relationship between the sideslip angle and vehicle stability. The β method is the most representative idea that has been put forward[15]. The influence of the sideslip angle on vehicle stability according to this method, is presented here.
   The β method explains the relationship between the sideslip angle and stability by analyzing the impact of the sideslip angle on the yaw moment and the lateral force. The relationship of the yaw moment with the lateral force and the slip angle can be obtained from the vehicle dynamics model. Taking the front tyres as the example, the lateral force curves when they move on the road with the friction coefficient of 0.8 and 0.2 are shown in Figure 3.25.
   As can be seen from Figure 3.25, on a low adhesion road the lateral force gets saturated even when the sideslip angle is small, and the lateral force provided is much smaller than that on the high adhesion road. The simulation results of the impact of the sideslip angle on the yaw moment and the lateral force are shown in Figures 3.26 and 3.27.

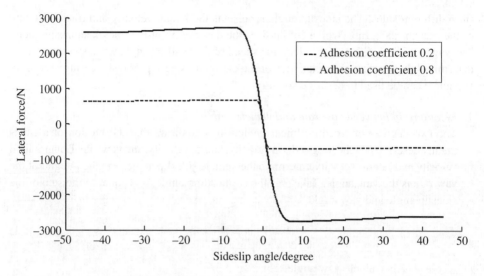

Figure 3.25   Front wheel lateral force curve.

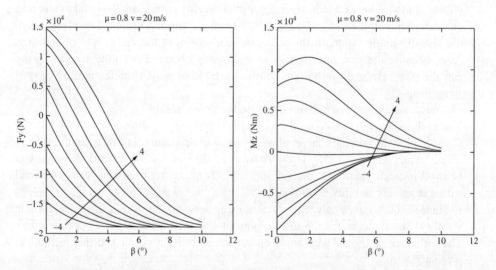

Figure 3.26   The impact of the sideslip angle with an adhesion coefficient of 0.8, at 20 m/s.

The β method shows that the vehicle tends to be stable when the yaw moment is positive or increases with the vehicle sideslip angle; otherwise, the car may lose stability. It can be seen from Figure 3.26 that a positive angle produces a positive yaw moment and lateral force, and a negative angle produces a negative yaw moment and lateral force when the sideslip angle is zero. However, the increment of the yaw moment and the lateral force decreases with the increase of the front wheel turning angle, which indicates that the tyre forces tend to enter the saturated zone as the

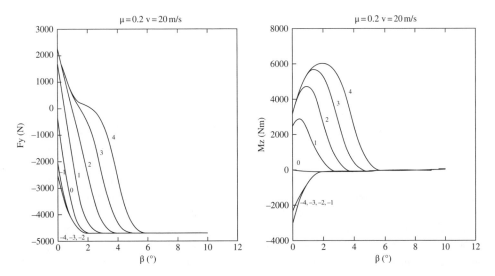

Figure 3.27 The impact of the sideslip angle with an adhesion coefficient of 0.2, at 20 m/s.

sideslip angle becomes larger. The vehicle yaw moment increases at first and then decreases along with the increase of the sideslip angle, and it gets close to zero when the sideslip angle is relatively large. The lateral force decreases with the increase of the sideslip angle and reaches a peak value when the sideslip angle is considerably large, which means that even if the driver turns the steering wheel, the yaw moment cannot be produced. That is why the vehicle is difficult to manipulate under large sideslip angle conditions.

Figure 3.27 shows the relationship of the yaw moment and the lateral force when the adhesion coefficient is low. It can be seen that as the adhesion coefficient between the tyre and the road's surface decreases, the yaw moment reaches zero more rapidly with the increase of the sideslip angle, and the lateral force gets saturated more quickly too. Therefore, the lower the adhesion coefficient, the more difficult it is for the driver to attain the yaw moment by turning the steering wheel. The lower the adhesion coefficient, the greater the impact of the sideslip angle on the vehicle stability, and the maximum sideslip angle needed to keep the vehicle stable is smaller. Therefore, when the vehicle moves on a low adhesion road, the sideslip angle should be strictly limited in order to maintain stability.

The conclusion is that dangerous situations cannot be avoided completely by only controlling the yaw rate, especially when the vehicle is running on a low adhesion road. Thus, in addition to considering the rolling of the vehicle body, the degree of deviation from the current track should also be considered so, the sideslip angle must be controlled. As the sideslip angle cannot be measured directly by the sensors on the vehicle, the estimation method has to be used, although the accuracy may be affected to some extent. In practical VSC systems, the range of the sideslip angle is generally set to avoid an unexpected intervention caused by inaccurate estimations.

3.5.5   Control System Structure

In order to meet the needs of vehicle stability control, VSC systems generally adopt a hierarchical control structure[1,14], which is shown in Figure 3.28. The control system gets the inputs from the sensors installed on the vehicle, and the motion parameter estimator. This is due to the motion parameters (such as the longitudinal vehicle speed, the sideslip angle, and the road adhesion coefficient) which are difficult to measure by sensors directly; therefore, those parameters need to be estimated based on the measured signal. Taking the deviation between the desired motion and the actual motion as the input, the upper motion controller is responsible for calculating the nominal yaw moment M_{YawNo} needed to eliminate the motion deviation. The lower actuator controller distributes the braking force on each wheel according to the nominal yaw moment and calculates the nominal braking pressure P_{WhlNo} of each wheel. Thus, the brake pressure is controlled by the lower controller. The required forces can then be applied to the vehicle through the interaction between the tyre and the road. There are inner and outer feedback control loops in the hierarchical control system. The outer feedback control loop provides the nominal yaw moment for the inner feedback loop, and the inner feedback control loop adjusts the tyre forces by the nominal braking pressure.

3.5.6   The Dynamics Models

Theoretical research on VSC usually needs the establishment of models such as the non-linear vehicle model, the linear half vehicle model, and the hydraulic system model. The nonlinear vehicle model is the controlled objective, and it must contain a nonlinear tyre

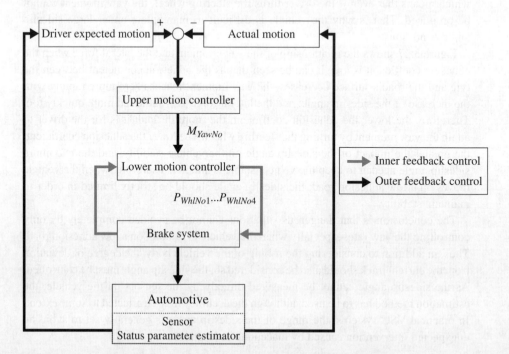

Figure 3.28   The hierarchical control system in VSC systems.

model to study the impact of the lateral force on the body motion. The linear half vehicle model provides the desired values of the motion parameters for reference. Since VSC controls the body motion through the hydraulic brake system, the complete hydraulic brake system model and the wheel braking model are essential.

1. *The desired value calculation model*

   The two degrees of freedom linear model is widely used as the desired value calculation model in VSC systems because it is simple and contains most of the important parameters describing the lateral motion like the body mass, front and rear tyres cornering stiffness, and wheelbase (Figure 3.29).

   The dynamics equations are

$$\left(k_f + k_r\right)\beta + \frac{1}{v_x}\left(l_f k_f - l_r k_r\right)r - k_f \delta_f = m\left(\dot{v}_y + v_x r\right) \tag{3.22}$$

$$\left(l_f k_f - l_r k_r\right)\beta + \frac{1}{v_x}\left(l_f^2 k_f + l_r^2 k_r\right)r - l_f k_f \delta_f = I_z \dot{r} \tag{3.23}$$

where δ_f is the front wheel turning angle, I_z is the inertia moment of the vehicle about the z-axis, l_f, l_r are the distances between the C.G. and the front and rear axle, v_x, v_y are the longitudinal and lateral velocities of the C.G., r is the yaw rate, and k_f, k_r are the front and rear tyres' cornering stiffness.

The desired yaw rate β_d and the sideslip angle r_d can be calculated by (3.24) and (3.25)[16]:

$$\beta_d = \frac{\left[2l_r\left(l_f + l_r\right)k_f k_r - mv_x^2 l_f k_f\right]}{2l_r\left(l_f + l_r\right)^2 k_f k_r - mv_x^2\left(l_f k_f - l_r k_r\right)}\delta_f \tag{3.24}$$

$$r_d = \frac{2\left(l_f + l_r\right)k_f k_r v_x}{2l_r\left(l_f + l_r\right)^2 k_f k_r - mv_x^2\left(l_f k_f - l_r k_r\right)}\delta_f \tag{3.25}$$

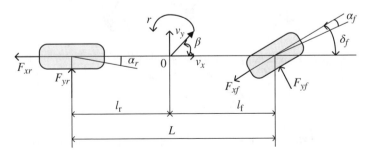

Figure 3.29 Two degrees of freedom linear vehicle model.

2. *The actual motion model*

There are many methods of establishing the dynamics model of a vehicle, amongst which modeling by formulating the equations and by using software are commonly used. As an example, a seven degree of freedom car model built by formulating the equations is presented. A car can be simplified as a seven degree of freedom model which includes longitudinal, lateral, and yaw motion of the body and rotational movements of the four wheels, as shown in Figure 3.30.

The dynamics equations of the body are:

$$m\left(\dot{v}_x - v_y r\right) = F_{xfl}' + F_{xfr}' + F_{xrl} + F_{xrr} \tag{3.26}$$

$$m\left(\dot{v}_y + v_x r\right) = F_{yfl}' + F_{yfr}' + F_{yrl} + F_{yrr} \tag{3.27}$$

$$I_z \dot{r} = l_f\left(F_{yfl}' + F_{yfr}'\right) - l_r\left(F_{yrl} + F_{yrr}\right) + \frac{D_f}{2}\left(F_{xfl}' - F_{xfr}'\right) + \frac{D_r}{2}\left(F_{xrl} - F_{xrr}\right) \tag{3.28}$$

where

$$F_{xfl}' = F_{xfl}\cos\delta_f - F_{yfl}\sin\delta_f$$

$$F_{yfl}' = F_{xfl}\sin\delta_f + F_{yfl}\cos\delta_f$$

$$F_{xfr}' = F_{xfr}\cos\delta_f - F_{yfr}\sin\delta_f$$

$$F_{yfr}' = F_{xfr}\sin\delta_f + F_{yfr}\cos\delta_f$$

and where v_x is the longitudinal velocity, v_y is the lateral velocity, r is the yaw rate, δ_{fi} s the front angle, F_{xi}, F_{yi} ($i = fl, fr, rl, rr$) are the longitudinal forces and lateral forces of the wheels, m is the car mass, I_z is the inertia moment of the car about the z-axis, l_f, l_r are the distance between the C.G. and front and rear axles, and D_f, D_r are the front and rear tracks.

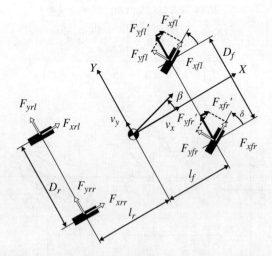

Figure 3.30 The seven degrees of freedom nonlinear dynamic model.

3.5.7 Setting of the Target Values for the Control Variables

1. *The ideal yaw rate*

   The ideal yaw rate is proportional to the steering wheel angle under certain vehicle speeds when the sideslip angle at the center of mass is small. The yaw rate gain relative to the steering wheel angle should be reduced appropriately when the vehicle speed increases to avoid the steering wheel being too sensitive at high speeds. Figure 3.31 shows the relationship between the ideal yaw rate and vehicle speed[17]. Each solid curve is with the same front wheel angle, and each dotted curve is the adhesion limitation of equal road adhesion coefficient. The ideal yaw rate is limited by the maximum force the road can provide.

   It is reasonable to use the two degrees of freedom linear model to calculate the expected motion parameters, but the influence of the road adhesion coefficient on them is not embodied in the model. In different road conditions, the lateral forces must satisfy the following constraint:

$$\left| a_y \right| \le \mu g \tag{3.29}$$

When the sideslip angle is very small, thus:

$$a_y \approx r v_x \tag{3.30}$$

So, the ideal yaw rate should also meet the following condition:

$$\left| r_d \right| \le \left| \frac{\mu g}{v_x} \right| \tag{3.31}$$

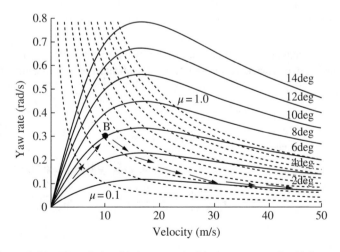

Figure 3.31 The relationship between the ideal yaw rate and vehicle speed.

2. *The stability region of the sideslip angle*

It can be seen that the sideslip angle threshold is closely tied to the adhesion coefficient through the above analysis by the β method. Because it is quite difficult to control the sideslip angle precisely, constructing a stable region of sideslip angle by using the sideslip angle and the yaw rate is a common practice. The boundary of the region can be described by the following equation:

$$\left| E_1 \dot{\beta} + \beta \right| \le \mu E_2 \tag{3.32}$$

where E_1 and E_2 are the constants of the stable boundary.

Equation (3.32) can be used as a control criterion of the sideslip angle, and it is also the basis for the yaw moment control algorithm design. The vehicle state can be considered to be stable when the inequality holds, while the vehicle will tend to oversteer when the inequality does not hold, which shows the necessity for the stability intervention.

3.5.8   Calculation of the Nominal Yaw Moment and Control

The upper and lower controllers are the core parts in the layered control system of a VSC (see Figure 3.28).

1. *Design of the upper controller*

When a vehicle is moving with an ideal yaw rate and the sideslip angle is within the stable range, the vehicle can be regarded as being in an ideal state. When the vehicle begins to lose stability, its motion state will be very different from the ideal one. At this time, the additional yaw moment ΔM is applied on the vehicle to stabilize the vehicle and make the actual vehicle state closer to the ideal. The upper controller calculates the value of the additional yaw moment ΔM, which consists of ΔM_r, produced by the yaw rate controller, and ΔM_β, produced by the sideslip controller, as shown in Figure 3.32.

Some control algorithms can be used in the yaw rate controller, such as PID control, feed forward and feedback combined control, optimal control, sliding-mode control,

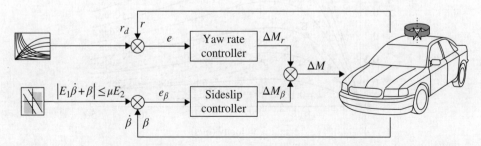

Figure 3.32   Logical diagram of the upper controller.

and fuzzy logic control. Because the sideslip angle of the vehicle is difficult to control, the simple but practical PD controller is normally used to stabilize the vehicle when it tends to be unstable.

$$\Delta M_\beta = k_{\beta p}\beta + k_{\beta d}\dot{\beta} \tag{3.33}$$

Then additional yaw moment can be obtained by the following equation:

$$\Delta M = \Delta M_r + \Delta M_\beta \tag{3.34}$$

2. *Design of the lower controller*

The function of the lower controller is to convert the additional yaw moment to the actual values of the variables, such as wheel cylinder pressure or the braking torque, so the controller can apply the brakes on a single or multiple wheels.

When braking on different wheels, the effectiveness of the additional yaw moments generated by either decreasing the lateral forces or by applying the brakes is different. The yaw moments produced by braking the front wheels is very different from the moments produced by braking the rear wheels. When steering, the directions of the additional yaw moments caused by the braking force of the outer front wheel and by the lateral force are the same, which is opposite to the turning direction. So this is the most effective way to correct oversteering, while braking the inner rear wheel is most suitable to correct understeering (Figure 3.33).

Since the key task of the VSC system is to help the driver to fulfill his turning purpose, the order made by the control strategy should not violate the driver's steering intention. So, the yaw moment needed must be calculated at any time. The steering

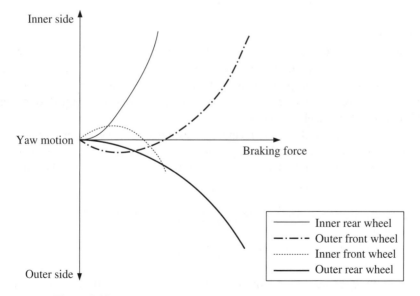

Figure 3.33 Yaw moment caused by braking on different wheels.

Table 3.2   The choice of the braked wheel.

$r - r_d$	δ_f	$\dot{\delta}_f$	Vehicle state	Braked wheel
+	+	+	oversteer	Front right
+	+	0	oversteer	Front right
+	+	−	oversteer	Front right
+	−	+		Without braking (no)
+	+	0	understeer	Rear right
+	−	0	understeer	Rear right
+	0	+	oversteer	Front right
−	+	+	understeer	Rear left
−	+	0	understeer	Rear left
−	+	−		Without braking (no)
−	−	+	oversteer	Front left
−	−	0	oversteer	Front left
−	−	−	oversteer	Front left
−	0	+	understeer	Rear left
−	0	−	oversteer	Front left

angle δ_f, the steering wheel velocity $\dot{\delta}_f$, and the difference between the actual yaw rate
and the ideal yaw rate ($e = r - r_d$) are used to judge the vehicle motion state. The
decision of which wheel should be braked can then be made. Table 3.2 shows the detail
strategy for a single wheel braking method. The positive sign of parameters is pointed
to the left.

There are two main methods that the lower controller can employ to implement the
control, the slip ratio control and the direct braking torque control[18,19].

3. *Slip ratio control*
   To establish the slip ratio control, two steps should be followed. First, the relationship
   of the yaw moment and the slip ratio needs to be determined and the change of the slip
   ratio of the wheel which is to be braked calculated according to the additional yaw
   moment. Second, the change to the slip ratio controller must be inputted to implement
   the control of the wheel.

   Take the single wheel braking method, for example. In the event of a vehicle losing
   stability, the VSC system needs to apply the brakes on the front left wheel to produce a
   yaw moment to make the vehicle stable. According to the tyre models, the longitudinal
   and lateral forces on this wheel can be expressed as:

$$F_{xfl} = \mu_x\left(S\right) \cdot F_z \tag{3.35}$$

$$F_{yfl} = \mu_y\left(S\right) \cdot F_z \tag{3.36}$$

where $\mu_x(S)$, $\mu_y(S)$ are the functions of the longitudinal and lateral adhesion coefficient,
and F_z is the vertical load on the wheel.

The yaw motion produced by the variation of the front left wheel slip ratio can be obtained by the following equation:

$$\Delta M = \sqrt{\left(\frac{T}{2}\right)^2 + l_f^2} \cdot \cos\left(\arctan\frac{2l_f}{T} - \delta_f\right) \cdot \frac{\partial F_{xfl}}{\partial S_0} \cdot \Delta S$$
$$-\sqrt{\left(\frac{T}{2}\right)^2 + l_f^2} \cdot \sin\left(\arctan\frac{2l_f}{T} - \delta_f\right) \cdot \frac{\partial F_{yfl}}{\partial S_0} \cdot \Delta S \tag{3.37}$$

where ΔM is the additional yaw moment, S_0 is the slip ratio of the front left wheel at the moment the controller starts, which can be calculated by the speed signal from the wheel velocity sensor.

Based on equation (3.37), the variation range of the slip ratio ΔS can be fixed according to the control values of the yaw moment, and the increment of the wheel cylinder pressure ΔP can then be calculated. For convenience, if a PD control algorithm is adopted, ΔP can be obtained by the following equation:

$$\Delta P = k_{Pd}\Delta S(i) + k_{Pd}\left(\Delta S(i) - \Delta S(i-1)\right) \tag{3.38}$$

When the driver doesn't step on the brake pedal, the target pressure P_{whlNo} is the increment of the wheel cylinder pressure $\Delta P(P_{whlNo} = \Delta P)$. When the driver applies the brakes, suppose the pressure in wheel cylinder is P_0, thus the target pressure P_{whlNo} is calculated by equation (3.39)

$$P_{whlNo} = P_0 + \Delta P \tag{3.39}$$

4. *Braking torque control*
Braking torque control is different from the slip ratio control. Instead of using the slip ratio controller, the additional yaw moment is converted to the control quantity of the wheel cylinder pressure directly. The principle of the cylinder pressure calculation is: (a) to change the calculated additional yaw moment ΔM to the variation of the longitudinal forces of one-side wheels; and (b), to determine the wheel cylinder pressure change by using the wheel motion model. Taking the one-side wheels braking model as an example, the additional yaw moment can be expressed by the longitudinal forces in Equation (3.40)[16]

$$\Delta M = \frac{1}{2}F_{xfr}D_f + \frac{1}{2}F_{xrr}D_r \tag{3.40}$$

Because the front and rear wheels on the same side are controlled uniformly, the longitudinal forces of the wheels on the same side are approximately equal. If F_d is the expectation of the longitudinal tyre force, $F_{xfr} = F_{xrr} = F_d$, then equation (3.40) can be rewritten as:

$$\Delta M = \frac{1}{2}F_d\left(D_f\cos\delta_f + D_r\right) \tag{3.41}$$

The wheel motion model is presented below[20]:

$$I_w \frac{d\omega}{dt} = T_t - T_b - Fr_0 \tag{3.42}$$

In equation (3.42), ω is obtained from the wheel velocity sensor, and $T_t = 0$ when braking (during normal running conditions, it can be calculated if the engine torque is known). The longitudinal tyre force can be expressed as:

$$F = -\frac{1}{r_0}\left(I_w \frac{d\omega}{dt} + T_b\right) \tag{3.43}$$

To the drum brake:

$$T_b = CP_w \tag{3.44}$$

So, the increment of the longitudinal tyre force is:

$$F_d = -\frac{1}{r_0}\left(I_w \frac{d\omega}{dt} + CP_w\right) \tag{3.45}$$

where $C = A_w u_b R_b$, I_w is the wheel's inertia, T_b is the brake torque, T_t is the driving torque, r_0 is the wheel radius, ω is the wheel's angular velocity, P_w is the target pressure of the wheel cylinder, A_w is the brake shoe area, u_b is the friction coefficient of the brake shoe, and R_b is the distance between the wheel center and the brake shoe.

5. *Simulation and analysis*

The simulation was done under the conditions that the car moved on two different roads with the adhesion coefficient of 0.9 and 0.4 at the speed of 120 km/h and 60 km/h respectively. The input steering wheel angle took the form of an amplitude increasing sine wave. The curve of the front wheel turning angle is shown in Figure 3.34. The simulation results are shown in Figures 3.35–3.40. Figure 3.35 to Figure 3.37 illustrate the yaw rate, sideslip angle, and the phase variation when the road adhesion coefficient is 0.9.

The simulation results are shown in Figures 3.38–3.40, when the road adhesion coefficient is 0.4.

From the simulation results, it can be concluded that the VSC can effectively keep the vehicle body stable. Without a VSC system, with the increase of the front wheel turning angle, the vehicle tends to be unstable. When the front wheel turning angle is big enough, the vehicle yaw rate and sideslip angle increase sharply. At this time, the vehicle will spin and deviate from the expected track significantly, and lose control completely. In the uncontrolled vehicle, as the vehicle speed is fixed during simulation, the yaw rate and sideslip angle have the tendency to increase indefinitely after the vehicle becomes unstable. But with the help of the VSC, the vehicle yaw rate and sideslip angle meet the expectations of the driver quite well. As can be seen from the phase plane diagrams shown in Figure 3.37 and Figure 3.40, the vehicle is in a stable state.

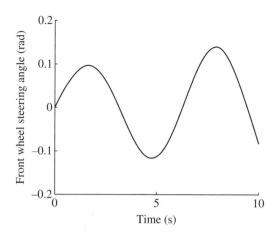

Figure 3.34 Front wheel steering angle.

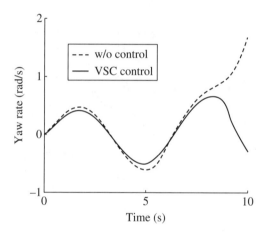

Figure 3.35 Yaw rate.

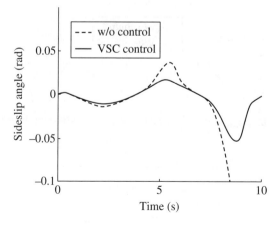

Figure 3.36 Sideslip angle.

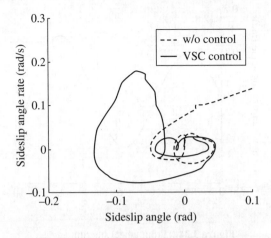

Figure 3.37   The phase diagram of the sideslip angle and sideslip angular velocity.

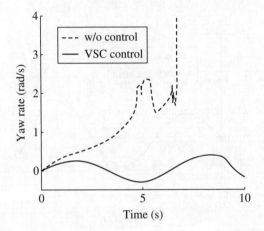

Figure 3.38   Yaw rate.

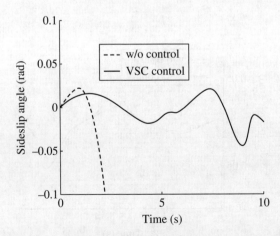

Figure 3.39   Sideslip angle.

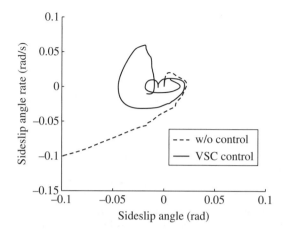

Figure 3.40 The phase diagram of the sideslip angle and sideslip angular velocity.

Appendix

Vehicle parameters	Symbol	Unit	Value
Vehicle mass	m	kg	1500
Front/rear wheel to centroid distance	l_f/l_r	m	1.219/1.252
Inertia of pitching moment	I_y	$kg \cdot m^2$	2436.4
Brake pedal force	F_b	N	220
Brake master cylinder diameter	D_m	m	0.02222
Transmission ratio of brake pedal mechanism	i_b	-	4.45
Boosting ratio booster	B	-	3.9
Brake disc working radius	D_f	m	0.104
Brake disc braking efficiency factor	C_f	-	0.8
Front wheel cylinder piston diameter	D_{wf}	m	0.054
Number of single clamp brake cylinders	n		1
Brake master cylinder efficiency	η_p	-	0.75
Brake drum working radius	D_r	m	0.100
Brake drum braking efficiency factor	C_r	-	2.15
Rear wheel cylinder piston diameter	D_{wr}	m	0.01746
Sprung mass centroid height	h	m	0.6
Front/rear unsprung centroid height	h_1/h_2	m	0.24/0.26

References

[1] Rajamani R. Vehicle Dynamics and Control. Springer US, 2006.

[2] Yu Z S. Automotive Theory, 4th edition. Mechanical Industry Press, Beijing, 2009.

[3] Li C M. Automotive Chassis Electronic Control Technology. Mechanical Industry Press, Beijing, 2004.

[4] Lu Z X. Automotive ABS, ASR and ESP Maintenance Illustrations. Electronic Industry Press, Beijing, 2006.

[5] Chu C B. Study on automotive chassis systems hierarchical coordination control. PhD thesis, Hefei University of Technology, Hefei, 2008.

[6] Wang D P, Guo K H, Gao Z H. Automotive drive anti-slip control technology. Automotive Technology, 1997, 4: 22–27.

[7] Zhang C B, Wu G Q, Ding Y L, et al. Study on control methods of automotive drive anti-slip. Automotive Engineering, 2000, 22(5): 324–328.

[8] Xiong X G. Study on control law of automotive drive anti-slip control system. Master's degree thesis, Hefei: Hefei University of Technology, 2010.

[9] McLellan D R, Ryan J P, Browalski E S, Heinricy J W. Increasing the safe driving envelope - ABS, traction control and beyond. Proceedings of the Society of Automotive Engineers, 92C014, 1992: 103–125.

[10] van Zanten A T, Erhardt R, Pfaff G. VDC, the Vehicle Dynamics Control System of Bosch. SAE Technical Paper, 950759.

[11] NHSTA. Federal Motor Vehicle Safety Standards No.126: Electronic Stability Control Systems. http://www. nhtsa.dot.gov/, 2007.

[12] van Zanten A T. Bosch VSC: 5 Years of Experience. SAE Technical Paper, 2000-01-1633.

[13] Nishio A, Tozu K, Yamaguchi H, Asano K, et al. Development of Vehicle Stability Control System Based on Vehicle Sideslip Angle Estimation. SAE Technical Paper 2001-01-0137.

[14] Wang D P, Guo K H. Study on the control principle and strategy of vehicle dynamics stability control. Journal of Mechanical Engineering, 2000, 36(3): 97–99.

[15] Hac A. Evaluation of Two Concepts in Vehicle Stability Enhancement Systems. Proceedings of 31st ISATA, Program Track on Automotive Mechatronics and Design, Düsseldorf, Germany, 1998, 205–213.

[16] Wang Q D, Zhang G H, Chen W W. Study on control of vehicle dynamics stability based on Sliding Mode Variable Structure. China Mechanical Engineering, 2009, 20(5): 622–625.

[17] Liu X Y. Study on vehicle stability control based on direct yaw moment control. PhD thesis, Hefei University of Technology, Hefei, 2010.

[18] Koibuchi K, Yamamoto M, Fukada Y, Inagaki S. Vehicle Stability Control System in Limit Cornering by Active Brake. SAE Technical Paper, 960487.

[19] Tseng H E, Ashrafi B, Madau D, et al. The development of vehicle stability control at Ford. IEEE/ASME Transactions on Mechatronics, 1999, 4(3): 223–234.

[20] You S H, Hahn J O, Cho Y M, Lee K I. Modeling and control of a hydraulic unit for direct yaw moment control in an automobile. Control Engineering Practice, 2006, 14(9): 1011–1022.

4

Vertical Vehicle Dynamics and Control

4.1 Vertical Dynamics Models

4.1.1 Introduction

The complex environment of vehicles means that the frequency range of vibration of the components, which can affect the comfort of drivers and passengers, is quite wide. Usually, NVH features (i.e., noise, vibration, and harshness) are used to characterize comfort. Generally, the range of vibration frequency can be divided into: 0–15 Hz for rigid motion, 15–150 Hz for vibration and resonance, and above 150 Hz for noise and scream.

Normally, a typical range of the resonance frequency can be divided as follows:

- the vehicle body is in the range of 1–1.5Hz (the damping ratio ζ is approximate 0.3);
- the wheel jumping is in the range of 10–12Hz;
- for the drivers and passengers it is in the range of 4–6Hz;
- for the powertrain mount it is in the range of 10–20Hz;
- for the structure it is over 20Hz; and
- for the tyre it is in the range of 30–50Hz and 80–100Hz.

This chapter mainly introduces the vertical dynamics which involves optimizing the design of parameters for a suspension system. This means that designers should coordinate the contradictory performance indexes such that the optimal performance of a suspension system can be achieved. To achieve this, one should ensure that the dynamic models reflect the actual working environment of vehicles, which largely depends on the complexity, the objects, and the demanded precision.

Integrated Vehicle Dynamics and Control, First Edition. Wuwei Chen, Hansong Xiao, Qidong Wang, Linfeng Zhao and Maofei Zhu.
© 2016 John Wiley & Sons Singapore Pte. Ltd. Published 2016 by John Wiley & Sons, Ltd.

Since vehicles are complex vibrating systems with multiple degrees of freedom, theoretically these systems will approach reality as the degrees of freedom increase. Therefore, it is natural to model the vehicles as various 3-dimensional models with dozens of degrees of freedom. From the point of view of comfort, the human–seat system is included in the systems when considering the responses to body vibration. For the road model, the systems include the input signals from the four wheels which are caused by road irregularities. These signals are different but they relate to each other. In addition, for vehicles which have low body stiffness and a long wheelbase, the bending and twist of the vehicle frame and body needs to be considered. If the effects from a powertrain mounting system need to be analyzed, one can separate this part from the vehicle body and consider the degrees of freedom relative to the vehicle body in the vertical direction. These methods will make the dynamic model closer to reality.

The increase of degrees of freedom indicates that more parameters are needed in the calculations. However, for some parameters, such as the mass, moment of inertia, and stiffness, etc., it is difficult to obtain precise values. Besides, the efficiency of the calculations is slowed down as the calculation precision increases. Thus, one should simplify the dynamic models of the vehicles by systematically considering the relative factors, such that the models are simple and can reflect reality.

4.1.2   Half-vehicle Model

Assuming that the surface of the road is flat, which means the difference of the road's surface by the two sides of vehicles can be ignored, one can establish a half-vehicle model with four degrees of freedom, as shown in Figure 4.1. For a half-vehicle model, following the dynamics equivalence, the vehicle body with mass m_2 and moment of inertia I can be decomposed into m_{2f}, m_{2r} and m_{2c}, which are, respectively, the mass of the front wheel, rear

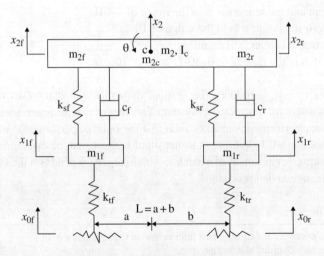

Figure 4.1   4-DOF half-vehicle model.

wheel, and center of mass. They satisfy the following conditions:

$$\begin{cases} m_{2f} + m_{2r} + m_{2c} = m_2 \\ m_{2f}a - m_{2r}b = 0 \\ I_c = m_{2f}a^2 + m_{2r}b^2 = m_c\rho_y^2 \end{cases} \tag{4.1}$$

where ρ_y is the radius of gyration around the horizontal axis y. Thus one has:

$$\begin{cases} m_{2f} = m_2 \dfrac{\rho_y^2}{aL} \\ m_{2r} = m_2 \dfrac{\rho_y^2}{bL} \\ m_{2c} = m_2 \left(1 - \dfrac{\rho_y^2}{ab} \right) \end{cases} \tag{4.2}$$

Here, $\varepsilon = \dfrac{\rho_y^2}{ab}$ is the distribution coefficient of the sprung mass. One has $m_{2c} = 0$, when $\varepsilon = 1$. Statistically, the value of ε for most vehicles ranges from 0.8 to 1.2, which can be viewed as close to 1. for $\varepsilon = 1$, the motions for m_{2f}, m_{2r} in the vertical direction are mutually independent. In this case, the vertical displacements x_{2f} and x_{2r} respectively on the front and rear wheel do not couple, and the main frequency is the same frequency as for the front and rear parts.

In Figure 4.1, a, b are the distances from the center of mass to the front/rear axle; m_s is the vehicle mass; m_{1f}, m_{1r} are the front/rear wheel mass; I_c is the moment of inertia of the body around the transverse axle; k_{sf} and c_f are the front suspension stiffness/damping coefficient; k_{sr} and c_r are the rear suspension stiffness/damping coefficient; k_{tf} and k_{tr} are the front/rear tyre stiffness coefficient; x_{0f} and x_{0r} are road roughness of the front/rear tyre contact; x_2 is the vertical displacement of the center of mass of the body; θ is the pitch angle; x_{2f} and x_{2r} are the vertical displacements of the front/rear end; x_{1f} and x_{1r} are vertical displacements of front/rear tyre; and c is the center of mass.

The motion equations of a 4-DOF half-vehicle model shown in Figure. 4.1 can be written as:

$$m_{1f}\ddot{x}_{1f} + c_f(\dot{x}_{1f} - \dot{x}_{2f}) + k_{sf}(x_{1f} - x_{2f}) + k_{tf}(x_{1f} - x_{0f}) = 0$$

$$m_{1r}\ddot{x}_{1r} + c_r(\dot{x}_{1r} - \dot{x}_{2r}) + k_{sr}(x_{1r} - x_{2r}) + k_{tr}(x_{1r} - x_{0r}) = 0$$

$$m_2\ddot{x}_2 - c_f(\dot{x}_{1f} - \dot{x}_{2f}) - k_{sf}(x_{1f} - x_{2f}) - c_r(\dot{x}_{1r} - \dot{x}_{2r}) - k_{sr}(x_{1r} - x_{2r}) = 0$$

$$I_c\ddot{\theta} + a(c_f(\dot{x}_{1f} - \dot{x}_{2f}) + k_{sf}(x_{1f} - x_{2f})) - b(c_r(\dot{x}_{1r} - \dot{x}_{2r}) + k_{sr}(x_{1r} - x_{2r})) = 0$$

When the pitch angle θ is small, the following approximations can be reached:

$$x_{2f} = x_2 - a\theta$$

$$x_{2r} = x_2 + b\theta$$

The matrix form of the equations of motion of the 4-DOF half-car model is:

$$[M]\ddot{X}+[C]\dot{X}+[K]X=[K_t]X_0 \tag{4.3}$$

where $[M] = \text{diag}[m_2, I_c, m_{1f}, m_{1r}]$ is the mass matrix. The damping matrix is:

$$[C]=\begin{bmatrix} C_f+C_r & -C_fa+C_rb & -C_f & -C_r \\ -C_fa+C_rb & C_fa^2+C_rb^2 & C_fa & -C_rb \\ -C_f & C_fa & C_f & 0 \\ -C_r & -C_rb & 0 & C_r \end{bmatrix}$$

The suspension stiffness matrix is:

$$[K]=\begin{bmatrix} k_{sf}+k_{sr} & -k_{sf}a+k_{sr}b & -k_{sf} & -k_{sr} \\ -k_{sf}a+k_{sr}b & k_{sf}a^2+k_{sr}b^2 & k_{sf}a & -k_{sr}b \\ -k_{sf} & k_{sf}a & k_{sf}+k_{tf} & 0 \\ -k_{sr} & -k_{sr}b & 0 & k_{sr}+k_{tr} \end{bmatrix}$$

The tyre stiffness matrix is:

$$[K_t]=\begin{bmatrix} 0 & 0 \\ 0 & 0 \\ k_{tf} & 0 \\ 0 & k_{tr} \end{bmatrix}$$

The road's surface input matrix is:

$$X_0 = \begin{bmatrix} x_{0f} & x_{0r} \end{bmatrix}^T$$

The system output matrix is:

$$X = \begin{bmatrix} x_2,\theta,x_{1f},x_{1r} \end{bmatrix}^T$$

It is necessary to use the full-vehicle model with seven degrees of freedom for studying the roll input from the road. The modeling process is similar to the half-vehicle model, except that the equations need to consider the factors for describing the roll motion of the body and the vertical motion of the other two tyres, which will be illustrated later.

4.2   Input Models of the Road's Surface

The roughness of the road's surface is the main disturbance into a vehicle system. In order to predict how the vehicle will respond to the input from the road's surface, it is necessary to describe the features of the road's surface correctly. The measurement of a road's surface roughness and the data processing, which needs specialized equipment, is important but time-consuming and not economical. This is the reason why this method has limited applications in engineering. Therefore, if a realistic road surface input model is established, it will prove a good foundation to the study of dynamic response and control.

4.2.1   Frequency-domain Models

From the study of the measurement data of a road's surface roughness, it shows that when the vehicle speed is constant, the road's roughness follows a Gaussian probability distribution, which is a stationary and ergodic stochastic process with a mean of zero. The power spectrum density (PSD) function and the variance of the road's surface can be used to describe its statistical properties. Since the velocity power spectral density is constant, which satisfies the definition and statistical properties of the white noise, it can be used to fit the time domain models of the road's surface roughness after the appropriate transformation. In the 1950s, people began to study the power spectrum of the road's surface, and used it to evaluate the quality of the road's surface and vehicle vibration responses. Until now, different methods for expressing the power spectrum of the road's surface have been proposed. In 1984, the international standards organization (ISO) proposed the draft to describe the road's surface roughness in the file ISO/TC108/SC2N67. According to this, China also drew up the measurement data report of the road's surface spectrum of the mechanical vibration (GB7031-2005). Both of the two reports suggested the fitting formula for the road's surface power spectrum $G_q(n)$ is[1]:

$$G_q\left(n\right) = G_q\left(n_0\right)\left(\frac{n}{n_0}\right)^{-p} \tag{4.4}$$

where n is the spatial frequency (m^{-1}), which is the inverse of the wavelength λ and means the numbers of waves per meter; $n_0 = 0.1$ is the referred frequency (m^{-1}); $G_q(n_0)$ is the coefficient of the road's surface irregularity (m^3), which is the value of the road's surface PSD under the referred frequency n_0; and p is the frequency exponent, which is the slope of the line of the log-log coordinates, and decides the frequency structure of the road's surface PSD.

In addition, the root-mean-square values of various road surfaces to describe its intensity, or the mean power of the road's surface stochastic exciting signals are defined as:

$$\sigma_q = \sqrt{\frac{1}{T}\int_0^T x^2\left(t\right)dt} \tag{4.5}$$

Table 4.1   Standard of road's surface irregularities grade.

Road surface grade	$G_q(n_0)/(10^{-6}m^3)$ $n_0 = 0.1m^{-1}$ Geometric mean	$\sigma_q/(10^{-3})$ $0.011m^{-1}<n$ $<2.83\,m^{-1}$ Geometric mean	Road surface grade	$G_q(n_0)/(10^{-6}m^3)$ $n_0 = 0.1m^{-1}$ Geometric mean	$\sigma_q/(10^{-3})$ $0.011m^{-1}<n$ $<2.83m^{-1}$ Geometric mean
A	16	3.81	E	4096	60.90
B	64	7.61	F	16384	121.80
C	256	15.23	G	65536	243.61
D	1024	30.45	H	262144	487.22

The two files mentioned above grade the road's surface power spectrum to eight levels and set the geometric mean of $G_q(n_0)$ of each grade. The frequency exponent of each graded road's surface spectrum is $p = 2$. The geometric mean of the root-mean-square values of the corresponding road's surface irregularities is also given, which is in the range $0.011m^{-1} <n< 2.83m^{-1}$, as shown in Table 4.1. Based on this, people usually use the speed PSD, $G_{\dot{q}}(n)$, and the acceleration PSD, $G_{\ddot{q}}(n)$, to describe the statistical characteristics of the road's surface irregularities.

The relations between them are:

$$G_{\dot{q}}(n) = (2\pi n)^2 G_q(n) \tag{4.6}$$

$$G_{\ddot{q}}(n) = (2\pi n)^4 G_q(n) \tag{4.7}$$

and when $p = 2$:

$$G_{\dot{q}}(n) = (2\pi n_0)^2 G_q(n_0) \tag{4.8}$$

Then, the amplitude of the road's surface power spectrum is a constant in the whole frequency range, which means it is white noise. It is convenient to do the calculations and analysis since the amplitude is only related to the irregularity coefficient $G_q(n_0)$.

In order to systematically study the characteristics of the road's surface, the vehicle speed should be considered. According to the speed u, the PSD of the spatial frequency $G_q(n)$, is translated to the power spectral density with the time frequency $G_q(f)$.

Assuming there is a vehicle with speed u, meaning the equivalent time frequency $f = un$, the relation between the PSD of the time and the spatial frequency is:

$$G_q(f) = \frac{1}{u} G_q(n) \tag{4.9}$$

The PSD of the time frequency is:

$$G_q(f) = G_q(n_0) n_0^2 \frac{u}{f^2} \qquad (4.10)$$

The relation of the PSD of the time frequency between the vertical velocity roughness $\dot{q}(t)$ and the acceleration roughness $\ddot{q}(t)$ can be obtained as follows:

$$G_{\dot{q}}(f) = 4\pi^2 G_q(n_0) n_0^2 u \qquad (4.11)$$

$$G_{\ddot{q}}(f) = 16\pi^4 G_q(n_0) n_0^2 u f^2 \qquad (4.12)$$

For the spatial frequency used for statistic analysis, the range is $0.011 m^{-1} < n < 2.83\ m^{-1}$. When the preferred speed $u = 10\sim30$m/s, it yields a time frequency range of $f = 0.33\sim28.3$Hz. The frequency range can effectively cover the natural frequency of the sprung mass (1~2Hz) and non-sprung mass (10~15Hz).

4.2.2   Time Domain Models

A road's surface PSD is a statistical magnitude of a certain road's surface roughness. The reconfiguration of the road's surface profile is not unique for a given road's surface PSD, and the obtained road's surface function is just a sample function of an equivalent road's surface profile, in a given road's surface spectrum, at a certain speed. To reconfigure the time domain model from the known road's surface spectrum, two conditions must be satisfied: (1) the road process is a steady stochastic process; (2) the process is ergodic. The basic idea of reconfiguration is to abstract the stochastic fluctuation of the road's surface profile to a white noise which satisfies certain conditions, then transform it to fit the time domain model of the road's surface stochastic roughness.

There are many methods of generating a time domain model of the road's surface profile for a stationary Gaussian stochastic process, such as wave-filter white noise, random sequence generation, harmonics superposition, AR (ARMA), and inverse Fast Fourier Transform. The wave-filter white noise method is widely used since it has a clear physical meaning and convenient calculations. In addition, the road's surface model parameters can be easily determined in terms of the data of the road's surface PSD and vehicle speeds.

According to what was mentioned above, when the vehicle is running at a uniform speed u, due to $\omega = 2\pi f$, equation (4.10) can be written as:

$$G_q(\omega) = (2\pi)^2 G_q(n_0) n_0^2 \frac{u}{\omega^2} \qquad (4.13)$$

when $\omega \to 0$, $G_q(\omega) \to \infty$.

Therefore, considering the lower cut-off angular frequency ω_0, the PSD can be written as:

$$G_q(\omega) = (2\pi)^2 G_q(n_0) n_0^2 \frac{u}{\omega^2 + \omega_0^2} \tag{4.14}$$

Equation (4.14) can be regarded as the responses from a first-order linear system under a white noise excitation. According to the random vibration theory, it is known that:

$$G_q(\omega) = |H(\omega)|^2 S_w \tag{4.15}$$

where $H(\omega)$ is the frequency response function; S_w is the PSD of a white noise $w(t)$ usually with $S_w = 1$.

Hence,

$$H(\omega) = \frac{2\pi n_0 \sqrt{G_q(n_0) \cdot u}}{\omega_0 + j\omega} \tag{4.16}$$

The equation of the time-domain expression of the road's surface roughness is[2]:

$$\dot{q}(t) = -2\pi n_{00} u q(t) + 2\pi n_0 \sqrt{G_q(n_0)} u w(t) \tag{4.17}$$

where n_{00} is the lower cut-off spatial frequency, usually 0.011m^{-1}; and $w(t)$ is a Gaussian white noise with zero mean.

4.3   Design of a Semi-active Suspension System

A suspension system is a very important component of any vehicle. It flexibly connects the vehicle body and the axles, bears the forces between the tyres and body, absorbs the shock from the road surface, and protects the vehicle from unwanted vibrations. The suspension plays a key role in the riding comfort and handling stability. Therefore, suspension design is one of the most important problems that automobile engineers focus on.

Suspension systems can be divided into passive suspension, semi-active suspension, and active suspension, according to the different control modes used. Most vehicles still implement traditional passive suspension systems. With the increase of vehicle speed and requirements of energy savings, eco-friendly technologies, safety and comfort, increasing demands are being made on the performance of suspension systems. The structure and main parameters of a passive suspension system cannot self-regulate under various speeds and road conditions; thus, it is impossible to reach the desired performance under some conditions. It is also limited to improve the performance of a passive suspension system through parameter optimization. That is why the current research into suspension systems focuses on electronic controlled suspension (ECS). Generally, ECS is divided into semi-active suspension and active suspension. Active suspension uses a force generator (or actuator) to displace the spring and damper of a traditional passive suspension. The actuator is normally hydraulic or pneumatic, which can generate the corresponding acting forces according to the control signals. In a vehicle suspension system, the semi-active suspension mainly uses the variable damping or other variable energy consumption components. For example, the damping force of a double-acting shock absorber can be changed through the diameter of the orifice in the piston, and the diameter of the orifice is determined by the real-time feedback control. Another kind of semi-active suspension is a magneto-rheological (MR) damper. The MR fluid in the damper is a liquid material which has rheological properties that can be changed with a magnetic field. Usually, with the increase of the magnetic field intensity, the viscosity of the fluid is increased. The dissipative force generated by the damper can be controlled through the electromagnetic field intensity. Here, the semi-active suspension system is addressed, and the active suspension system will be introduced later in this chapter.

4.3.1 Dynamic Model of a Semi-active Suspension System

The half-vehicle dynamic model with passive suspension has already been given before. Changing the passive damping force to a variable damping force U_f and U_r, the half-vehicle dynamic model with a semi-active suspension is obtained, as shown in Figure 4.2.

The indexes of ride comfort mainly consist of the vertical vibration acceleration of the vehicle body, pitching angular acceleration, suspension dynamic travel, and tyre dynamic load. The vehicle body acceleration is the main evaluation index for ride comfort. Suspension dynamic travel involves the limiting travel, and the mismatch will increase the probability of hitting the limiting blocks which will reduce ride comfort. The dynamic load between the tyres and the road's surface influences the adhesive effect which relates to driving safety. In order to achieve the desired performance of a vehicle, it is necessary to comprehensively choose and evaluate the parameters of a suspension system when the ride comfort analysis is made. So, in the design of a semi-active suspension system, the vertical

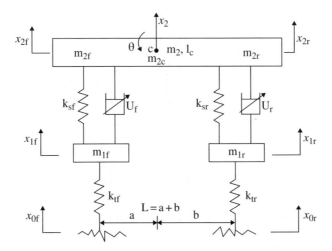

Figure 4.2 The half-vehicle model with a semi-active suspension.

body acceleration $\ddot{x}_2$, suspension dynamic travel f_d and tyre dynamic load F_d are chosen as the output variables of the system.

A half-vehicle model with a semi-active suspension is shown in Figure 4.2. If the distribution coefficient of a sprung mass is close to 1, the vertical vibration of the front and rear sprung masses will be almost independent. Then, it can be simplified to a quarter-vehicle model which is shown in Figure 4.3. Here, the semi-active suspension system is achieved by controlling the damping force of a shock absorber. Thus, the dynamic model of a shock absorber can be described by the damping force U.

The equations of motion for a quarter-vehicle model shown in Figure 4.3 are written as:

$$\begin{cases} m_1\ddot{x}_1 - k_s(x_2 - x_1) + k_t(x_1 - x_0) - c_s(\dot{x}_2 - \dot{x}_1) + U = 0 \\ m_2\ddot{x}_2 + c_s(\dot{x}_2 - \dot{x}_1) + k_s(x_2 - x_1) - U = 0 \end{cases} \tag{4.18}$$

The total damping force of a semi-active suspension can be regarded as the common damping force $c_s(\dot{x}_2 - \dot{x}_1)$ plus the controllable damping force U, which can be expressed as:

$$F = c_s\left(\dot{x}_2 - \dot{x}_1\right) + U$$

The state variables are:

$$X = \begin{bmatrix} x_1 - x_0 & x_2 - x_1 & \dot{x}_1 & \dot{x}_2 \end{bmatrix}^\mathrm{T}$$

For simplicity, take $x_0 = w(t)$, which represents the road's speed input as white noise. The state equation of the system can be written as:

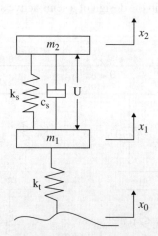

Figure 4.3 The quarter-vehicle model with a semi-active suspension, where m_1 is the unsprung mass; m_2 is the sprung mass; k_t is the tyre stiffness coefficient; k_s is the spring stiffness coefficient; c_s is the damping coefficient; U is the controllable force; x_2 is the sprung mass travel; x_1 is the non-sprung mass travel; and x_0 is the road excitation input.

$$\dot{X} = AX + Bu + Gw \qquad (4.19)$$

Taking the sprung mass acceleration $\ddot{x}_2$ as the output variable, one has the output equation:

$$Y = CX + Du \qquad (4.20)$$

where

$$A = \begin{bmatrix} 0 & 0 & 1 & 0 \\ 0 & 0 & -1 & 1 \\ \dfrac{-k_t}{m_1} & \dfrac{k_s}{m_1} & \dfrac{-C_s}{m_1} & \dfrac{C_s}{m_1} \\ 0 & \dfrac{-k_s}{m_2} & \dfrac{C_s}{m_2} & \dfrac{-C_s}{m_2} \end{bmatrix} \quad B = \begin{bmatrix} 0 \\ 0 \\ \dfrac{-1}{m_1} \\ \dfrac{1}{m_2} \end{bmatrix}$$

$$C = \begin{bmatrix} 0 & \dfrac{-k_s}{m_2} & \dfrac{C_s}{m_2} & \dfrac{-C_s}{m_2} \end{bmatrix}$$

$$D = \begin{bmatrix} \dfrac{1}{m_2} \end{bmatrix} \quad G = \begin{bmatrix} -1 \\ 0 \\ 0 \\ 0 \end{bmatrix}$$

4.3.2 Integrated Optimization Design of a Semi-active Suspension System

Active and semi-active suspensions have been widely applied and developed in the automotive industry as computers, and electronic and hydraulic servo techniques have improved. At the same time as the development of control theories, plenty of control methods have been applied to active and semi-active suspension systems, such as optimum control, preview control, adaptive control, neural network control and fuzzy logic control. Usually, the optimization theory is first applied to design the mechanical structure parameters of an active or semi-active suspension system. Then, some control strategies are applied to design the controller. These methods divide the procedure of a mechanical control system into two parts. Although both of them are applied to the optimization design process, it's actually hard for an active or semi-active suspension to reach the desired effect in practice. In fact, from the control point of view, the mechanical structure parameters, which are optimized by just considering motional or static conditions, are not always optimal. In the same way, the control parameters designed by just considering the system's stability and dynamics are not optimal either. There are complex and coupled relationships between the mechanical structure system and the control system, and this relationship should be considered in order to obtain global optimal parameters. Figure 4.4 gives the concept of an integrated design for a mechanical/control system. As shown in Figure 4.4, the control target and the control system

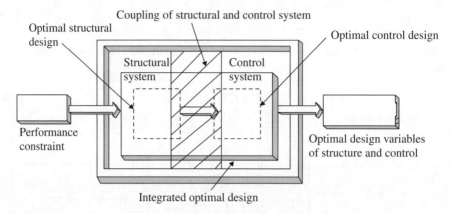

Figure 4.4   The integrated design of the controlled target and the control system.

are interrelated. It is more reasonable to optimize the two systems at the same time rather than to do optimization separately in order to get better results[3].

Thus, traditional design methods cannot coordinate the performance between the mechanical and the control systems. It is not a global optimal design, and it is necessary to use a global optimal design which considers both the mechanical structure parameters and the control parameters in order to get the ideal performance from a system.

The integrated design of the mechanical structure parameters and the control parameters started from the late 1980s and early 1990s of last century. For example, Japanese scholar Asada proposed an integrated optimization method of mechanical structure parameters and control parameters about single-link and two-link manipulators successively[4–5]; American scholar Anton adopted the approach of recursive experiments to integrated optimization of the mechanical structure and control parameters[6]. The stability and robustness study about the integrated optimization methods of the structure and control parameters has been done by Alexander, Savant, Shi and other scholars[7–10]. Here, based on a semi-active suspension system, one would be aimed at doing some exploratory work about the integrated optimization design of the structure and control parameters for engineering applications.

4.3.3   The Realization of the Integrated Optimization Method[11–12]

As shown in Figure 4.5, this optimization method adopts a cyclic structure, mainly due to the need to estimate whether the mechanical structure parameters and control parameters of the system are optimized simultaneously. If the optimal results are not reached, the mechanical structure parameters and control parameters should be modified. Then they should be checked again until both of them reach the best configuration.

When a vehicle is running, the vibration caused by the road's roughness excitation is a stochastic and uncertain disturbance, which makes the response characteristics of a suspension system become very complicated. Based on this, the genetic algorithm (GA)

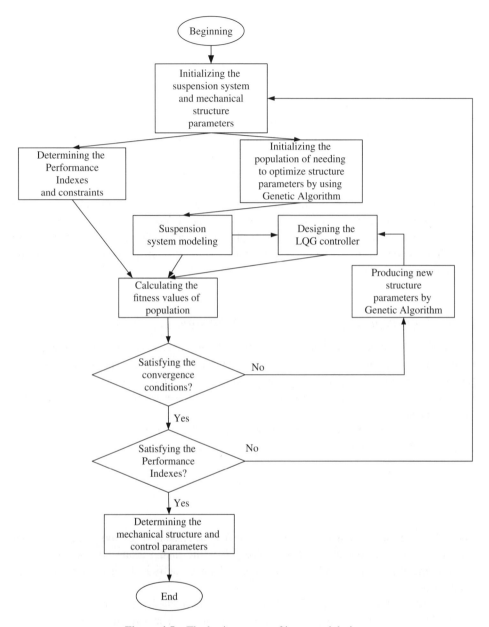

Figure 4.5   The basic process of integrated design.

and the LQG control method of the integrated optimization design for a suspension system
is applied, and the specific steps are as follows:

1. Determine the structure parameters, structure models, performance indexes and constraint conditions of a suspension system.

2. Initialize the population to optimize the structure parameters by using the genetic algorithm, and to establish the mathematical model of a suspension system.
3. Design a controller for the population by using the LQG control strategy, and calculate the fitness values.
4. Judge whether the system meet the convergence conditions. If not, recalculate by using the genetic algorithm until it meets the convergence conditions.
5. Determine whether to meet the performance indexes of a suspension system. If the conditions are met, determine the structure and control parameters of a suspension system, and end the optimization process. Otherwise, go back to step (1), reinitialize the structure and control parameters of a suspension system and recalculate them.

The structure parameters, models, performance indexes, and constraint conditions of a suspension system depend on specific problems. The details are not shown here. The following sections introduce the mathematical models, the implementation of the genetic algorithm, the design of a LQG controller for a semi-active suspension system, and the simulation results.

4.3.4 Implementation of the Genetic Algorithm

The genetic algorithm assimilates the evolution of the "survival of the fittest" of a natural system, which provides a robust searching method in a complex spatial domain. The genetic algorithm is carried out by using simple calculations, and does not need any restrictive assumptions (such as function continuity, derivative existence, and unimodal, etc.) for the searching space.

For a quarter-vehicle model, if the requirements for riding comfort and driving safety are considered, the performance index function J can be written as

$$J = \sqrt{\frac{w_1 J_1^2 + w_2 J_2^2 + w_3 J_3^2}{w_1 + w_2 + w_3}} \qquad (4.21)$$

where J_1 is the total variance of the body's vertical acceleration responses; J_2 is the total variance of the tyre dynamic load responses; J_3 is the total variance of the suspension dynamic travel responses; and w_1–w_3 are the influence coefficients of each performance index for J.

Let the fitness function $f = 1/J$. The design goal is to find the parameters which minimize the performance index J and maximize the function f. The structure parameters of a suspension system needing optimization are the suspension stiffness k_s, tyre stiffness k_t, and unsprung mass m_1, as shown in the following steps.

1. *Parameter coding*
   A binary code is used where each parameter is coded by a different length respectively, and then the parameters are cascaded to form a chromosome string. For any solving parameter x_i, the binary code with k bit length is used, and the upper and lower bound of

x_i is a and b, respectively. If M_i is the corresponding binary code, the relationship between M_i and x_i is as follows:

$$x_i = b + M_i \left[(a-b)/(2^k - 1) \right]$$

2. *Generate the initial population*
   Use a stochastic method to generate $N = 100$ individuals as the initial population.
3. *Selection*
   Use the roulette method to select; i.e., when the fitness value of the i-th individual of the population is f_i, the probability that it selects is $P_{si} = f_i \Big/ \sum_{i=1}^{N} f_i$. The larger the fitness value of the individual, the more the chance of being selected, and vice versa.
4. *Crossover*
   Make the crossover rate $P_c = 0.6$ and use multiple-point (3 points) crossover. First, match stochastically the individuals of the populations; then set stochastically the cross-over point K_{C1}, K_{C2} and K_{C3} in each matched individual within a fixed range; finally, the matched individuals exchange information in the corresponding location.
5. *Mutation*
   Take a small probability for the mutation probability $P_m = 0.1$, use a single-point muta-tion (which means to pick up a part of the individuals stochastically), and then inverse it.
6. *Assessing the values of fitness functions*
   If the values of the fitness function of continuing 15 generations do not change, the mutation probability P_m with 10% speed is increased. If the values of the fitness function still do not change for a continuing 5 generations after increasing the mutation proba-bility, then it converges and the genetic algorithm ends; otherwise, turn to step (7).
7. *Convergence condition*
   The error of the fitness function $\Delta f = f_{max} - f_{min}$. When $\Delta f < 0.05$, the algorithm converges, and the cycle of the genetic algorithm ends; otherwise, turn to step (3).

4.3.5   LQG Controller Design

Here, the LQG (linear quadratic Gauss) controller based on an observer is used. For the model as shown in equation (4.18), let $w(t)$ as the road speed excitation (white noise that acts on the system), and $v(t)$ as the measurement noise. If these signals are a Gauss process with zero mean, that is:

$$E(w) = E(v) = E(wv^T) = 0;\ E(ww^T) = Q_0;\ E(vv^T) = R_0$$

The LQG optimal problem is to design a control input u and minimize the quadratic performance index:

$$L = \lim_{t_f \to \infty} E \left\{ \int_0^{t_f} \begin{bmatrix} X^T & u^T \end{bmatrix} \begin{bmatrix} Q & N_c \\ N_c^T & R \end{bmatrix} \begin{bmatrix} X \\ u \end{bmatrix} \right\} \qquad (4.22)$$

where Q and R are the weighting matrices for state variables and control variables respectively; N_c is the weighting matrix for the two variables; and t_f is the end time.

Generally, equation (4.22) can be simply shown as:

$$L = \lim_{t_f \to \infty} E\left(\int_0^{t_f} X^T Q X + u^T R u \right) \qquad (4.23)$$

According to the separation theorem and the Kalman filtering principle, the state equation of the optimal estimator can be obtained as follows:

$$\hat{X} = A\hat{X} + Bu + K_l \left(y - C\hat{X} \right) \qquad (4.24)$$

The gain K_l of the filter is:

$$K_l = PC^T R_0^{-1} \qquad (4.25)$$

Here, P satisfies the following Raccati equation:

$$A^T P + PA - PC^T R_0^{-1} CP + GQ_0 G^T = 0 \qquad (4.26)$$

Then, the estimated state $\hat{X}$ is used to be substituted into X, and to design the optimal control as:

$$u = -K_c X \qquad (4.27)$$

where

$$K_c = R^{-1} B^T P_c \qquad (4.28)$$

Here, P_c satisfies the following Riccatic algebra equation:

$$A^T P_c + P_c A - P_c B R^{-1} B^T P_c + Q = 0 \qquad (4.29)$$

4.3.6   Simulations and Result Analysis

Some parameters for the integrated optimization simulation are shown in Table 4.2. When considering the performance of a suspension system, the values $k_s \in [8000, 18000]$, $k_t \in [80000, 180000]$, $m_1 \in [30,40]$, $w_1 = 0.65$, $w_2 = 0.25$, and $w_3 = 0.1$ are taken for the simulation. Assuming a vehicle is running on a level B road with a medium speed v = 20m/s, and the road's surface roughness coefficient is $G_q(n_0) = 256 \times 10^{-6} m^2 / m^{-1}$, the root mean square value $\sigma = 0.317(m/s)$ of the stochastic excitation signals of the road speed can be obtained. Using MATLAB to do the simulation, the results are shown in Table 4.2 and Figures 4.7–4.9.

From Table 4.2, it is clear that, after the integrated optimization method, the suspension stiffness k_s increases, the tyre stiffness k_t significantly decreases and the unsprung mass m_1

stays almost unchanged. Theoretically, the unsprung mass has little effect on the riding comfort so, in the integrated optimization process, the unsprung mass m_1 need not be the optimal target. The adherence between the tyre and road's surface can be significantly increased and the driving safety improved through reducing the tyre stiffness k_t. The increase of the suspension stiffness k_s decreases the suspension dynamic travel and ensures the limit travel.

The stochastic exciting signals of road speed are shown in Figure 4.6. The convergence curves of the evaluation index *J* are shown in Figure 4.7. In the integrated optimization process of the mechanical structure parameters and control parameters, the evaluation index *J* converges to a stable minimum value by using the genetic algorithm after a certain number of cycles of selection, crossover, and mutation. From Figure 4.8(a), it is clear that, by using the LQG controller for a semi-active suspension, the integrated optimization system can reduce the acceleration response of the sprung mass and improve the riding comfort, when compared with the non-integrated optimization system (i.e., the semi-active suspension system with given structure parameters to design the LQG controller). As shown in Figures 4.8(b) and (c), riding comfort and driving safety have been increased, and both the tyre and suspension dynamic travel are improved after the integrated optimization control for a semi-active suspension with the LQG controller is implemented. Similarly, from the frequency response curves of the sprung mass acceleration, as shown in Figure 4.9, the

Table 4.2 Some physical parameters of the suspension system.

Before and after optimization	Structure parameters			
	Sprung mass m_2/kg	Unsprung mass m_1/kg	Suspension stiffness k_s/N.m^{-1}	Tyre stiffness k_t/ N.m^{-1}
Before integrated optimization	300	34	14000	120000
After integrated optimization	300	32	15200	100670

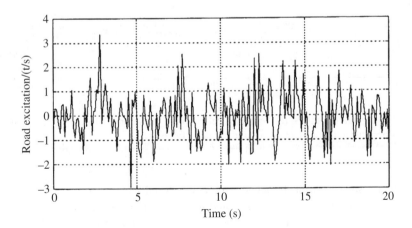

Figure 4.6 Stochastic exciting signals of road speed.

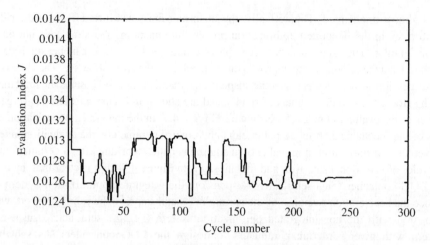

Figure 4.7 Convergence curves of the evaluation index *J.*

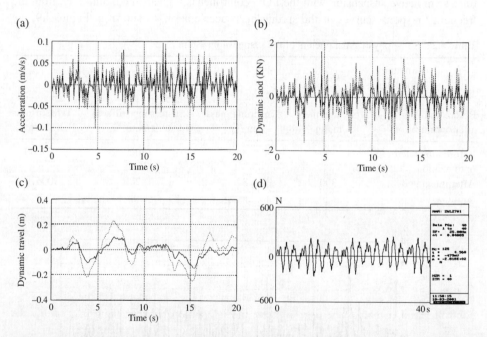

Figure 4.8 Simulation results. (a) Vertical acceleration of the sprung mass. (b) Tyre dynamic load. (c) Suspension dynamic travel. (d) Controlled damper force.

acceleration response in the frequency domain 1–50rad/s of the integrated optimization system has improved significantly compared with the non-integrated optimization system. Here the solid line in the figures represents the integrated optimization, and the dotted line represents the non-integrated optimization.

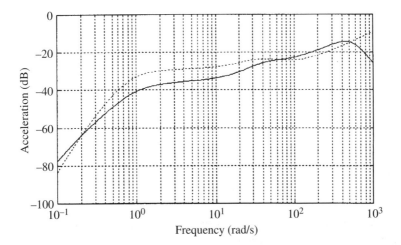

Figure 4.9 Frequency response of the sprung mass' vertical acceleration.

Table 4.3 Comparison of the simulation results (RMS).

Indexes	Method 1			Method 2		
Velocity	σ_{F_d}/N	$\sigma_{\ddot{x}_2}$/m/s^2	$\sigma_{x_1-x_2}$/m	σ_{F_d}/N	$\sigma_{\ddot{x}_2}$/m/s^2	$\sigma_{x_1-x_2}$/m
v = 20m/s	168.1779	0.8987	0.0411	344.0955	1.7045	0.0459
v = 30m/s	170.5303	0.9077	0.0634	349.2768	1.7282	0.0741

Table 4.3 compares the data of the tyre dynamic load, body acceleration, suspension dynamic travel, and other indexes of the suspension system under different velocities. Method 1 is the integrated optimization semi-active suspension with the LQG controller, and method 2 is the non-integrated optimization semi-active suspension with the LQG controller. It is obvious that when using method 1 each index is better than when using method 2.

Through the simulation results of the mechanical structure and control parameters of a quarter-vehicle model, it can be seen that the integrated optimization method has a significant effect on improving riding comfort and driving safety. The integrated optimization method effectively reaches the performance of the mechanical structure and the controller and combines them to improve the interaction between them in a traditional optimization process. Thus, the designed suspension system can obtain the global optimum.

4.4 Time-lag Problem and its Control of a Semi-active Suspension[13–14]

The time-lag phenomenon widely exists in vehicle control systems, and is one of the main factors that causes system instability and decreases performance. Due to the great influence on performance in a suspension system, the time-lag problem is one of the hotspot studies of semi-active suspension systems. In addition, time-lag mainly affects the low-frequency characteristics of a suspension system which is more sensitive to the human body. Then again, it will have a strong impact on riding comfort and handling stability. Therefore, it is necessary to consider the time-lag effect on a semi-active suspension system and adopt some measures to weaken the adverse effect of time-lag on a suspension system.

4.4.1   Causes and Impacts of Time-lag

Currently, the active and semi-active suspension systems and other control systems are mainly composed of signal detection devices, controllers, actuators, and controlled objects. The sources of time-lag in a control system are:

1. Transmission time-lag of the measured signals from the sensors to the controller;
2. Time-lag caused by the calculation of the control law;
3. Time-lag of transmitting the control signals to the actuator;
4. Time-lag caused by the response time of the actuator;
5. Time-lag caused by the time required to establish the control.

It is thus clear that time-lag mainly exists in three steps of the control process. First, it exists in the signal detection channel of a sensor, such as the response time-lag and transmission time-lag of a sensor. Second, it exists in the disturbance channel of a controlled target, for example, the required time that a driver exerts steering wheel angle or steering force until the front wheel starts to turn. Finally, it exists in the control input channel of an actuator, such as the required time from a step motor, direct current motor, solenoid valve and an actuator to receive instructions to start an action. The time-lag in a vehicle system in different locations has different effects on the performance of a control system, and it is necessary to analyze the location of the time-lag effect on a control system.

Supposing that there are different time-lags τ_1, τ_2 and τ_3 in the disturbance channel of a controlled target, the detection channel of the sensor signals, and the control input channel of an actuator respectively. Then, a closed loop control system with pure time-lag is shown in Figure 4.10.

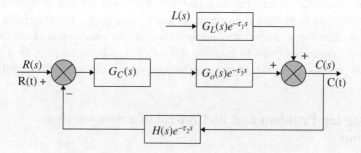

Figure 4.10   Closed loop system with time-lag.

The transfer function $G_{CR}(s)$ and $G_{CL}(s)$ of a closed loop system with the time-lag under the given desired output and the external disturbances can be seen in Figure 4.10.

$$\begin{cases} G_{CR}(s) = \dfrac{C(s)}{R(s)} = \dfrac{G_C(s)G_0(s)e^{-\tau_3 s}}{1+G_C(s)G_0(s)H(s)e^{-(\tau_2+\tau_3)s}} \\[3mm] G_{CL}(s) = \dfrac{C(s)}{L(s)} = \dfrac{G_L(s)e^{-\tau_1 s}}{1+G_C(s)G_0(s)H(s)e^{-(\tau_2+\tau_3)s}} \end{cases} \qquad (4.30)$$

where $G_C(s)$ is the controller's transfer function; Go (s) is the transfer function of the input channel of the controlled target; $H(s)$ is the transfer function of the sensor measure channel; $G_L(s)$ is the transfer function of the disturbance channel of the controlled target; R(t) is the desired output signal of the system; and C(t) is the actual output signal of the system.

Both of their characteristic equations can be expressed as:

$$1+G_C(s)G_0(s)H(s)e^{-(\tau_2+\tau_3)s} = 0 \qquad (4.31)$$

From equations (4.30) and (4.31), it is known that the time-lag locations impacting the performance of a control system are as follows.

1. *The disturbance channel of a controlled target*
   The pure lag factor $e^{-\tau_1 s}$, which is generated by the time-lag τ_1 that exists in the disturbance channel of a controlled target, appears in the numerator of the transfer function $G_{CL}(s)$. Compared with a no-lag control process, the control output C(t) under the action of the external disturbance L(t) will need to delay a pure lag time τ_1. Because the characteristic equation (4.31) of a closed-loop system does not contain the pure lag factor $e^{-\tau_1 s}$, the time-lag τ_1 of the disturbance channel has no influence on the stability of the control system. This makes the control output C(t) of the whole transient process of the control system delay a pure-lag time τ_1.
2. *The detection signal channel of sensors*
   The time-lag τ_2 that exists in the detection channel of sensors appears in the characteristic equation (4.31). It affects the stability of the control system but has no influence on the real-time value of the control output C(t).
3. *The control input channel of an actuator*
   The pure lag factor $e^{-\tau_3 s}$ which is generated by the time-lag τ_3 of the input channel of an actuator not only affects the stability of a closed-loop system, but it also makes the output C(t) delay a pure lag τ_3.

4. *All three channels*

If the control input channel, signal detection channel, and other links of the closed-loop system have a time-lag, then the pure lag factor will appear in the characteristic equations of a closed-loop system. In addition, the time-lag of two channels has an influence of a pure lag additive action on the stability of the closed-loop system.

It is thus clear that although the time-lag in the disturbance channel of the controlled target does not affect the system stability, it has great relevance with the real-time display of the output signals of the system. When time-lag exists in the control input channel and signal detection channel of a chassis control system, the stability of the closed-loop system is affected by the total lag $(\tau_2 + \tau_3)$ of adding both lags. Therefore, in the vehicle control process, it is necessary to compensate or control the time-lag of each link, to weaken the negative impact on the control performance, and to avoid the unexpected instability of the system.

4.4.2   Time-lag Variable Structure Control of an MR (Magneto-Rheological) Semi-active Suspension

For a semi-active suspension system, Magneto-rheological Semi-active Suspension (MSAS) has been extensively studied and has certain applications. It uses MR dampers as actuators to change the values of the damping force. The control system adjusts the exciting current that is given to the magnetic coil in the damper according to the road's surface excitation and suspension status, so as to change the magnetic field intensity of the coil. It also makes the apparent viscosity of the Magneto Rheological Fluid (MRF) in the damping channel change. Thus, continuous adjustment of the damping force can be achieved. The response process of the MSAS involves measuring the vehicle status and road signals, calculating the control effect and achieving the controllable damping force. This will generate a larger time-lag, impacting the control performance of the system if the response time is longer in some links. At present, these methods are used to solve the time-lag impact on the control performance of a semi-active suspension system:

1. Smith pre-evaluation compensation method. In the control channel with pure lag of the system, a parallel compensator is used to eliminate the effect of pure lag to the control process.
2. Critical time-lag values of the control system are calculated to eliminate the effect of time-lag by using time delay feedback links.
3. The controller is designed directly from the dynamic equations with time-lag.

Although the first and second methods can improve the effect of the time-lag, they mainly compensate for the time-lag after the controller has been designed. Because they do not consider the effect of time-lag in controller design, it's hard to ensure the stability of the control system.

The third method is used first to build up the dynamic equations with time-lag, then the effect of time-lag in controller design is considered in advance, so it is generally suitable and easy to ensure the stability of the control system. Thus, it is the main method used to solve the time-lag nowadays.

4.4.2.1   Time-lag test for an MR damper

As a key component of an MSAS system, the performance of an MR damper can directly affect the control performance of a semi-active suspension. The dynamic performance of an MR damper not only includes the maximal damping force, the maximal travel, and controllable range of the damping force, but also the dynamic response time of the MR damper. Here, the dynamic response time for a self-developed MR damper is tested, so reference to the control strategy design of a semi-active suspension with time-lag is provided.

The identification methods of time-lag include: putting a step signal to the input end of the target with time-lag, then start timing from the moment the step signal is added in, and to end timing when the responses from the output of the target with time-lag can be observed. This period of time τ is the time-lag of the target. According to this identification method, the self-developed MR damper (as shown in Figure 4.11(a)) is tested, with an amplitude of 25mm and frequency of 0.04Hz with the sine wave load by using the MR damper rig (Figure 4.11(b)). Here, the piston rod of the MR damper is in constant motion with a speed of 4mm/s. In the stretching travel of the damper, the transient current is among 0–0.5, 0–1.0, 0–1.5, 0–2.0 A. In the test, it takes the transient process from one steady state of the damping force to another, which is caused by the transient current. The test results under different transient current conditions are shown in Figure 4.12.

By the tests results shown in Figure 4.12, a time-lag table of the MR damper can be obtained. Table 4.4 shows that the response lag of an MR damper has the following characteristics:

1. The dynamic response time of an MR damper is less affected by the transient range of the current.
2. The response time when the transient current increases is significantly less than when it decreases. This is mainly caused by the remanence in the magnetic circuit of the MR damper, and the rapid decline of the damping force is blocked by the remanence field.
3. The response lag of the MR damper is no more than 28ms, and it has a faster dynamic response speed. In order to further improve the performance of the control system, the effect of time-lag in control strategy design should be considered.

(a)   (b)

Figure 4.11   MR damper and the test rig. (a) Self-developed MR damper. (b) Test rig for an MR damper.

(a)   (b)

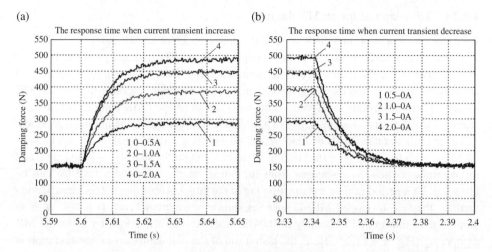

Figure 4.12   Response time of MR damper under different transient current conditions.

Table 4.4   The time-lag of the MR damper with transient current.

	The range of transient current							
Response lag	0–0.5A	0–1.0A	0–1.5A	0–2.0A	0.5–0A	1.0–0A	1.5–0A	2.0–0A
τ/ms	20.7	22.4	20.3	21.6	26.4	25.6	27.7	26.3

4.4.2.2   MSAS model

According to the nonlinear Bingham model and the measured results of the indicator diagram of an MR damper, the following mathematical model describes the characteristics of an MR damper:

$$\begin{cases} F(t) = c_s v + F_{MR}(t)\operatorname{sgn}(v) \\ F_{MR}(t) = \bar{a}_1 I^2(t) + \bar{a}_2 I(t) + \bar{a}_3 \end{cases} \tag{4.32}$$

where $F(t)$ is the damping force of the MR damper; c_s is the viscous damping coefficient; sgn is the sign function; $F_{MR}(t)$ is the coulomb damping force; v is the velocity of the piston rod; $\bar{a}_1, \bar{a}_2, \bar{a}_3$ are experimental constants ($\bar{a}_1 \neq 0$); and I is the current of the MR damper coil.

With the development of modern computer technologies and high-performance microprocessors, the time-lag generated by the signal measuring, processing, and control law calculation is very small. So, the response lag of an MR damper is considered, and a 2-DOF quarter car model of a semi-active suspension with the MR lag is built, similar to equation (4.18). The dynamic equations are:

$$\begin{cases} m_2\ddot{x}_2 + k_s(x_2 - x_1) + c_s(\dot{x}_2 - \dot{x}_1) + F_{MR}(t-\tau)\operatorname{sgn}(\dot{x}_2 - \dot{x}_1) = 0 \\ m_1\ddot{x}_1 - k_s(x_2 - x_1) + k_t(x_1 - x_0) - c_s(\dot{x}_2 - \dot{x}_1) - F_{MR}(t-\tau)\operatorname{sgn}(\dot{x}_2 - \dot{x}_1) = 0 \end{cases} \tag{4.33}$$

The parameters in the equations have the same meaning as in equation (4.18).

Defining $x = [x_2 \ x_2 - x_1 \ \dot{x}_2 \ \dot{x}_2 - \dot{x}_1]^T$ as the state variables of the system, the output variables are $y = [\ddot{x}_2 \ x_2 - x_1 \ x_1 - x_0]^T$, $u(t) = F_{MR}(t)\text{sgn}(v)$, $w(t) = x_0(t)$. Thus, the state equation and output equation can be derived as:

$$\begin{cases} \dot{x} = Ax + Bu(t-\tau) + Ew(t) \\ y = Cx + D_1 u(t-\tau) + D_2 w(t) \end{cases} \qquad (4.34)$$

where

$$A = \begin{pmatrix} 0 & 0 & 1 & 0 \\ 0 & 0 & 0 & 1 \\ 0 & -k_s/m_2 & 0 & -c_s/m_2 \\ k_t/m_1 & -k_s/m_1 - k_t/m_1 - k_s/m_2 & 0 & -c_s/m_1 - c_s/m_2 \end{pmatrix}$$

$$B = \begin{pmatrix} 0 & 0 & -1/m_2 & -1/m_2 - 1/m_1 \end{pmatrix}^T \quad E = \begin{pmatrix} 0 & 0 & 0 & -k_t/m_1 \end{pmatrix}^T$$

$$C = \begin{pmatrix} 0 & -k_s/m_2 & 0 & -c_s/m_2 \\ 0 & 1 & 0 & 0 \\ 1 & -1 & 0 & 0 \end{pmatrix} \quad D_1 = \begin{pmatrix} 1/m_2 & 0 & 0 \end{pmatrix}^T \quad D_2 = \begin{pmatrix} 0 & 0 & -1 \end{pmatrix}^T$$

4.4.2.3 Designing a variable structure controller with time-lag

In the above-mentioned modeling process, there is a time-lag τ in the control input channel $u(t)$ of the control system (4.34) for considering the response lag of an MR damper. The time-lag existing in the control input channel will affect the control effect of the entire MSAS and even make the system unsteady. In order to overcome the time-lag problem, an effective treatment is to consider the control input lag when the controller is designed. The goal is to design a controller which not only can make the closed-loop system steady, but can also make the system reach the expected performance when there is a time-lag.

Sliding mode control (SMC) is a nonlinear control method which differs with conventional control methods due to the inherent discontinuity of control. It uses a special sliding mode control method and makes the system's state move along a sliding surface by switching the control variables in order to reach the expected target. Thus, the sliding mode surface has an indifference to the parameter perturbation of the system and external disturbances. The SMC method is used to design an MSAS controller with its control input being the time-lag. The stability analysis and critical lag calculation of an MSAS with time-lag is carried out in the following section.

1. *Sliding mode of system*

   There is a control input lag in the equation (4.34) which increases the difficulty of designing a sliding mode. When the switching is carried out, in order to make the error smaller and avoid serious chattering phenomenon, the switching functionality is constructed as follows:

$$s = Gx + \int_{t-\tau}^{t} GBu(t-\xi)d\xi + \Gamma \tag{4.35}$$

G is the constant 1×4 matrix which meets the condition that the inverse of matrix GB exists. It is obvious then that matrix G exists. Γ is a sliding mode compensator, and has the following mathematical model:

$$\dot{\Gamma} = -G(A + BK)x \tag{4.36}$$

Here, K is the unknown constant 1×4 matrix which is related to the input lag τ. Under the condition that the generalized sliding mode condition $s\dot{s} < 0$. is satisfied, one can use the exponential approach law to improve the dynamic quality of the motion, and obtain the derivative of s with respect to time by using a saturation function to replace the sign function, which can be expressed as:

$$\dot{s} = GBu - GBKx + GEw = -\tilde{\alpha} f_{\text{sat}}(s) - \tilde{\beta}s \tag{4.37}$$

where $\tilde{\alpha}$ is the approaching velocity; $\tilde{\beta}$ is the reaching velocity; and $f_{\text{sat}}(s)$ is the saturation function.

The equivalent SMC law can be obtained from the following:

$$u = Kx - (GB)^{-1}\left(GEw + \tilde{\alpha} f_{\text{sat}}(s) + \tilde{\beta}s\right) \tag{4.38}$$

The system's trajectory reaches the sliding manifold under control law (4.38), the sliding model equation of motion on the sliding manifold is:

$$\dot{x} = Ax + BKx(t-\tau) + E(w(t) - w(t-\tau)) = Ax + BKx(t-\tau) + E\delta \tag{4.39}$$

where

$$\delta = w(t) - w(t-\tau)$$

2. *Judging the system's stability*
   Theorem   If there are 4×4 symmetric and positive definite matrixes L, N, S; a 1×4 matrix M; and a 4×4 matrix J, then the following matrix inequalities hold:

$$\begin{pmatrix} \Phi & BM - J & E & \tau LA^{\text{T}} \\ M^{\text{T}}B^{\text{T}} - J^{\text{T}} & -S & 0 & \tau M^{\text{T}}B^{\text{T}} \\ E^{\text{T}} & 0 & 0 & \tau E^{\text{T}} \\ \tau AL & \tau BM & \tau E & -\tau L \end{pmatrix} < 0 \tag{4.40}$$

$$\begin{pmatrix} N & J \\ J^{\text{T}} & L \end{pmatrix} \geq 0 \tag{4.41}$$

where $\Phi = LA^{\mathrm{T}} + AL + \tau N + J + J^{\mathrm{T}} + S$ for any $0 \leq \tau \leq \bar{\tau}$, $\bar{\tau}$ is the critical time-lag of the system.

If $K = ML^{-1}$ is taken, the sliding mode equation (4.39) will be asymptotically stable. For details, see reference[13].

The inequalities (4.40) and (4.41) are linear matrix inequalities (LMIs) of matrices L, M, N, J, S and T, so the asymptotic stability of the sliding mode (4.39) is transformed into the feasibility of LMIs, which can be solved by the solver **feasp** in MATLAB LMI toolbox.

3. *Critical time-lag of system stability*

   Applying the theorem stated above, the maximum permissible lag of the MSAS can be gained, which is the critical time-lag $\bar{\tau}$ of the system keeping it steady. It can be obtained by solving the following optimization problem:

$$\max_{L, M, N, J, S, T} \tau$$

$$s.t.\ L > 0, N > 0, S > 0, \begin{pmatrix} N & J \\ J^{\mathrm{T}} & L \end{pmatrix} \geq 0 \qquad (4.42)$$

$$\begin{pmatrix} \Phi & BM - J & E & \tau LA^{\mathrm{T}} \\ M^{\mathrm{T}}B^{\mathrm{T}} - J^{\mathrm{T}} & -S & 0 & \tau M^{\mathrm{T}}B^{\mathrm{T}} \\ E^{\mathrm{T}} & 0 & 0 & \tau E^{\mathrm{T}} \\ \tau AL & \tau BM & \tau E & -\tau L \end{pmatrix} < 0$$

In reference[13], the optimization problem (4.42) can also be transformed into the following problem to minimize the generalized eigenvalues.

$$\min \sigma$$

$$s.t.\ \begin{pmatrix} N & J \\ J^{\mathrm{T}} & L \end{pmatrix} \geq 0, L > 0,\ N > 0,\ S > 0 \qquad (4.43)$$

$$\begin{pmatrix} LA^{\mathrm{T}} \\ M^{\mathrm{T}}B^{\mathrm{T}} \\ E^{\mathrm{T}} \end{pmatrix} L^{-1} \begin{pmatrix} LA^{\mathrm{T}} & M^{\mathrm{T}}B^{\mathrm{T}} & E^{\mathrm{T}} \end{pmatrix} < -\sigma \begin{pmatrix} \Phi & BM - J & E \\ M^{\mathrm{T}}B^{\mathrm{T}} - J^{\mathrm{T}} & -S & 0 \\ E^{\mathrm{T}} & 0 & 0 \end{pmatrix}$$

Using the solver **gevp** in MATLAB LMI toolbox to obtain the optimization formula (4.43), a globally optimal solution σ^* would be reached. Then, the critical time-lag $\bar{\tau} = 1/\sigma^*$ of the MSAS can be obtained.

4.4.3   Simulation Results and Analysis

By using the SMC (named controller **I**) with the time-lag, the MSAS simulation mentioned before is carried out. The results are compared with the traditional SMC (named controller **II**) without the time-lag. Some parameters used in the simulation are shown in Table 4.5.

Table 4.5 Simulation parameters.

Parameters	Values	Unit
k_s	20	kN.m^{-1}
k_t	180	kN.m^{-1}
m_2	320	kg
m_1	40	kg
c_s	1317	N.s.m^{-1}
$\bar{a}_1$	18.98	/
$\bar{a}_2$	51.22	/
$\bar{a}_3$	314.77	/

In order to make the system's state trajectory reach the sliding mode surface quickly and without a large chatter, the parameters $\tilde{a} = 0.15$, $\hat{\beta} = 34$ are selected after repeating the adjustment. Thus, the critical time-lag $\bar{\tau} = 32$ms is reached by solving equation (4.43), which is close to the actual time-lag (<28ms) of the MR damper tested before. The value of $\mathbf{K} = (-198.14\ 182.06\ 19.87\ -3.67)$ can also be obtained by the solver **feasp**. To make sure that $\mathbf{GB}$ is reversible after repeatedly doing adjustments, $\mathbf{G} = (14.28\ -1.25\ 0\ -35.6)$ is determined. Then, the actual control effect of the MSAS is better.

4.4.3.1   Simulation of the Bump Road Input

Assume a vehicle is running along a bump road at a speed of 30 km/h. The length of the bumb is 0.1m, and the height is 0.06m. The mathematical model of the bump road can be described by the following equation:

$$x_0(t) = \begin{cases} \dfrac{\tilde{a}}{2}\left(1 - \cos\left(\dfrac{2\pi u_c}{l}\right)\right) & 0 \leq t \leq \dfrac{l}{u_c} \\ 0 & t > \dfrac{l}{u_c} \end{cases} \tag{4.44}$$

where $\tilde{a}$ is the bump height; l is the bump length; u_c is the vehicle speed.

The simulation curves of the body vertical acceleration, suspension dynamic deflection, and tyre dynamic load by using controller **I** and **II**, and the passive suspension under the bump road excitation conditions, are shown in Figures 4.13–4.15. The specific simulation results are shown in Table 4.6. As shown in the figures and table, by using controller **I**, the peak responses of the body acceleration, suspension dynamic deflection, and tyre dynamic displacement are improved by 56.15%, 38.71% and 26.9% respectively, compared with the passive suspension. As well, the settling time of the transient process is shorter. The peak values of the controller **II** without considering time-lag improved by 25.02%, 20.04%, 23.35% respectively, compared with the passive suspension. Due to a 28ms time-lag in the control channel, it will be out-of-step with the control in the simulation.

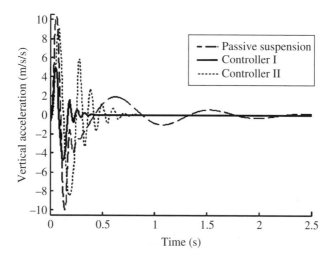

Figure 4.13   Body vertical acceleration.

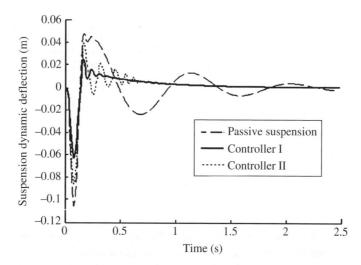

Figure 4.14   Suspension dynamic deflection.

4.4.3.2   Simulation of the Road's Surface Stochastic Input

Assume a vehicle is running along a grade **B** road at a uniform speed of 70km/h, and the filtering white noise is the stochastic road input. The road's roughness coefficient is 64×10^{-6} m³/cycle, and the low cut-off frequency is 0.1Hz. The simulation results of three different modes of stochastic road's surface are shown in Figures 4.16–4.18.

Figure 4.16 shows that the optimal damper force calculated by controller **II** cannot be achieved in real time, due to the 28ms time-lag in the control channel. Therefore, it is difficult to reduce vehicle vibration. The response lag of the MR damper of the controller **I** is

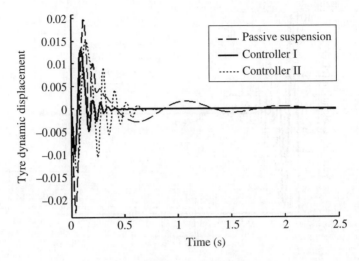

Figure 4.15   Tyre dynamic displacement.

Table 4.6   Simulation results of a bump road input.

Parameters	Control methods	Peak responses	Settling time t/s
Body vertical acceleration $\ddot{x}_2$/m.s^{-2}	Passive suspension	10.43	1.52
	Controller **I**	5.02	0.37
	Controller **II**	8.57	0.68
Suspension dynamic deflection $x_2 - x_1$/cm	Passive suspension	10.23	1.78
	Controller **I**	6.27	1.15
	Controller **II**	8.18	1.43
Tyre dynamic displacement $x_1 - x_0$/cm	Passive suspension	1.97	1.68
	Controller **I**	1.44	0.45
	Controller **II**	1.51	0.72

considered in advanced, and the input time-lag in the MSAS control channel is compensated reasonably, such that the body vertical acceleration is restrained effectively.

In Figure 4.17, controller **I** makes the suspension dynamic deflection achieve a better effect compared with controller **II**, when the time-lag is considered.

In Figure 4.18, it is clear to see that the time-lag has a greater effect on the body vibration. Controller **I** makes the peak values of the first resonance decrease by 44.16% at 28ms lag, compared with the passive suspension; controller **II** makes the peak values of the first resonance decrease by 18.18%. This is consistent with the simulation results of the bump input. It is obvious that the existence of time-lag greatly influences the ride comfort.

In order to analyze the influence of the control effect under different control input lag, through changing time-lags in simulations, the root mean square (RMS) values of the body vertical acceleration, suspension dynamic deflection, and tyre dynamic displacement are obtained under different time-lags, as shown in Table 4.7 and Table 4.8.

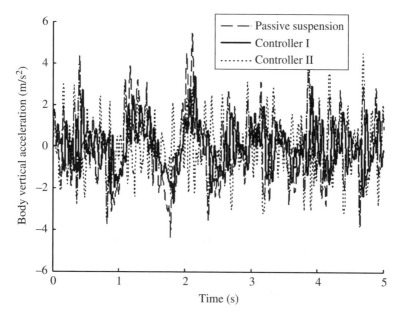

Figure 4.16 Body vertical acceleration.

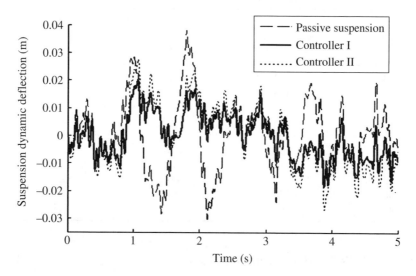

Figure 4.17 Suspension dynamic deflection.

In Table 4.7 and Table 4.8, the RMS values of the body vertical acceleration, suspension dynamic deflection, and tyre dynamic displacement, obtained by using controller **I** and controller **II**, are increased gradually with the increase of the time-lag in the control input channel. But controller **II** increases more, and this leads to controller **II** being unstable and divergent when the time-lag gets larger. This reflects that the control quality rapidly deteriorates with the increase of the time-lag in the control input channel.

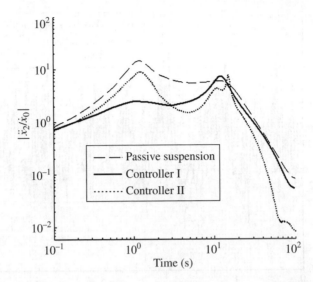

Figure 4.18 Amplitude-frequency characteristics of the body acceleration.

Table 4.7 RMS values of vehicle response by controller I.

Time-lag τ/ms	Body acceleration $\ddot{x}_2$ / m.s^{-2}	Suspension dynamic deflection $x_2 - x_1$ / cm	Tyre dynamic displacement $x_1 - x_0$ / cm
0	0.374	0.656	0.192
5	0.403	0.685	0.198
10	0.421	0.711	0.204
15	0.458	0.741	0.207
20	0.492	0.780	0.213
25	0.507	0.787	0.219
30	0.514	0.225	0.225
35	0.856	1.007	0.349
40	1.545	1.137	0.408
45	2.793	1.392	0.577

4.4.4 Experiment Validation

To verify the validity of the control strategy, an experimental study is performed on a 4-channel vibration rig with a car implementing the MSAS. The experimental setup is shown in Figure 4.19. Several cases were tested in the experiments. Controller **I** and controller **II** separately considered two types of road inputs (bump road and stochastic road), and the results are compared with those from the passive suspension. The vehicle parameters used in the tests are almost the same as with the simulation data.

The test results of the bump road input are shown in Figure 4.20. The peak values of the body vertical acceleration decrease from 11.6m/s^2 to 4.98 m/s^2. The peak values of the tyre

Table 4.8 RMS values of the vehicle response by controller II.

Time-lag τ/ms	Body acceleration $\ddot{x}_2$ / m.s^{-2}	Suspension dynamic deflection $x_2 - x_1$ / cm	Tyre dynamic displacement $x_1 - x_0$ / cm
0	0.353	0.637	0.178
5	0.423	0.721	0.204
10	0.472	0.777	0.216
15	0.591	0.803	0.217
20	0.714	0.821	0.224
25	0.911	0.935	0.315
30	1.209	0.910	0.331
35	1.515	1.118	0.352
40	1.883	1.469	0.497
45	3.661	1.988	0.612

Figure 4.19 Experimental setup.

dynamic load decrease remarkably and the settling time significantly shortens. These results are gained under the conditions of the bump road input and carried out by controller **I**, in which the time-lag is considered. For body vertical acceleration and tyre dynamic load carried out by controller **II** (without time-lag), the improved effect is less evident compared with controller **I**. The results show that the time-lag has a great influence on the control quality of the system, and simulations come up with the same conclusions.

Under stochastic road inputs with an amplitude of 10mm, and a signal bandwidth of 0.1–20Hz, the test results are shown in Table 4.9.

From Table 4.9, it is clear that the RMS value of the body acceleration is reduced from 0.89m/s^2 to 0.46m/s^2 by using controller **I**. Compared with the passive suspension, it is improved by 48.3%, and the RMS value of the tyre dynamic load is reduced from 0.704KN to 0.427KN, which is improved by 26.9%. By using controller **II**, the RMS

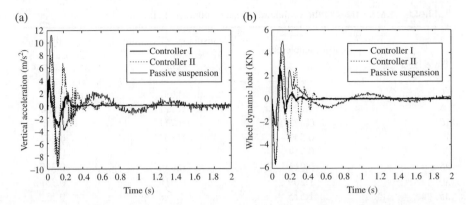

Figure 4.20 Experimental results of a bump road input. (a) Body acceleration. (b) Tyre dynamic load.

Table 4.9 RMS values comparison between simulation and experiment results.

Control mode	Body acceleration $\ddot{x}_2$/m.s^{-2}		Tyre dynamic load F_d/KN	
	Experiment results	Simulation results	Experiment results	Simulation results
Controller **I**	0.46	0.51	0.427	0.484
Controller **II**	0.77	0.83	0.512	0.575
Passive suspension	0.89	1.02	0.584	0.606

value of the tyre dynamic load is improved by only 12.3%, as the time-lag leads to it being out-of-step with the control. The experimental results are close to the simulation results, which fully verify that the stability of the MSAS can be improved further by using the sliding mode control method and considering time-lag. It also shows that the lag influence on the control quality can be reduced, and good vibration performance of the MSAS can be obtained.

Of course, this method can be applied to other control systems of vehicles with time-lag, due to the SMC strategy considering the influences of time-lag. In addition, this method starts directly from the time-lag equations of motion of the system, and thus can be widely used.

4.5   Design of an Active Suspension System[15–16]

Many advances have been made in active suspension and control theory nowadays. There are many control strategies for a 7-DOF full-vehicle model and a 4-DOF half-vehicle model, such as H_∞ control, H_2 control, L_2 control, adaptive control, and optimal control. Generally speaking, H_∞ control strategy can solve the robust stability problem of a controlled target, and H_2 control strategy lets the controlled targets have a better dynamic performance. Combining H_2/H_∞ to build a hybrid control strategy, the problems of robust stability and optimal dynamic performance of a controlled target can be solved with better results. For a half-vehicle model, the riding comfort is related to the body vertical acceleration, pitching angular acceleration, and suspension dynamic deflection. The handling stability is related to the front and rear tyre dynamic loads. Based upon this consideration, and also considering how to reduce the complexity of a controller, a multi-objective optimization control strategy for an active suspension system is introduced in this section. Choosing H_∞ norm as the robust performance index of a high-order unmodeled, H_2 norm as the time-domain LQG performance index of perturbation action, and choosing weighting matrices to set the frequency-domain performance index of a suspension system, a multi-objective H_2/H_∞ hybrid controller based on LMI (Linear Matrix Inequality) is designed. Simulation results show that the performance indexes by using the multi-objective H_2/H_∞ hybrid control method under the premise of the system with unmodeled robust stability, the handling stability and riding comfort of a vehicle can be improved effectively compared with the H_2 control or H_∞ control.

4.5.1   The Dynamic Model of an Active Suspension System

The 12-order state equation of a half-vehicle model of an active suspension, as shown in Figure 4.21, can be represented as:

$$\dot{x} = Ax + Gw + Bu \tag{4.45}$$

and the state variables are:

$$x = \begin{bmatrix} \dot{x}_2 & \dot{\theta} & x_{2f} - x_{1f} & x_{2r} - x_{1r} & \dot{x}_{1f} & \dot{x}_{1r} & x_{1f} - x_{0f} & x_{1r} - x_{0r} & q_1 & q_2 & v_1 & v_2 \end{bmatrix}^T \tag{4.46}$$

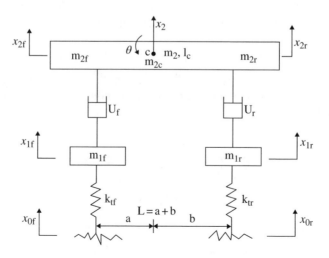

Figure 4.21   Half-vehicle model of an active suspension.

the control variables are:

$$u = \begin{bmatrix} i_1 & i_2 \end{bmatrix}^T \tag{4.47}$$

the disturbance variables are:

$$w = \begin{bmatrix} \dot{x}_{0f} & \dot{x}_{0r} & t_p \end{bmatrix}^T \tag{4.48}$$

where $\dot{x}_2$ is the body vertical velocity, $\dot{\theta}$ is the pitching angle, $x_{2f} - x_{1f}$, $x_{2r} - x_{1r}$ are the front and rear suspension dynamic deflections respectively, $\dot{x}_{1f}$, $\dot{x}_{1r}$ are the front and rear vertical velocities respectively, $x_{1f} - x_{0f}$, $x_{1r} - x_{0r}$ are the front and rear tyre dynamic displacements respectively, q_1, q_2 are the hydroelectric velocities of the front and rear servo-valves respectively, and v_1, v_2 are the hydroelectric volume of the front and rear servo-valves respectively. The control input $u = \begin{bmatrix} i_1 & i_2 \end{bmatrix}^T$ is the current of the front and rear servo-valves. The disturbance $w = \begin{bmatrix} \dot{x}_{0f} & \dot{x}_{0r} & t_p \end{bmatrix}^T$ is the road's surface speed excitation of the front and rear tyres and handling torque respectively.

4.5.2   Design of the Control Scheme

An active suspension system with an uncertainty of high-order unmodeled is considered, as shown in Figure 4.22.

In Figure 4.22, we are measuring the output $y = \begin{bmatrix} \dot{x}_2 & \dot{\varphi} & x_{2f} - x_{1f} & x_{2r} - x_{1r} \end{bmatrix}^T$; control input $u = \begin{bmatrix} i_1 & i_2 \end{bmatrix}^T$; performance evaluation $z = \begin{bmatrix} \ddot{x}_2 & \dot{\varphi} & x_{2f} - x_{1f} & x_{2r} - x_{1r} & x_{1f} - x_{0f} & x_{1r} - x_{0r} \end{bmatrix}^T$; and road disturbance input $w = \begin{bmatrix} \dot{x}_{0f} & \dot{x}_{0r} & t_p \end{bmatrix}^T$.

Here, $G_0(s)$ is a nominal active suspension model of a 4-DOF half-vehicle, and a real model is $G(s) = G_0(s)(I + \Delta(s))$, when the uncertainty Δ of the dynamic unmodeled is considered.

Assuming $\Delta(s) = diag[\Delta_1(s) \quad \Delta_2(s)]$; where $\Delta_1(s)$ is the high-order unmodeled part of a servo-valve attached to the front suspension, and $\Delta_2(s)$ is the high-order unmodeled part of a servo-valve attached to the rear suspension. According to reference [15], the upper bound functions of $\Delta_1(s)$ and $\Delta_2(s)$ are

$$W_s(s) = \frac{3s^2 + 132s + 5800}{s^2 + 132s + 17400} \tag{4.49}$$

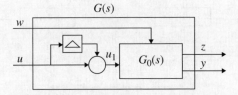

Figure 4.22   Diagram of controlled target.

The weighting coefficient matrix S_w is introduced to improve the system robustness of anti-disturbance, the weighting coefficient matrix S_z and the weighting function matrix $W_1(s)$ are introduced to improve the evaluation index z of the system. Considering that the system has a maximum uncertainty of $\Delta_m(s) = diag[W_s(s) \quad W_s(s)]$, so as to enhance the robustness of dynamic unmodeled part. For the unmodeled part $\Delta(s)$, let $\Delta_1(s) = a_1 W_s(s), \Delta_2(s) = a_2 W_s(s),$ $|a_1| \leq 1, |a_2| \leq 1,$ and the weighting function matrix $W_1(s) = diag(W_{11}(s) \quad W_{12}(s) \quad 1 \quad 1 \quad 1 \quad 1).$ Thus, the weighting function of the body vertical acceleration is $W_{11}(s)$, and the weighting function of the pitching angular acceleration is $W_{12}(s)$.

The human body is most sensitive to vertical vibration in the frequency range from 4Hz to 12.5Hz, and to horizontal vibration in the frequency range from 0.5Hz to 2Hz, according to the ISO2631-1(1997). Therefore, we choose $W_{11}(s) = \dfrac{s^2 + 30.16s + 1421}{s^2 + 15.08s + 1421}$ to increase the weight of the vertical vibrations in the 4~12.5Hz zone, and $W_{12}(s) = \dfrac{s^2 + 7.54s + 88.83}{s^2 + 3.77s + 88.83}$ to increase the weight of the horizontal vibrations in the 0.5~2Hz zone. In addition, we choose a weighting coefficient matrix as follows:

$$S_z = diag(0.03, 0.01, 8, 8, 0.1, 0.1)$$

$$S_w = diag(0.05, 0.05, 1.5 \times 10^4)$$

The control system of the generalized controlled target is shown in Figure 4.23, and the simplified form is shown in Figure 4.24.

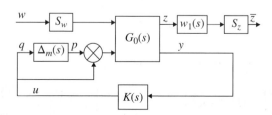

Figure 4.23 Control block diagram of the generalized controlled target.

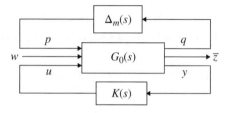

Figure 4.24 Simplified block diagram of the control system.

The transfer function of the generalized controlled target is:

$$G_g(s) = diag\left(S_z W_1(s), I\right) G_0(s) diag\left(S_w, I\right) \qquad (4.50)$$

From Figures 4.23 and 4.24, the state equations of a generalized controlled target can be obtained.

$$\begin{cases} \dot{x}_g = A x_g + B_1 w_1 + B_2 u \\ \dot{z}_1 = C_1 x_g + D_{11} w_1 + D_{12} u \\ z_2 = C_2 x_g + D_{21} w_1 + D_{22} u \\ y = C_3 x_g + D_3 w_1 \end{cases} \qquad (4.51)$$

where $w_1 = [p \quad w]^T$, $z_1 = q$, $z_2 = \bar{z}$, x_g are the state variables of a generalized controlled target, and A, B_1, B_2, C_1, C_2, C_3, D_{11}, D_{12}, D_{21}, D_{22}, D_3 are the coefficient matrices.

Channel T_{qp}: The transfer function from p to q. Consider the disturbance of p, which is generated by the uncertain link $\Delta_m(s)$, and use the H_∞ norm to evaluate T_{qp} to get better robustness.

Channel $T_{\bar{z}w}$: The transfer function from w to $\bar{z}$. Consider the road disturbance w, which is a stochastic white noise signal, and use the H_2 norm to evaluate $T_{\bar{z}w}$ to get a better dynamic performance index.

4.5.3   Multi-objective Mixed H_2/H_∞ Control

Design the output feedback controller $K(s)$ for a generalized controlled target. The equations of the controller $K(s)$ are:

$$\begin{cases} \dot{x}_c = K x_c + L y \\ u = M x_c + N y \end{cases} \qquad (4.52)$$

The transfer function matrix of the controller is:

$$K(s) = \frac{u(s)}{y(s)} = M(sI - K)^{-1} L + N = \begin{bmatrix} K_{11}(s) & K_{12}(s) & K_{13}(s) & K_{14}(s) \\ K_{21}(s) & K_{22}(s) & K_{23}(s) & K_{24}(s) \end{bmatrix} \qquad (4.53)$$

The control inputs are:

$$i_1 = K_{11}(s)\dot{z}_s + K_{12}(s)\dot{\theta} + K_{13}(s)z_{st1} + K_{14}(s)z_{st2}$$
$$i_2 = K_{21}(s)\dot{z}_s + K_{22}(s)\dot{\theta} + K_{23}(s)z_{st1} + K_{24}(s)z_{st2}$$

Design the controller to solve the matrices K, L, M and N.

1. *Two control schemes of a multi-objective mixed H_2/H_∞ control*

Control scheme I: Design an output feedback controller to satisfy the following conditions:

(a) The closed-loop system is stable;

(b) $\left\| T_{qp} \right\|_{\infty} < \gamma$;

(c) $\min \left\| T_{\bar{z}w} \right\|_2$

Control scheme II: Design an output feedback controller to satisfy the following conditions:

(a) The closed-loop system is stable;

(b) $\left\| T_{qp} \right\|_{\infty} < m_1$

(c) $\left\| T_{\bar{z}w} \right\|_2 < n_1$

(d) $\min \left(a \left\| T_{qp} \right\|_{\infty}^2 + b \left\| T_{\bar{z}w} \right\|_2^2 \right)$

Adjust the weights of the norm of the two channels $\|T_{qp}\|$ and $\|T_{\bar{z}w}\|$ in the H_2/H_{∞} mixed control, and to satisfy the condition: $\min(a\|T_{qp}\|_{\infty}^2 + b\|T_{\bar{z}w}\|_2^2)$, so that a better performance of robust stability and dynamic performance can be obtained.

2. *Mixed control algorithm of H_2/H_{∞}*

The mixed control algorithm of H_2/H_{∞} based on LMI can be expressed as:

1. The suboptimal control problem of H_2/H_{∞} is solvable if and only if there are symmetric matrices X, Y, Z, and meet the condition that $X - Y^{-1} = ZZ^T$ and the following three LMIs:

$$
\begin{bmatrix}
A^T X + XA + JC_3 + \left(JC_3\right)^T & XB_1 + JD_3 & \left(C_1 + D_{12}NC_3\right)^T \\
\left(XB_1 + JD_3\right)^T & -\gamma I & \left(D_{11} + D_{12}ND_3\right)^T \\
C_1 + D_{12}NC_3 & D_{11} + D_{12}ND_3 & -\gamma I
\end{bmatrix} < 0
$$

$$
\begin{bmatrix}
AY + YA^T + B_2 F + \left(B_2 F\right)^T & B_1 + B_2 ND_3 & \left(C_1 Y + D_{12}F\right)^T \\
\left(B_1 + B_2 ND_3\right)^T & -\gamma I & \left(D_{11} + D_{12}ND_3\right)^T \\
C_1 Y + D_{12}F & D_{11} + D_{12}ND_3 & -\gamma I
\end{bmatrix} < 0
$$

$$
\begin{bmatrix}
\alpha I & C_2 Y + D_{22}F & C_2 + D_{22}NC_3 \\
\left(C_2 Y + D_{22}F\right)^T & Y & I \\
\left(C_2 + D_{22}NC_3\right)^T & I & X
\end{bmatrix} \geq 0
$$

2. There are full column rank matrices U and V, and make $X - Y^{-1} = UU^T$, $V = (I - YX)U^{-T}$ and $F = NCY + MV^T$, $J = XBN + UL$

3. There is a $Q_{21} = (A + B_1 NC_3)^T + X(A + B_1 NC_3)Y + JC_3 Y + XB_1 F$, and

$$
\begin{bmatrix} Q_{31} & Q_{32} & \cdots & Q_3 \end{bmatrix} =
$$

$$
\begin{bmatrix}
\left(B_1 + B_2 ND_3\right)^T & \left(XB_1 + JD_3\right)^T & \cdots\cdots & -\gamma I & \left(D_{11} + D_{12}ND_3\right)^T \\
C_1 Y + D_{12}F & C_1 + D_{12}NC_3 & \cdots\cdots & D_{11} + D_{12}ND_3 & -\gamma I
\end{bmatrix}
$$

4. The coefficient matrix of equation (4.52) is:

$$\begin{bmatrix} K & L \\ M & N \end{bmatrix} = \begin{bmatrix} U^{-1}\left(Q_{32}^{T}Q_{3}^{-1}Q_{31} - Q_{21}\right)V^{-T} & U^{-1}\left(J - XB_{1}N\right) \\ \left(F - NC_{3}Y\right)V^{-T} & -D_{22}^{-1}D_{21}D_{3}^{-1} \end{bmatrix}$$

4.5.4   Simulation Study

Take $a_1 = 1$, $a_2 = 1$ in simulation. In control scheme **I**, suppose that $\left\|T_{qp}\right\|_{\infty} = 0.7$, thus, min $\left\|T_{\bar{z}w}\right\|_{2} = 0.952$ can be obtained. Then, take $a = 0.5$, $b = 0.5$, $m_1 = 1$, $n_1 = 1$ in the control scheme **II**, thus, $\left\|T_{qp}\right\|_{\infty} = 0.657$ and $\left\|T_{\bar{z}w}\right\|_{2} = 0.483$ can be obtained. **ass1**, **ass2**, **ass3**, **ass4** and **pss** denote the H_{∞} control, H_2 control, multi-objective mixed H_2/H_{∞} control scheme **I**, and control scheme **II**, and passive suspension respectively. Figures 4.25–4.28 are the response curves under the disturbance which is the given pulse signal excitation at a vehicle speed of 10km/h, and a pulse signal amplitude of 0.06. In Figures 4.25 and 4.26, the performance curves obtained by scheme **II** are small, and the convergence speed is fast. Thus, the overall performance indexes are better. In Figures 4.27 and 4.28, compared with the passive suspension system, the dynamic deflections of the front and rear suspensions by using H_{∞} control, H_2 control, and control scheme **II** are dropped considerably. In Figure 4.27, the peak values of the front suspension deflection from large to small are **pss**, **ass4**, **ass1**, and **ass2** control strategy. In Figure 4.28, the peak values of the rear suspension deflection, from large to small, are: **pss**, **ass4**, **ass1**, and **ass2** control strategy. But the curves of multi-objective mixed H_2/H_{∞} control scheme **II** have a fast convergence speed and smaller oscillations overall. Figures 4.29 and 4.30 give the response curves under the excitation of a random white noise. The white noise has zero mean and a variance of 0.01. Under the

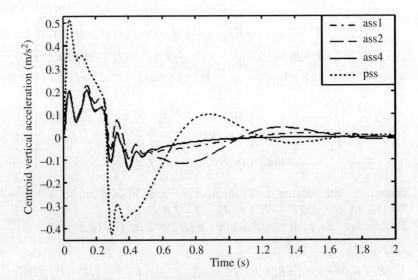

Figure 4.25   Vertical acceleration responses.

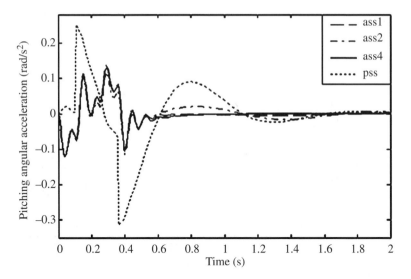

Figure 4.26 Pitching angular acceleration responses.

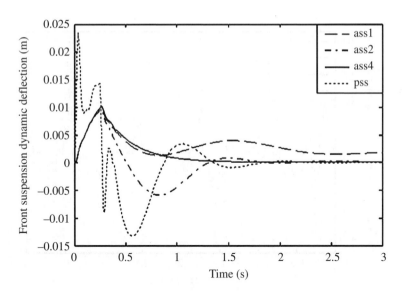

Figure 4.27 Responses of the front suspension dynamic deflection.

excitation of white noise, the vertical acceleration of the car's body and the pitching angular acceleration can obtain a better control effect using the control scheme **II**. It is clear to see from Figure 4.31 that the control scheme **II** can greatly reduce the gain of the vertical acceleration from the front tyre to the center of the body mass in the frequency range of 4–12Hz. In Figure 4.32, the control scheme **II** can greatly reduce the gain of the pitching angular acceleration from the front tyre to the body in the frequency range of 1–2Hz.

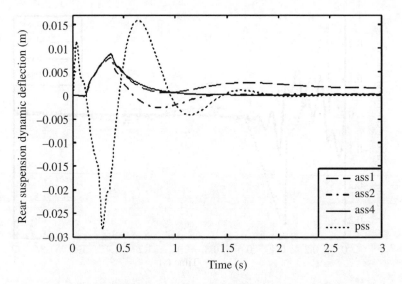

Figure 4.28 Responses of the rear suspension dynamic deflection.

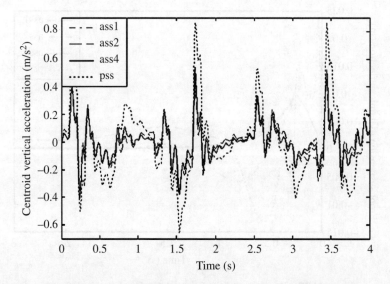

Figure 4.29 Vertical acceleration responses of the center of mass.

In Figures 4.33 and 4.34, the gains of the front and rear suspension dynamic deflections, from large to small, are: passive suspension, H_2 control, control scheme **II**, and control scheme **I**, respectively. In Figures 4.35 and 4.36, the front and rear tyre dynamic loads, from large to small, are: passive suspension, H_2 control, control scheme **I**, and control scheme **II**, respectively. It is clear that control scheme **II** can obtain a better vehicle handling stability.

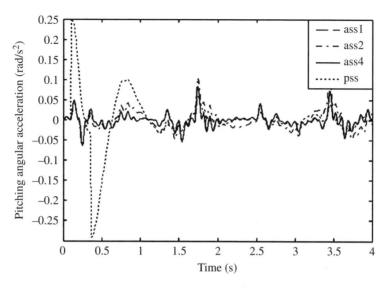

Figure 4.30   Pitching angular acceleration response of the center of mass.

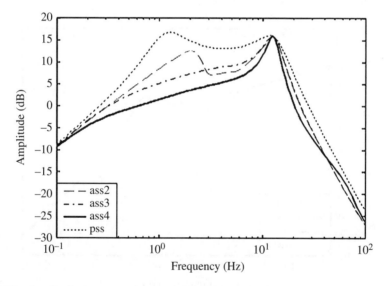

Figure 4.31   Amplitude-frequency characteristics of the transfer function of the vertical acceleration from the front tyre-road to the center of mass.

4.6   Order-reduction Study of an Active Suspension Controller[17-19]

In order to improve vehicle riding comfort and handling stability, various advanced control strategies have been used for the design of active suspension control systems in recent years. The orders of an active suspension model can be more than 20 according to different

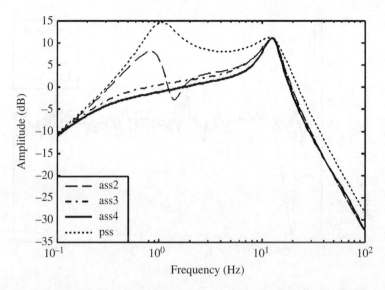

Figure 4.32 Amplitude-frequency characteristics of the transfer function of the pitching angular acceleration from the front tyre-road to the center of mass.

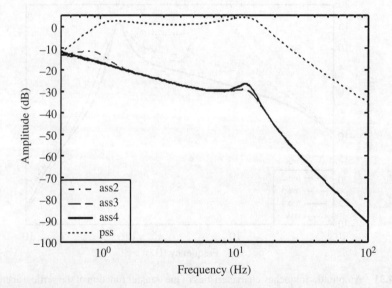

Figure 4.33 Amplitude-frequency characteristics of the transfer function of the front suspension deflection from the front tyre-road input.

degrees of freedom. Considering the human sensitivity frequency range, the orders of a generalized system by using weighting frequencies will be higher, and the orders of a controller by using H_∞, H_2, LQG strategies, etc., will be higher than a generalized system. A high order controller brings great difficulty to engineering applications, increases the complexity, and reduces the real-time and reliability of a control system.

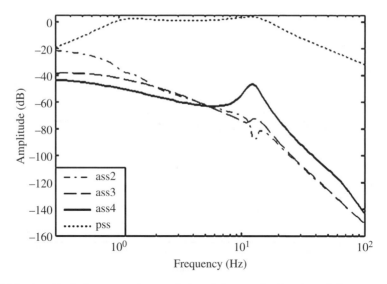

Figure 4.34 Amplitude-frequency characteristics of the transfer function of the rear suspension deflection from the rear tyre-road input.

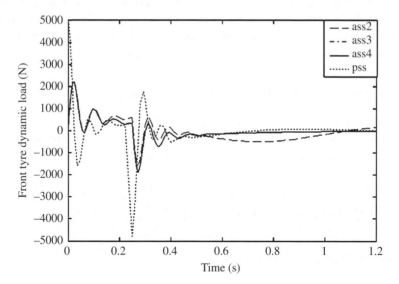

Figure 4.35 Front tyre dynamic load.

Thus, reducing the orders of a controller as much as possible under the conditions of keeping the performance of a closed loop control system is still unknown. There are usually two kinds of research methods investigating this aspect: one method is to reduce the order of a high order controlled target and design a controller according to the reduction model; the second is to reduce the order of a designed high order controller, such as by using the controller reduction method based on LMI to design a low order controller of an active

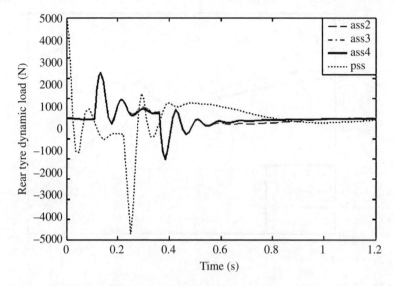

Figure 4.36   Rear tyre dynamic load.

suspension system. This section introduces how to design an H_∞ controller based on a 7-DOF full vehicle model, and design a controller by the Optimal Hankel-norm Reduction (OHNR) method. It shows that the Hankel reduction method can reach a better control effect compared with the other two kinds of reduction methods.

4.6.1   Full Vehicle Model with 7 Degrees of Freedom

Figure 4.37 is a full vehicle model with 7 degrees of freedom, which considers the vertical, pitch, and roll motion of a sprung mass and the vertical motion of non-sprung mass.

The vertical motion equation of the center of body mass is:

$$m_2\ddot{x}_2 = c_{sA}(\dot{x}_{1A} - \dot{x}_{2A}) + k_{sA}(x_{1A} - x_{2A}) + c_{sB}(\dot{x}_{1B} - \dot{x}_{2B}) + k_{sB}(x_{1B} - x_{2B}) + c_{sC}(\dot{x}_{1C} - \dot{x}_{2C}) +$$
$$k_{sC}(x_{1C} - x_{2C}) + c_{sD}(\dot{x}_{1D} - \dot{x}_{2D}) + k_{sD}(x_{1D} - x_{2D}) + f_A + f_B + f_C + f_D$$

$$(4.54)$$

The body pitching motion equation is:

$$I_p\ddot{\theta} = \left[c_{sC}(\dot{x}_{1C} - \dot{x}_{2C}) + k_{sC}(x_{1C} - x_{2C}) + c_{sD}(\dot{x}_{1D} - \dot{x}_{2D}) + k_{sD}(x_{1D} - x_{2D}) + f_C + f_D\right]b -$$
$$\left[k_{sA}(x_{1A} - x_{2A}) + c_{sB}(\dot{x}_{1B} - \dot{x}_{2B}) + k_{sB}(x_{1B} - x_{2B}) + c_{sA}(\dot{x}_{1A} - \dot{x}_{2A}) + f_A + f_B\right]a$$

$$(4.55)$$

The body roll motion equation is:

$$I_r\ddot{\varphi} = \left[c_{sA}(\dot{x}_{1A} - \dot{x}_{2A}) + k_{sA}(x_{1A} - x_{2A}) - c_{sB}(\dot{x}_{1B} - \dot{x}_{2B}) - k_{sB}(x_{1B} - x_{2B}) + (f_A - f_B)\right]t +$$
$$\left[c_{sC}(\dot{x}_{1C} - \dot{x}_{2C}) + k_{sC}(x_{1C} - x_{2C}) - c_{sD}(\dot{x}_{1D} - \dot{x}_{2D}) - k_{sD}(x_{1D} - x_{2D}) + (f_C - f_D)\right]t$$

$$(4.56)$$

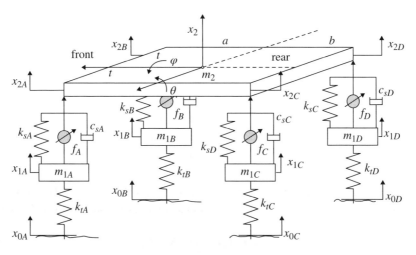

Figure 4.37   Diagram of a 7-DOF full vehicle model.

The vertical motion equations of four non-sprung masses are:

$$m_{1A}\ddot{x}_{1A} = k_{tA}\left(x_{0A} - x_{1A}\right) + k_{sA}\left(x_{2A} - x_{1A}\right) + c_{sA}\left(\dot{x}_{2A} - \dot{x}_{1A}\right) - f_A \tag{4.57}$$

$$m_{1B}\ddot{x}_{1B} = k_{tB}\left(x_{0B} - x_{1B}\right) + k_{sB}\left(x_{2B} - x_{1B}\right) + c_{sB}\left(\dot{x}_{2B} - \dot{x}_{1B}\right) - f_B \tag{4.58}$$

$$m_{1C}\ddot{x}_{1C} = k_{tC}\left(x_{0C} - x_{1C}\right) + k_{sC}\left(x_{2C} - x_{1C}\right) + c_{sC}\left(\dot{x}_{2C} - \dot{x}_{1C}\right) - f_C \tag{4.59}$$

$$m_{1D}\ddot{x}_{1D} = k_{tD}\left(x_{0D} - x_{1D}\right) + k_{sD}\left(x_{2D} - x_{1D}\right) + c_{sD}\left(\dot{x}_{2D} - \dot{x}_{1D}\right) - f_D \tag{4.60}$$

The state variables of the system are:

$$\boldsymbol{x} = \begin{bmatrix} x_2 & \dot{x}_2 & \varphi & \dot{\varphi} & \theta & \dot{\theta} & \dot{x}_{1A} & \dot{x}_{1B} & \dot{x}_{1C} & \dot{x}_{1D} & x_{1A} & x_{1B} & x_{1C} & x_{1D} \end{bmatrix}^T$$

The disturbance input is:

$$\boldsymbol{w} = \begin{bmatrix} x_{0A} & x_{0B} & x_{0C} & x_{0D} \end{bmatrix}^T$$

The control input is:

$$\boldsymbol{u} = \begin{bmatrix} f_A & f_B & f_C & f_D \end{bmatrix}^T$$

The system output is:

$$\boldsymbol{z} = \begin{bmatrix} \ddot{x}_2 & \ddot{\varphi} & \ddot{\theta} & x_{2A} - x_{1A} & x_{2B} - x_{1B} & x_{2C} - x_{1C} & x_{2D} - x_{1D} & f_A & f_B & f_C & f_D \end{bmatrix}^T$$

The measuring output is:

$$y = \begin{bmatrix} \ddot{x}_2 & \dot{\varphi} & \dot{\theta} \end{bmatrix}^T$$

The state space model of an active suspension system is:

$$\begin{cases} \dot{x} = Ax + B_1 w + B_2 u \\ z = C_1 x + D_{11} w + D_{12} u \\ y = C_2 x + D_{21} w + D_{22} u \end{cases} \tag{4.61}$$

Take $G_0 = C(sI - A)^{-1} B + D$. This model is the minimal realization model of a state space with 14 orders. The corresponding state space model of a passive suspension system is:

$$G_{0p} = C_1 (sI - A)^{-1} B_1 + D_{11} \tag{4.62}$$

4.6.2   Controller Design

In the design of an H_∞ output feedback controller for the target $G_0(s)$, the diagram of the output feedback of a closed-loop control system is shown in Figure 4.38.

The goal of H_∞ control is to design a controller $K(s)$, which can make the interior of a closed-loop system stable and minimize $\|T_{zw}(s)\|_\infty$, where $T_{zw}(s)$ is the transfer function from the disturbance input w to the controlled output z.

Since,

$$\begin{bmatrix} z \\ y \end{bmatrix} = G_0(s) \begin{bmatrix} w \\ u \end{bmatrix} = \begin{bmatrix} G_{11}(s) & G_{12}(s) \\ G_{21}(s) & G_{22}(s) \end{bmatrix} \begin{bmatrix} w \\ u \end{bmatrix}$$

Thus,

$$T_{zw}(s) = G_{11} + G_{12} K (I - G_{22} K)^{-1} G_{21} \tag{4.63}$$

The weighting transfer functions are:

$$W_1 = \frac{s^2 + 314.2s + 987}{s^2 + 43.98s + 987} \quad \text{(vertical motion direction)}$$

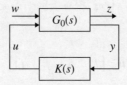

Figure 4.38 H_∞ output feedback.

$$W_2 = \frac{s^2 + 50.27s + 25.72}{s^2 + 7.037s + 25.72} \quad \text{(horizontal motion direction)}$$

The weighting transfer function matrix of the controlled output is:

$$W_{perf} = diag\begin{pmatrix} W_1 & W_2 & W_2 & 1 & 1 & 1 & 1 & 1 & 1 & 1 & 1 \end{pmatrix}$$

The weighting coefficient matrix of the disturbance input is:

$$S_w = diag\begin{pmatrix} 0.0001 & 0.0001 & 0.0001 & 0.0001 \end{pmatrix}$$

The weighting coefficient matrix of the controlled output is:

$$S_z = diag\begin{pmatrix} 6 & 5 & 5 & 25 & 25 & 25 & 25 & 0.022 & 0.022 & 0.022 & 0.022 \end{pmatrix}$$

The diagram of the H_∞ weighted control is shown in Figure 4.39.

According to the H_∞ control theory, in order to effectively suppress the disturbance input, it should have $\left\| T_{\overline{zw}}(s) \right\|_\infty = \left\| S_z \cdot W_{perf} \cdot T_{zw}(s) \cdot S_w \right\|_\infty$ to reach the minimum. Now the controller design problem can be converted into the H_∞ standard design containing the weighting function of a generalized system. By using the MATLAB/LMI toolbox, the controller $K = C_k(sI - A_k)^{-1}B_k + D_k$ can be solved.

4.6.3   Controller Order-reduction

4.6.3.1   Application of the Optimal Hankel-norm Reduction (OHNR) Method

By calculations, the order of the H_∞ controller $K(s)$ is 20, $K(s)$ is stable, and (A_k, B_k, C_k, D_k) is the minimal realization of $K(s)$. Thus, the reduction order controller $G_i(s)$ can be obtained as follows by selecting the orders of the controller K(s) from order 5 to 10. The results are shown in Table 4.10.

In Table 4.10, E_i is the difference of the transfer function matrix between an ith-order controller and a full-order controller. $\left\| T_{\overline{izw}} \right\|_\infty$ is the H_∞ norm of the transfer function matrix from the disturbance output to the controlled input in the closed-loop system which is composed of an ith-order controller and the generalized target. $\left\| T_{izw} \right\|_\infty$ is the H_∞ norm of the transfer function matrix from the disturbance output to the controlled input in a closed-loop system which is composed of an ith-order controller and the original controlled target. It is

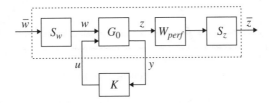

Figure 4.39 H_∞ weighted control.

clear that the increase of error rate for all indexes is larger from 10th to 9th-order and from 9th to 8th-order.

In order to further determine the dimensions of order-reduction controllers, the 5th–10th and 20th-order controllers are implemented to an active suspension system, respectively, to observe the control effect. The simulation results show that the 10th-order controllers have almost the same control effect than the full-order controller. The control effect of the 5–7th-order controllers obviously worsens, and the control effect of 8th and 9th-order controllers is almost the same. Figure 4.40 shows the closed-loop amplitude-frequency response curves of the vertical acceleration when different order controllers are used, with the front tyre-road travel input to the center of the body mass.

Hence, in order to reduce the orders of the controllers as much as possible without lossing the control effect of a closed-loop system, reducing the 20th to an 8th-order controller is appropriate.

Table 4.10 Comparison of the OHNR results of a 20 order H_∞ controller.

The order of controller i	$\lambda_i = \|E_i\|_H$	$\dfrac{\lambda_i}{\|K(s)\|_H}$	$\beta_i = \|E_i\|_\infty$	$\dfrac{\beta_i}{\|K(s)\|_\infty}$	$\gamma_i = \|T_{i\overline{z}w}\|_\infty$	$\dfrac{\gamma_i - \gamma_{20}}{\gamma_{20}}$	$\eta_i = \|T_{izw}\|_\infty$	$\dfrac{\eta_i - \eta_{20}}{\eta_{20}}$
10	360.4	4.8%	438.1	3.1%	0.8503	3.1%	83796	34.6%
9	817.6	10.8%	1019.9	7.3%	0.8434	2.3%	56810	-8.7%
8	1267.5	16.8%	2257.5	16.1%	0.8629	4.6%	100640	61.7%
7	1373.6	18.2%	2454.9	17.5%	0.8666	5.1%	118900	91.0%
6	1540.7	20.4%	2031.1	14.4%	0.8479	2.8%	85922	38.0%
5	1700.8	22.5%	2095.7	14.9%	0.8470	2.7%	89639	44.0%

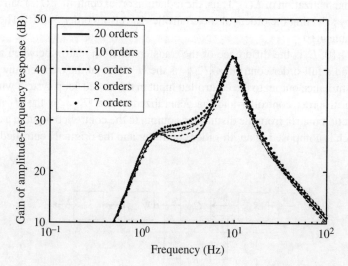

Figure 4.40 Amplitude-frequency characteristics of the vertical acceleration by using different controllers.

4.6.3.2   Other Order-reduction Methods

Modal truncation (MT) and balanced truncation (BT) methods are usual reduction methods for a linear time invariant system.

1. *Modal truncation method*
   Truncate the mode which has the lesser effect on the dynamic response of a system, such as high frequency modes.
       According to the MT method, a 20th-order controller would be reduced. In the order-reduction process, it can be found that when the order of a controller was reduced to the 6th-order, the control effect become worse. Figure 4.41 shows the closed-loop amplitude-frequency response curves of the front tyre-road travel to the body vertical acceleration, by using different order controllers based on the MT method.
2. *Balanced truncation method*
   Keep the important modes and remove the unimportant modes of the system. The importance of the modes is reflected by the corresponding Hankel singular values.
       According to the Hankel singular values designed in the previous 20th-order controller, it can be seen that the singular values are larger before the 9th order. By using the BT method to a full-order controller for reduction, the control effect worsens when orders are reduced to the 8th-order. Figure 4.42 shows the amplitude-frequency response curves of the front tyre-road travel to the body vertical acceleration by using different order controllers based on the BT method.

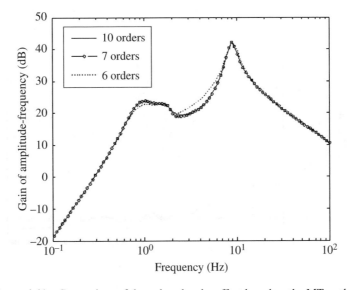

Figure 4.41   Comparison of the reduced-order effect based on the MT method.

4.6.3.3   Comparison of Different Reduction Methods

The use of the MT, BT, and OHNR methods to reduce a 20th-order H_∞ controller to an 8th-order controller and their comparison will be studied in this section.

The difference of the Maximal Singular Value (MSV) is compared in Figure 4.43, which is an 8th-order controller and a full-order controller designed by different reduction methods. It can be found that the MSV of the MT method is much larger than the other two

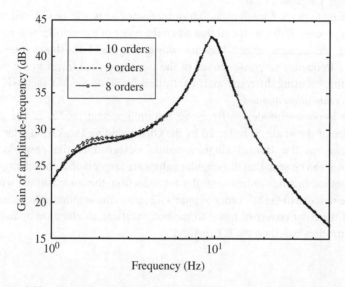

Figure 4.42   Comparison of the reduced-order effect based on the BT method.

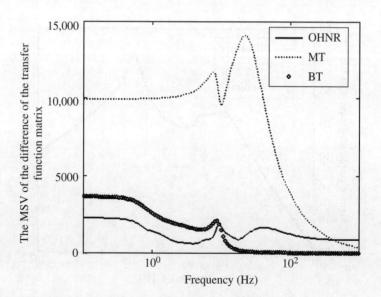

Figure 4.43   Comparison of MSV of E_8.

methods within the whole frequency bands as a result of the MT method truncating a lot of modes. The MSV of the BT method is larger than the OHNR method within a low frequency band and tends to zero in a larger than 100Hz frequency band. The important frequency band of designing a suspension controller is in 0.5–12.5Hz. The MSV of the OHNR method is the smallest of the three methods, and it means that using the OHNR method in this frequency band to get a reduction controller is closer to a 20th-order controller.

The corresponding index values of the MT method are much larger in each norm index, which are listed in Table 4.11; the $\left\|E_8\right\|_H$ and $\left\|E_8\right\|_\infty$ of the BT method are larger than the OHNR method.

In order to further compare the BT and OHNR methods, Figure 4.44 shows the amplitude–frequency response curves of the front tyre-road travel to the body vertical acceleration by using an 8th-order controller based on the two kinds of methods. It is clear that the performance of the closed-loop system is better within 4–10Hz by using the OHNR method, as shown in Figure 4.44.

4.6.4 Simulation Analysis

The above analysis shows that the OHNR method can get a reduced-order controller which is closer to a full-order controller. In order to further analyze the reduction effect, a

Table 4.11 Comparison of the effect of different reduced-order methods.

Reduced-order methods	$\left\|E_8\right\|_H$	$\left\|E_8\right\|_\infty$	$\left\|T_{8\overline{zw}}\right\|_\infty$
MT	7570.7	14072	0.9677
BT	2292.3	3652.3	0.8386
OHNR	1267.5	2257.5	0.8633

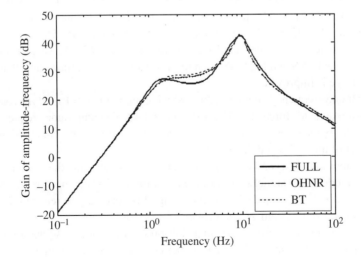

Figure 4.44 Comparison of the reduction effect of the BT and OHNR methods.

Table 4.12 Some vehicle physical parameters.

Parameters	Symbols	Dimensions	Values
Distance from CG to front axle	a	m	1.4
Distance from CG to rear axle	b	m	1.7
Half-track width	t	m	0.45
Sprung mass	m_2	kg	1500
Unsprung mass	$m_{1A}, m_{1B}, m_{1C}, m_{1D}$	kg	59
Damper coefficient: front	$c_{sA}(c_{sB})$	$N \cdot s / m$	1000
Damper coefficient: rear	$c_{sC}(c_{sD})$	$N \cdot s / m$	1100
Front suspension stiffness	$k_{sA}(k_{sB})$	N/m	35000
Rear suspension stiffness	$k_{sC}(k_{sD})$	N/m	38000
Pitch moment of inertia	I_P	$kg \cdot m^2$	2160
Roll moment of inertia	I_r	$kg \cdot m^2$	460
Tyre stiffness	$K_{tA}, K_{tB}, K_{tC}, K_{tD}$	N/m	190000

simulation analysis was carried out for four systems: a Reduction Active Suspension System (RASS) composed of an 8th-order controller based on the OHNR method and an original active suspension; a Full-order Active Suspension System (FASS) composed of a 20th-order controller; an original active suspension; and a Passive Suspension System (PSS). The simulation results are shown in Figures 4.45–4.48. Some vehicle parameters used in the simulations are shown in Table 4.12.

From Figure 4.45, it is clear that the FASS and RASS are similar in the amplitude-requency curves within a vertical vibration of 4~12Hz and an angular vibration of 1~2Hz. The amplitudes are largely decreased compared with the PSS, thus playing better with vibration isolation.

In the whole frequency ranges, the amplitude-frequency characteristics of the RASS and the FASS are almost the same as shown in Figure 4.46. Within the frequency range 1~12Hz, which is between the natural frequency of the body (about 1~2Hz) and the natural frequency of the tyres (about 8–12Hz), the amplitudes of the FASS and the RASS are lower than the PSS. When a vehicle runs under a pulse excitation (Figure 4.47), the suspension dynamic deflections of the FASS and the RASS can reach steady state in a short time. Thus, the riding performance is improved.

The amplitude-frequency characteristics, which contain the left front tyre-road travel to the suspension actuator force, are shown in Figure 4.48. The amplitude-frequency characteristics of the other tyres to each actuator force are similar. The curves show that the control effect of the FASS and the RASS is similar.

From the above analysis and calculation examples, it is clear that the established 7-DOF full-vehicle model with an active suspension is a minimal state space realization model with 14 orders. In order to apply it to engineering, it is necessary to reduce the order for a weighted generalized system H_∞ controller with a 20th-order. The closed-loop control performance of a reduced-order controller (8th-order controller) designed by the Hankel norm optimal index is similar with a 20th-order controller. Many simulations have shown that, for a vehicle implementing an active suspension with a reduced-order controller and a

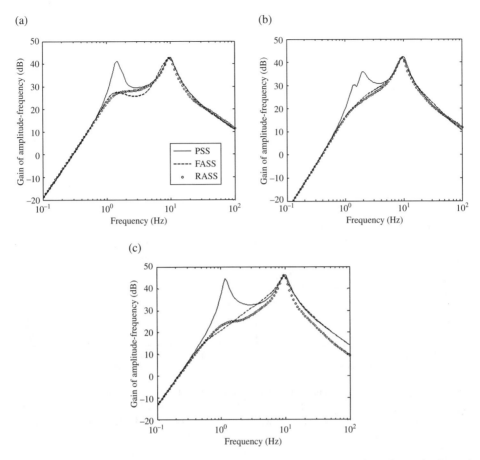

Figure 4.45 Amplitude-frequency characteristics of the left front tyre-road travel x_{0A} to body accelerations $\ddot{x}_2$, $\ddot{\theta}$, $\ddot{\varphi}$. (a) Amplitude-frequency characteristics of $\ddot{x}_2 \sim x_{0A}$ (b) Amplitude-frequency characteristic of $\ddot{\theta} \sim x_{0A}$. (c) Amplitude-frequency characteristics of $\ddot{\varphi} \sim x_{0A}$.

full-order controller, the frequency domain performances of the body vertical acceleration, pitching angular acceleration, and rolling angular acceleration are similar, and they are better than a passive suspension. In the usual frequency band, the suspension dynamic deflection for a reduced-order controller and a full-order controller is almost the same, and it is better than a passive suspension within 1~12Hz. Both the reduced-order control system and the 20th-order control system improve the riding comfort. Using an 8-order controller to replace a 20th-order controller not only has a better control performance, but also brings greater convenience to engineering practices.

The above study results involve just one method. By using the controller reduction method of the frequency weighted left factorization, the controller order can also decrease greatly. The closed-loop control performance loses less when a 7th-order controller is used to replace a full 20th-order controller. There many research results regarding this, and it is beyond the scope this book to show all the details here.

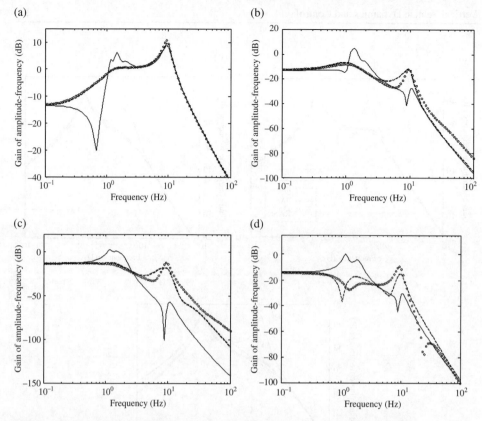

Figure 4.46 Amplitude-frequency characteristics of the left front tyre-road travel x_{0A} to the suspension dynamic deflection. (a) Amplitude-frequency characteristics of $(x_{2A} - x_{1A}) \sim x_{0A}$. (b) Amplitude-frequency characteristics of $(x_{2B} - x_{1B}) \sim x_{0A}$. (c) Amplitude-frequency characteristics of $(x_{2C} - x_{1C}) \sim x_{0A}$. (d) Amplitude-frequency characteristics of $(x_{2D} - x_{1D}) \sim x_{0A}$.

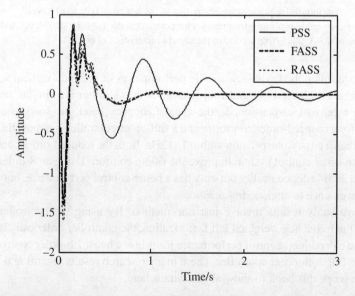

Figure 4.47 Time domain responses of the left front suspension dynamic deflection $x_{2A} - x_{1A}$ when 4 tyres getting pulse excitation with amplitude $1cm$ and width $0.1s$.

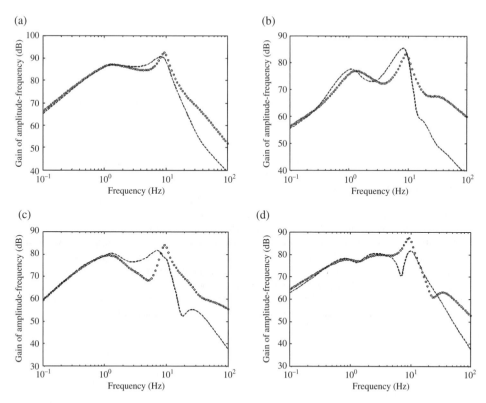

Figure 4.48 Amplitude-frequency characteristics of the left front tyre-road excitation to actuator force. (a) Amplitude-frequency characteristics of $f_A \sim x_{0A}$. (b) Amplitude-frequency characteristics of $f_B \sim x_{0A}$. (c) Amplitude-frequency characteristics of $f_C \sim x_{0A}$. (d) Amplitude-frequency characteristics of $f_D \sim x_{0A}$.

References

[1] Yu Z S. Automobile Theory. Beijing: China Machine Press, 2009.

[2] Yu F, Lin Y. Vehicle System Dynamics. Beijing: China Machine Press, 2005.

[3] Gu Z Q, Ma K G, Chen W D. Vibration Active Control. Beijing: National Defence Industry Press, 1997.

[4] Asada H, Park J H, Rai S. A control-configured flexible arm: Integrated structure/control design. Proceedings of the IEEE, International Conference on Robotics and Automation, Sacramento, California, 1991.

[5] Asada H, Park J H. Integrated structure/control design of a two-link non-rigid robot arm for high speed positioning.Proceedings of the IEEE, International Conference on Robotics and Automation, 1992.

[6] Pil A C, Asada H H. Integrated structure/control design of mechatronic systems using a recursive experimental optimization method. IEEE/ASME Transactions on Mechatronics, 1996, 1(3): 191–203.

[7] Mayzus A, Grigoriadis K. Integrated structural and control design for structural systems via LMIs. Proceedings of the IEEE, International Conference on Control Applications, Kohala Coast – Island of Hawaii, 1999.

[8] Savant S V, Asada H H. Integrated structure/control design based on model validity and robustness margin. Proceedings of the American Control Conference, San Diego, California, 1999.

[9] Shi G, Robert E S. An algorithm for integrated structure and control design with variance bounds. Proceedings of the 35th Conference on Decision and Control, Kobe, Japan, 1995.

[10] Kajiwara I, Nagamatsu A. Integrated design of structure and control system considering performance and stability. Proceedings of the IEEE, International Conference on Control Applications, Kohala Coast – Island of Hawaii, 1999.

[11] Wang Q R, Zhu W L, Chen W W. Integrated optimization of structure and control parameters for an automotive semi-active suspension system based on genetic algorithm and LQG control. Automotive Engineering, 2002, 24(3): 236–240.

[12] Chen W W, Wang Q R, Zhu W L. Integrated structure-control design for a suspension system based on generic algorithm and H_∞ control. Journal of Vibration Engineering, 2003, 16(2): 143–148.

[13] Zhu M F. Research on Decoupling Control Method of Vehicle Chassis and Time-lag Control of Chassis Key Subsystems. Doctoral dissertation, Hefei University of Technology, Hefei, 2011.

[14] Chen W W, Zhu M F. Delay-dependent H_2 / H_∞ control for vehicle magneto-rheological semi-active suspension. Chinese Journal of Mechanical Engineering, 2011, 24(6): 1028–1034.

[15] Fang M, Shi M G, Chen W W. Research on multi-objective and mixed H_2 / H_∞ control of a vehicle active suspension. Transactions of the Chinese Society for Agricultural Machinery, 2005, 36(3): 4–7, 18.

[16] Shi M G, Chen W W. Mixed H_2 / H_∞ control based on game theory and its application to the vehicle active suspension system. Control Theory & Applications, 2005, 22(6): 882–888.

[17] Wang H B, Fang M, Chen W W. A study on H_∞ controller order reduction for vehicle suspension system based on frequency-weighted left factorization. Automotive Engineering, 2006, 28(9): 812–816.

[18] Fang M, Wang H B, Chen W W. Order-reduction of H_∞ controller for the active suspension of a vehicle system. Control Theory & Applications, 2007, 24(4): 553–560.

[19] Wang J, Xu W L, Chen W W. Optimal Hankel-norm reduction of active suspension model with application in suspension multi-objective control. International Journal of Vehicle Design, 2006, 40(1/2/3): 175–195.

5

Lateral Vehicle Dynamics and Control

5.1 General Equations of Lateral Vehicle Dynamics[1–2]

In motion, vehicles modeled as a rigid body have six degrees of freedom. In this chapter, vehicles are assumed to only have planar motion parallel to the road's surface. But, considering the tyre cornering properties, we assume the following:

1. No vertical, pitch, or roll motion with respect to the y-axis and x-axis, respectively
2. For constant speed motion, no account is taken of the tangential force and aerodynamic effect
3. Ignoring the influence of the clearance, friction and deformation of the steering system, and taking the rotation angle of the front wheel as an input directly
4. The change of the tyre characteristics and the role of the aligning torque on the left wheel caused by load changes are disregarded.

Then, the vehicle is simplified as a "bicycle" model, as shown in Figure 5.1, which is the two degrees of freedom model with lateral and yaw motion.

The vehicle receives the driver's instructions, the front wheel is turned by an angle δ_f, the centrifugal force is generated at the centroid, which causes lateral reaction forces F_{y1} and F_{y2} on the front and rear, and the corresponding cornering angles α_1 and α_2, respectively. Thus, the direction of the front and rear wheel speeds u_1 and u_2 can be determined. According to the motion theorem of rigid bodies, the instantaneous center of rotation o' can be obtained. The turning radius R is the distance between o' and the centroid o. The speed of the centroid is $v_1 = \omega_r R$, in which ω_r is the yaw rate. The component of v_1 in the x-axis is:

$$u = v_1 \cos \beta \tag{5.1}$$

Integrated Vehicle Dynamics and Control, First Edition. Wuwei Chen, Hansong Xiao, Qidong Wang, Linfeng Zhao and Maofei Zhu.
© 2016 John Wiley & Sons Singapore Pte. Ltd. Published 2016 by John Wiley & Sons, Ltd.

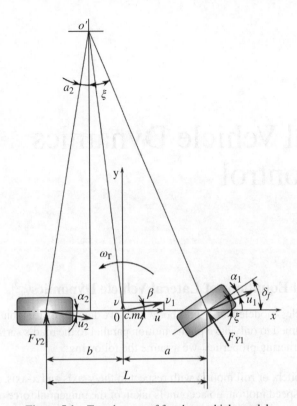

Figure 5.1 Two degrees of freedom vehicle model.

Here, β is the angle between the centroid speed and the longitudinal axis.

As β is very small, we can assume that $\cos \beta = 1$. Equation (5.1) can be written as follows:

$$u = v_1 = \omega_r R \tag{5.2}$$

The component of v_1 in the y-axis is:

$$v = v_1 \sin \beta, \beta = \frac{v}{v_1} = \frac{v}{u} \tag{5.3}$$

So that the centroid acceleration a_y in the y-axis is:

$$a_y = \dot{v} + u\omega_r \tag{5.4}$$

The differential equations of motion are derived from the force and moment equilibrium equations, as follows:

$$F_{y1} + F_{y2} = ma_y = m\left(\dot{v} + u\omega_r\right)$$
$$aF_{y1} - bF_{y2} = I_z\dot{\omega}_r \tag{5.5}$$

where F_{y1}, F_{y2} are the lateral reaction forces on the front and rear wheels, m is the vehicle mass, a, b are the horizontal distance between the front and rear axles and the vehicle centroid, and I_z is the moment of inertia of the vehicle body with respect to the z-axis.

The values of the lateral forces depend on the cornering stiffness and angle such as:

$$F_{y1} = K_{\alpha 1}\alpha_1 \qquad F_{y2} = K_{\alpha 2}\alpha_2$$

where $K_{\alpha 1}$, $K_{\alpha 2}$ are the cornering stiffnesses of the front and rear tyres, α_1, α_2 are the sideslip angles of the front and rear tyres.

The values of α_1, α_2, of the front and rear tyres can be obtained from the following geometric relationships:

$$\alpha_1 = \frac{a\omega_r}{u} + \beta - \delta_f \tag{5.6}$$

$$\alpha_2 = \beta - \frac{b\omega_r}{u} \tag{5.7}$$

Combining these last equations with equation (5.3), the following equations can be obtained:

$$\left(K_{\alpha 1} + K_{\alpha 2}\right)\beta + \frac{\omega_r}{u}\left(aK_{\alpha 1} - bK_{\alpha 2}\right) - K_{\alpha 1}\delta_f = m\left(\dot{v} + u\omega_r\right) \tag{5.8}$$

$$\left(aK_{\alpha 1} - bK_{\alpha 2}\right)\beta + \frac{\omega_r}{u}\left(a^2 K_{\alpha 1} + b^2 K_{\alpha 2}\right) - aK_{\alpha 1}\delta_f = I_z\,\dot{\omega}_r \tag{5.9}$$

With these two equations, the response of various operating conditions can be analyzed.

5.2 Handling and Stability Analysis

5.2.1 Steady State Response (Steady Steering)

If a step angle is put into the vehicle, its response is a constant driving circle, and the yaw rate is constant, which is $\omega_r = $ constant, $\dot{\omega}_r = 0$, $\dot{v} = 0$, when combined with equations (5.8) and (5.9), can get:

$$\left(K_{\alpha 1} + K_{\alpha 2}\right)\frac{v}{u} + \frac{\omega_r}{u}\left(aK_{\alpha 1} - bK_{\alpha 2}\right) - K_{\alpha 1}\delta_f = mu\omega_r$$

$$\left(aK_{\alpha 1} - bK_{\alpha 2}\right)\frac{v}{u} + \frac{\omega_r}{u}\left(a^2 K_{\alpha 1} + b^2 K_{\alpha 2}\right) - aK_{\alpha 1}\delta_f = 0 \tag{5.10}$$

Eliminating v in the above equations, ω_r can be obtained:

$$\frac{\omega_r}{\delta_f} = \frac{u/L}{1 + Ku^2} \tag{5.11}$$

where $K = \dfrac{m}{L^2}\left(\dfrac{a}{K_{\alpha 2}} - \dfrac{b}{K_{\alpha 1}}\right)$ is the stability factor, and L is the wheelbase.

In Germany, by using (EG) = KL, then

$$\frac{\omega_r}{\delta_f} = \frac{u}{L + (EG)u^2} \tag{5.12}$$

where (EG) is the steering gradient, and $\dfrac{\omega_r}{\delta_f}$ is the steady state yaw rate gain, also known as the steering sensitivity.

The values of the stability factor K have a great influence on stability. The following three cases, K = 0, K > 0, and K < 0, are analyzed.

(1) K = 0

By putting K = 0 into equation (5.11), $\dfrac{\omega_r}{\delta_f} = u/L$. The steering characteristic is similar to the one with no tyre sideslip. Yaw rate has a linear relationship with vehicle speed, and the slope is 1/L, which for the steering characteristic is also called neutral steering. The curve of $\dfrac{\omega_r}{\delta_f} \sim u$ is shown in Figure 5.2.

(2) K > 0

From equation (5.11), the yaw rate gain is smaller than the one for neutral steering. ω_r/δ_f is no longer a linear relationship with vehicle speed. The curve of $\omega_r/\delta_f \sim u$ is a line of the yaw gain below the neutral steering, and it grows to a certain point and then bends down.

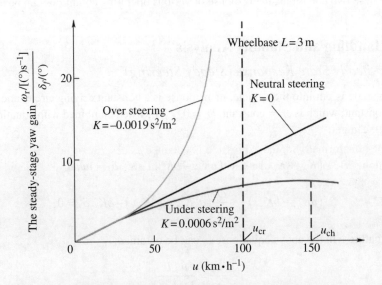

Figure 5.2   Vehicle steady yaw rate gain curves.

This steering characteristic is called understeering, and the greater K is, the lower the yaw rate gain curve is, and the greater the understeering is. According to the condition which the tangent line at the maximum of the curve parallels to the axis u (the derivative of u is zero), it can be calculated that, when $u_{ch} = 1/\sqrt{K}$, the vehicle steady yaw rate gain achieves a maximum value, and $\omega_r / \delta_f = u_{ch} / 2L$.

This maximum value is half of the yaw rate gain of the neutral steering vehicle whose wheelbase L is equal. At this time, u_{ch} is called the characteristic speed. That is to say, when the understeer increases, the characteristic speed u_{ch} reduces as K increases. In modern cars, the characteristic speed is designed between 65 and 100 km/h.

(3)   K < 0

For this case, the denominator of equation (5.11) is less than 1, and the yaw rate gain ω_r/δ_f is bigger than the neutral steering. With the vehicle speed increasing, the curve will bend upward (shown in Figure 5.2). This steering characteristic is called over-steering. The greater the absolute value of K, the greater the oversteering. Obviously, when $u_{cr} = \sqrt{(-1/K)}$, the steady state yaw rate gain tends to infinity (see Figure 5.2), where u_{cr} is called the critical speed, and the lower the critical speed, the greater the oversteering.

The physical meaning of the critical speed is that when the vehicle speed with over-steer characteristics reaches this value, the extremely small steering angles of the front wheel will cause a great yaw rate. This means that the car turns to face instability, and then a sharp steering, sideslip, or rollover will occur. This speed is called the critical speed. All modern cars are designed with understeer characteristics, so there is no real critical speed. Instead, the characteristic speed has an important practical significance, and it becomes an important performance index in vehicle design. From equation (5.11), if the lateral acceleration a_y is multiplied by the expression of stability factor K, then we get:

$$K = \frac{m}{L^2}\left(\frac{a}{K_{\alpha 2}} - \frac{b}{K_{\alpha 1}}\right) = \frac{1}{a_y L}\left(\frac{F_{y2}}{K_{\alpha 2}} - \frac{F_{y1}}{K_{\alpha 1}}\right) = \frac{1}{a_y L}(\alpha_1 - \alpha_2) \qquad (5.13)$$

In equation (5.13), since the sign of the lateral acceleration a_y is opposite to the sideslip angles of the front and rear tyres, and therefore the absolute values of the angles α_1 and α_2, and of the acceleration a_y are taken, the signs of the front and rear items in the brackets should be transformed. So, the expression is obtained as described above.

Equation (5.13) shows the relationship between the steering characteristics and the side-slip angle difference of the front and rear tyres as follows:

$$\left.\begin{array}{l} (\alpha_1 - \alpha_2) > 0, \ K > 0, \ \text{Understeering} \\ (\alpha_1 - \alpha_2) < 0, \ K < 0, \ \text{Oversteer} \\ (\alpha_1 - \alpha_2) = 0, \ K = 0, \ \text{Neutral steering} \end{array}\right\} \qquad (5.14)$$

Equation (5.14) shows that only when $\alpha_1 = \alpha_2$ can a neutral steering state be obtained.

Now, assuming that C_n can be found on the center line of the longitudinal axis, when the total lateral force is applied on this point, $\alpha_1 = \alpha_2$ can be obtained. The distance between C_n and the front and rear axles are a' and b', respectively. If $\alpha_1 = \alpha_2 = \alpha$, then:

$$F_{y1} = K_{\alpha 1}\alpha, \quad F_{y2} = K_{\alpha 2}\alpha \tag{5.15}$$

The resultant force is then:

$$F_y = F_{y1} + F_{y2} = \left(K_{\alpha 1} + K_{\alpha 2}\right)\alpha \tag{5.16}$$

According to moment equilibrium conditions, a' and b' can be calculated, respectively:

$$a' = \frac{F_{y2}L}{F_{y1} + F_{y2}} = \frac{K_{\alpha 2}}{K_{\alpha 1} + K_{\alpha 2}}L \tag{5.17}$$

$$b' = \frac{F_{y1}L}{F_{y1} + F_{y2}} = \frac{K_{\alpha 1}}{K_{\alpha 1} + K_{\alpha 2}}L \tag{5.18}$$

The vehicle steering characteristics can be determined by the ratio between $(a'-a)$ and wheelbase L. This ratio is also called the static margin (SM).

$$SM = \frac{a'-a}{L} = \frac{K_{\alpha 2}}{K_{\alpha 1} + K_{\alpha 2}} - \frac{a}{L} \tag{5.19}$$

When the turning point C_n coincides with the centroid $(a' = a)$:

$$SM = 0 \quad \text{Neutral steering } (\alpha_1 = \alpha_2)$$

When the centroid is before the neutral point $(a < a')$:

$$SM > 0 \quad \text{Understeering } (\alpha_1 > \alpha_2)$$

When the centroid is after the neutral point $(a > a')$:

$$SM < 0 \quad \text{Oversteering } (\alpha_1 < \alpha_2)$$

So, during the design, according to the tyre cornering stiffness $K_{\alpha 1}$, $K_{\alpha 2}$, the location (a', b') of C_n should be calculated first. And then, in the overall design, in order to ensure good steering characteristics, the relationship $a \leq a'$ must be satisfied by the centroid location (a, b).

5.2.2 Transient Response

First, equations (5.8) and (5.9) can be rewritten as follows:

$$A\beta + \frac{\omega_r}{u}B - K_{\alpha 1}\delta_f = m\left(\dot{v} + u\omega_r\right) \tag{5.20}$$

$$B\beta + \frac{\omega_r}{u}D - aK_{\alpha 1}\delta_f = I_z\dot{\omega}_r \tag{5.21}$$

where $A = K_{\alpha 1} + K_{\alpha 2}$, $B = (aK_{\alpha 1} - bK_{\alpha 2})$, $D = \left(a^2 K_{\alpha 1} + b^2 K_{\alpha 2}\right)$, $\beta = v/u$.

From equation (5.21), the following can be calculated:

$$\beta = \left(I_z \dot{\omega}_r - \frac{D}{u} \omega_r + \alpha K_{\alpha1} \delta_f \right) \Big/ B$$

Combining this with equation (5.20), and eliminating β, the differential equation of $\ddot{\omega}_r$ can be organized into the following:

$$m_0 \ddot{\omega}_r + h\dot{\omega}_r + c\omega_r = b_1 \dot{\delta}_f + b_0 \delta_f \qquad (5.22)$$

where $m_0 = muI_z$, $h = -[mD + I_z A]$, $c = muB + (AD - B^2)/u$, $b_1 = -muaK_{\alpha1}$, $b_0 = LK_{\alpha1}K_{\alpha2}$.
This is a second-order differential equation of forced vibration, and can be further rewritten as:

$$\ddot{\omega}_r + 2\omega_0 \xi \dot{\omega}_r + \omega_0^2 \omega_r = B_1 \dot{\delta}_f + B_0 \delta_f \qquad (5.23)$$

where $\omega_0^2 = C/m_0$, $B_1 = b_1/m_0$, $\xi = h/2\omega_0 m_0$, $B_0 = b_0/m_0$.
When the angle step input is applied to the front wheel, the front wheel angle can be expressed as:

$$t < 0, \delta_f = 0; \quad t \geq 0, \delta_f = \delta_0; \quad t > 0, \dot{\delta}_f = 0$$

Substituting this last expression into equation (5.23), when $t > 0$:

$$\dot{\omega}_r + 2\omega_0 \xi \dot{\omega}_r + \omega_0^2 \omega_r = B_0 \delta_0 \qquad (5.24)$$

Equation (5.24) is a second-order non-homogeneous differential equation with constant coefficients. Its general solution is equal to the sum of a special solution and a general solution of its corresponding homogeneous differential equation. Its special solution is:

$$\omega_r = \frac{B_0 \delta_0}{\omega_0^2} = \frac{u/L}{1 + Ku^2} \delta_0 = \left(\frac{\omega_r}{\delta} \right)_s \delta_0$$

The steady state yaw rate is $\omega_{r0} = \left(\dfrac{\omega_r}{\delta} \right)_s \delta_0$.

The taken Laplace transform is taken from equation (5.23) on both sides getting:

$$\left(s^2 + 2\xi\omega_0 s + \omega_0^2 \right) \omega_r(s) - B_1 \delta_0 = B_0 \delta_0 / s$$

$$\omega_r(s) = \frac{(B_1 s + B_0)\delta_0}{s\left(s^2 + 2\xi\omega_0 s + \omega_0^2 \right)} \qquad (5.25)$$

Considering $\xi = 1, \xi > 1, \xi < 1$, then the solution of equation (5.25) has three cases.
The roots of the characteristic equation are:

$$\xi < 1, \ s = -\xi\omega_0 \pm \omega_0 \sqrt{1 - \xi^2} i \quad \text{A pair of conjugate complex roots}$$
$$\xi = 1, \ s = -\omega_0 \qquad\qquad\qquad \text{A double root}$$
$$\xi > 1, \ s = -\xi\omega_0 \pm \omega_0 \sqrt{\xi^2 - 1} \quad \text{Two different real roots}$$

The general solution of the homogeneous equation is:

$$\xi < 1, \omega_r = Ce^{-\xi\omega_0 t}\sin(\omega_0\sqrt{1-\xi^2}\,t+\phi)$$
$$\xi = 1, \omega_r = (C_1+C_2)e^{-\omega_0 t}$$
$$\xi > 1, \omega_r = C_3 e^{\left(-\xi\omega_0+\omega_0\sqrt{\xi^2-1}\right)t} + C_4 e^{\left(-\xi\omega_0-\omega_0\sqrt{\xi^2-1}\right)}$$

where C, ϕ, C_1, C_2, C_3, C_4 are integration constants,which can be determined by the initial motion conditions.

When $\xi > 1$, it means that there is a large damping in the system, and the yaw rate response $\omega_r(t)$ is monotonically increasing. With the growth of time, ω_r approaches the steady state yaw rate. But, after the vehicle exceeds the critical speed, ω_r is divergent and tends to infinity, and the vehicle loses stability at this time.

When $\xi = 1$, called critical damping, the yaw rate response $\omega_r(t)$ is also monotonically increasing and approaches to a steady state yaw rate.

When $\xi < 1$, called small damping, the yaw rate response $\omega_r(t)$ is a reduction sine curve which converges to a steady state yaw rate.

As vehicles usually have small damping, the variation of the yaw rate response when $\xi < 1$ is the following

$$\omega_r(t) = \frac{B_0\delta_0}{\omega_0^2} + Ce^{-\xi\omega_0 t}\sin\left(\omega_0\sqrt{1-\xi^2}\,t+\phi\right)$$

Let $\omega = \omega_0\sqrt{1-\xi^2}$. The initial conditions for the motion are:

$$t = 0,\ \omega_r = 0,\ v = 0,\ \delta_f = \delta_0,\ \beta = 0$$

According to equation (5.9),

$$\dot{\omega}_r(0) = -\frac{aK_{\alpha 1}\delta_0}{I_z}$$

Equation (5.25) can also be written as rational fractions:

$$\omega_r(s) = \frac{A_1}{s} + \frac{A_2 s + A_3}{s^2 + 2\xi\omega_0 s + \omega_0^2} \tag{5.26}$$

Comparing equation (5.25) with equation (5.26), the following can be obtained:

$$\left.\begin{aligned} A_1 &= B_0\delta_0 / \omega_0^2 \\ A_2 &= -A_1 = -B_0\delta_0 / \omega_0^2 \\ A_3 &= B_0\delta_0\left(\frac{B_1}{B_0}\omega_0^2 - 2\xi\omega_0\right)/\omega_0^2 \end{aligned}\right\} \tag{5.27}$$

Equation (5.25) can be obtained from the Laplace inverse transform:

$$\omega_r(t) = L^{-1}\left[\omega_r(s)\right] = A_1 + A_2 e^{-\xi\omega_0 t}\cos\omega t + A_4 e^{-\xi\omega_0 t}\sin\omega t \tag{5.28}$$

where:

$$A_4 = \frac{A_3 - A_2\xi\omega_0}{\omega} = \frac{B_1\delta_0 - \dfrac{B_1\delta_0\xi}{\omega_0}}{\omega}$$

Equation (5.28) can be further rewritten as:

$$\omega_r(t) = \frac{B_0\delta_0}{\omega_0^2}\left[1 - A'e^{-\xi\omega_0 t} + \sin\left(\omega t + \Phi\right)\right] \tag{5.29}$$

where:

$$A' = \sqrt{1 + \left(\frac{A_4}{A_2}\right)^2} \qquad \text{and} \qquad \Phi = \arctan\left(A_2 / A_4\right)$$

The time history curves of the yaw rate response are shown in Figure 5.3, in which ξ has a great impact on the yaw rate response.

The greater the damping ratio ξ, the faster the attenuation of the yaw rate response. When the damping ratio ξ is small, the attenuation of the yaw rate response is slow and the overshoot is large. The following curves in Figure 5.3 show that the yaw rate finally tends to a steady yaw rate $B_0\delta_0 / \omega_0^2$.

Generally, the quality evaluation of the transient response comes from the undamped natural frequency ω_0, damping ratio ξ, overshoot, and other parameters.

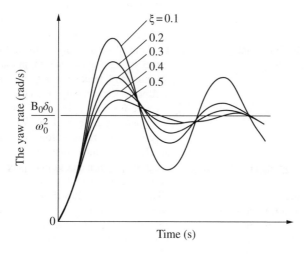

Figure 5.3   Yaw rate response time history.

5.2.3   The Frequency Response Characteristics of Yaw Rate

If the vehicle is a linear dynamic system, it has frequency retention. When the input is a sine function, the steady-state output is also a sine function with the same frequency. But the input/output amplitude and phase are different. The ratio of the input/output amplitude is a function of frequency f, called amplitude-frequency characteristics $A(f)$. The phase difference is also a function of f, called phase-frequency characteristics $\phi(f)$. Both of them are named as the frequency response characteristics, and can be represented unified by using the frequency response function $H(f)$.

5.3   Handling Stability Evaluations

Currently, handling stability is evaluated subjectively and experimentally by most automobile manufacturers.

5.3.1   Subjective Evaluation Contents

The evaluation contents of a vehicle handling stability analysis include static state evaluations and dynamic evaluations; the common evaluation items are depicted in Table 5.1.

5.3.2   Experimental Evaluation Contents

Currently, Chinese standard QC/T 480–1999 is implemented to evaluate handling stability. The criterion thresholds and evaluation of the handling and stability of a vehicle can be carried out by using six test evaluations, such as steady steering, transient response of the

Table 5.1   Subjective evaluations of handling stability.

Static evaluations	Steering wheel	Steering wheel size
		Adjusting angles of the steering wheel
	Steering in-situ	Slow and even steering force under idle speed
		Fast steering force under idle speed
Dynamic evaluations	Straight-running stability	Uniform speed running on smooth road surface
		Accelerating running on smooth road surface
		Running with braking on smooth road surface
		Uniform speed running on rough road surface
	Steering portability	Steering force under low-speed cornering
		Steering force under high-speed cornering
		Steering wheel rebound force with large lateral acceleration while cornering
	Aligning performance	Aligning performance at low speed
		Aligning performance at medium and high speeds
	Steady turn performance(fixed circle driving)	Understeering characteristics
		Roll characteristics
		Limited speed of tyre sideslip
	Transient performance	Body roll at high-speed lane change
		The steering response to high-speed lane change
		The tyre grip performance of high-speed lane change

steering wheel angle step input, transient response of the steering wheel angle pulse input, steering aligning performance, steering portability, and snaking tests. Information about the vehicle handling and stability is basically included in the evaluation of these projects. Of course, more companies carry out test evaluations with reference to advanced country standards.

5.4 Four-wheel Steering System and Control[3]

The sideslip angle of the body's centroid is mainly used to describe the problems of keeping track of the vehicle stability control. Although the centroid sideslip angle can be determined by the lateral and longitudinal forces applied to the vehicle, it is difficult to use the braking or the driving forces to directly control the vehicle lateral forces. Four-wheel steering is an important component of vehicle chassis active control technology, and also an important method of controlling vehicle stability. It can not only reduce the centroid sideslip angle, but also increase the tyre lateral force margin. It has the advantage of easily implementing track keeping and sideslip angle control, and is one of the developing directions of modern automotive technologies.

With the development of four-wheel steering vehicles, their control objectives are being deepened, and the control performance is being improved. In particular, it requires that the system can withstand the impact of vehicle parameter changes and maintain the desired steering characteristics, and also requires that the system still has a good response characteristic when the tyres reach the limit state. Under certain conditions, better control effects can be achieved by adaptive control and robust control. However, as the input values of the steering angle are very small under high lateral acceleration, it is difficult to accurately identify the vehicle real-time response parameters and, subsequently, design an adaptive control system and analyze the system's stability. In the application process of robust control, its ultimate controller design can be attributed to the solution of linear matrix inequalities, so it cannot be fully inclusive of the nonlinear error caused by the reduced tyre cornering stiffness and may make the control system unstable.

As the vertical and horizontal forces of the tyres and the road adhesion are interdependent and have a nonlinear relationship with the vertical loads, the nonlinear control algorithms should be used for the four-wheel steering. But, the uncertainty of the driving conditions and the environmental changes, robustness and adaptiveness should be required for the system.

Among them, the control algorithms based on neural network theory are effective methods for four-wheel steering nonlinear control. The nonlinear characteristics of the tyre/road contact at high lateral acceleration can be identified by using the neural network model. Through the online correction of the neural network weightings, the system has the adaptive capacity of withstanding vehicle parameter changes and maintaining the desired steering characteristics to accommodate for the nonlinear motion of the vehicles.

Based on the ideas expressed above, the hybrid control system of the yaw rate feedback and adaptive neural network is constructed. Through online correction of the neural network weightings, it has the adaptive capacity when the nonlinear control system is implemented. The virtual prototype model of the four-wheel steering vehicle is also constructed based on the software ADAMS/Car, and the co-simulation analysis in ADAMS and MATLAB environment is studied, so the effectiveness of the hybrid control system is verified.

5.4.1   Control Objectives of the Four-wheel Steering Vehicle

The rear tyre lateral force can be controlled directly and independently by a four-wheel steering vehicle, and the front and rear sideslip angles and lateral forces of the tyres can be changed simultaneously. Thus, the vehicle transient response performances and steering control capability can be improved. The control objectives can be summarized as follows:

1. Reducing the sideslip angle at the centroid
2. Achieving low speed with good maneuverability, high speed with good stability
3. In a certain frequency range, the gain and phase between the lateral acceleration, yaw rate, and steering angles with small changes
4. Reducing the phase difference between the lateral acceleration and yaw rate, and their respective phases
5. Achieving the desired steering characteristics
6. Withstanding the system parameter changes, maintaining the desired steering characteristics
7. When the tyres reach the limit state, they still have a good response.

5.4.2   Design of a Four-wheel Steering Control System

5.4.2.1   Yaw Rate Feedback Control of a 2-DOF Linear Model

The simplified 2-DOF linear model of a four-wheel steering vehicle is an important vehicle model that is still used in the various literature, which studies the dynamic characteristics of the four-wheel steering vehicle. Although the yaw rate and centroid sideslip angles determined by the 2-DOF linear vehicle model are only accurate in the linear region of the tyres, it is more stable to the vehicle, and it is relatively easy to grasp the steering characteristics for a driver. The model can be expressed as:

$$\begin{cases} mu\left(\dot{\beta}+\omega_r\right) = K_{\alpha1}\left(\beta+a\omega_r\,/\,u-\delta_f\right)+K_{\alpha2}\left(\beta-b\omega_r\,/\,u-\delta_r\right) \\ I_z\dot{\omega}_r = K_{\alpha1}a\left(\beta+a\omega_r\,/\,u-\delta_f\right)-K_{\alpha2}b\left(\beta-b\omega_r\,/\,u-\delta_r\right) \end{cases} \tag{5.30}$$

where δ_f, δ_r are the steering angles of the front and rear wheels respectively, and the other symbols are the same as before.

Comparing with the feed forward control, feedback control of the four-wheel steering system reduces the external interference more effectively. When the above linear model is used by the rear wheel steering control, the prominent features are the simple calculations and fast response. The control quantity depends on the steering angles of the front wheels and vehicle operating parameters, and it can be expressed as:

$$\delta_r = -c_1\delta_f + c_2 u\omega_r \tag{5.31}$$

where $c_1 = 1$, $c_2 = \dfrac{mb}{K_{\alpha1}L} + \dfrac{ma}{K_{\alpha2}L}$, L is the wheelbase, and a, b are the horizontal distances between the body mass center and the front and rear axles.

5.4.2.2   The Design of a Nonlinear Control System Based on the Neural Network Theory

Based on the feedback control scheme of a 2-DOF linear model, the tyre cornering force is considered to vary linearly with the sideslip angles, without considering the impact of non-linear factors such as tyres. When the sideslip angles are larger, the error of the linear model is larger. This cannot meet the higher control requirements of modern four-wheel steering vehicles. Based on this, the problem can be solved by using nonlinear control theory. The four-wheel steering control method based on artificial neural network is an effective way which considers the nonlinear dynamic characteristics of the vehicle and tyres. The neural network control system of the four-wheel steering system can be expressed as follows:

In Figure 5.4, y is the system output, including the yaw rate and the sideslip angles, and $y*$ is the reference system input. The network structure 6-10-2 is used in the identification neural network AN1.The six input variables are the yaw rate ω_r, centroid sideslip angle β, speed u, lateral acceleration a_y, and front, rear wheel steer angles δ_f and δ_r. The two output variables are the yaw rate ω_r and centroid sideslip angle β. The activation function of the hidden layer and output layer is the bipolar function. Considering the nonlinear effects of the tyres, the variation of the front steering angles with time by using a multi-body dynamic model for sampling needs a variety of modes. The rear steering angles can be a random mode. The calculation of data in each mode is saved as learning samples for later training.

The yaw rate feedback control shows that the output of the rear steering angles is related to the yaw rate, vehicle speed, and front steering angle. Therefore, the network structure 4-6-1 is used in the neural controller AN2. The four input variables are the yaw rate ω_r, speed u, front steering angle δ_f, and lateral acceleration a_y. The output variable is the rear steering angle δ_r. The activation function of the hidden layer and output layer is the bipolar function. The output of the rear steering angles is needed for proper gain calculation. Here, comparing with the yaw rate feedback control of a linear 2-DOF model, the input of AN2 has an additional parameter, the lateral acceleration. The purpose is to show the nonlinear

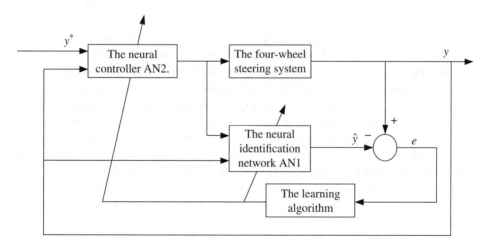

Figure 5.4   The neural network control system of a four-wheel steering vehicle.

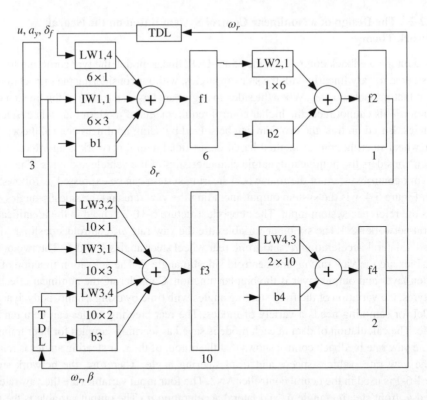

Figure 5.5   The topology of the neural network system of a four-wheel steering system.

characteristics and compensate the linear control error at large lateral acceleration. The offline training of AN1 and AN2 can be used to access weighting coefficients of each layer.

The neural controller AN2 is trained through the back-propagation algorithm. First, the neural identification network AN1 is required to do the error back-propagation. However, in order to facilitate the use of the function call method in MATLAB, the neural network toolbox for training is used, and the system is treated as a complex network. Therefore, AN1 and AN2 are combined into a composite neural network system of a four-wheel steering vehicle system, which is a composite network 4-6-1-10-2. The first layer of the network has 4 neurons, the second layer has 6 neurons, the third layer has only 1 neuron, and the output signals are the rear steering angles. The fourth layer has 10 neurons, and the output layer has 2 neurons. The output signals are the body yaw rate ω_r and sideslip angle β. With the exception of the input layer, the activation function of each layer is the bipolar function. Wherein, in addition to receive the steering angle signals of the rear wheel of the second layer, the third layer also accepts the feedback signals of the fourth layer. The four variables of the network are ω_r, u, δ_f, a_y, respectively, and the six variables of the second layer are δ_r, u, a_y, δ_f, ω_r, β. The topology of the neural network control system of the four-wheel steering system is shown in Figure 5.5.

Excluding the input layer, Figure 5.5 can be viewed as a 4-layer neural network where f_1, f_2, f_3, f_4 are the transfer function (activation function) of each layer of the network, respectively. In addition, b_1, b_2, b_3, b_4 are the offset (threshold values) of each layer neurons, respectively. Also, $IW_{1,1}$ is the network weight between the input layer and the first layer, and $IW_{3,1}$ is the network weight between the input layer and the third layer. The network weight $LW_{1,4}$ is between the fourth layer and the first layer of the network, $LW_{2,1}$ is the network weight between the first layer and the second layer, $LW_{3,2}$ is the network weight between the second layer and the third layer, $LW_{3,4}$ is the network weight between the fourth layer and the third layer, and $LW_{4,3}$ is the network weight between the third layer and the fourth layer. TDL is the time-lag, and the current time signal can be delayed several times.

Assuming AN2 has been trained, in order to train AN1, the output error of AN2 is back-propagated to AN1. The error signals shared with the layers of AN1 are used to adjust the weight matrix of the corresponding layer, until the index function of the network performance satisfies the requirements. In training, only the weight matrix of AN2 is modified. The weight matrix of AN1 represents the nonlinear dynamic characteristics of the vehicle and tyres. The trained property of the network object can be set to non-modification by using MATLAB, and the system identification is completely trained before. A network performance index J presented in equation (5.32) is the sum of the squares of the sideslip angle β. Where p is the sample size, i is the order number, and q is the number of the sampling.

$$J = \sum_{i=1}^{p} \sum_{n=1}^{q} \beta(n)^2 \tag{5.32}$$

5.4.2.3 A Hybrid Control System of the Yaw Rate Feedback and Neural Network

Nonlinear problems can be solved effectively by four-wheel steering neural network control systems. However, when the neural network is trained to be used offline, it is not adaptive. To make the system adaptive, this neural network control system can be trained offline to obtain network weights. After that, using online methods, the system has the adaptive capacity through the online learning correction. In order to facilitate the network with the weights which were adjusted online, the yaw rate feedback control and neural network control are combined. The hybrid control system is shown in Figure 5.6.

The next step is the online correction of AN2 to make the neural network adaptive. Methods are summarized as follows. When the output error between AN1 and the vehicle model exceeds the limit value, the input and output variables of the neural network in this condition are trained as a new sample, and the weight coefficient matrix of AN1 is set immutably. The weight coefficient matrix of AN2 is corrected online, and the correct neural controller is used for control calculations at the next time. Thus, AN1 is adaptive to the changes of the controlled target, and the neural network control system has the function of self-learning/self-tuning while online. It can overcome the various errors caused by parameter variation, not only solving nonlinear problems, but also by making the system adaptive.

The amount of actual control from the hybrid control system is determined by the sum of the output of the feedback controller and neural controller. This reflects the nonlinear compensation from the neural controller to the vehicle yaw rate feedback control, which is based on the linear 2-DOF model. Meanwhile, the effectiveness of the four-wheel steering control system is guaranteed by the feedback controller, and the learning tasks of the neural

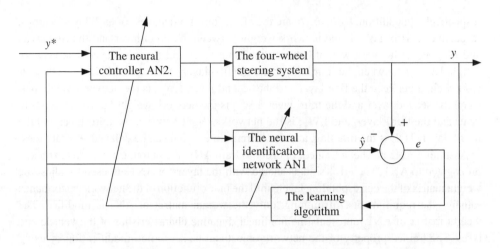

Figure 5.6 The four-wheel steering hybrid control system.

network are reduced. This makes the workload of the online self-learning/self-tuning reduce, and the nonlinear adaptive control achievable.

5.4.3 Multi-body Dynamics Modeling of a Four-wheel Steering Vehicle

Currently, the virtual prototyping software ADAMS has been widely used for modeling, solving, and visualizating vehicle dynamics technology. But due to functional limitations of the ADAMS software, the CAR module does not include the templates of a rear-wheel steering system. To make the rear wheel section with a steering function, a secondary development is created based on the existing rack and pinion steering templates. The rear-wheel steering subsystem is composed by the MacPherson rear suspension; and then the four-wheel steering vehicle model is assembled. The subsystems include the MacPherson front (rear) suspensions, front (rear) rack and pinion steering systems, braking system, wheels, body and so on.

In the ADAMS/CAR module, the rear-wheel steering angle can be controlled to achieve the four-wheel steering by adjusting the steering-gear angular displacement. Meanwhile, the displacement of the steering-gear angular displacement can be determined by the above control signals of the rear-wheel steering angles. Considering the four-wheel steering has certain requirements for the delay characteristics of the hydraulic system, the large delay of the hydraulic system will have an effect on the control stability of the four-wheel steering. Through theoretical analysis and experimental research into the hydraulic system, the delay characteristics of the hydraulic control system can be treated as the composition of the throttle and pure delay characteristics. To simplify the analysis, ignoring the throttle characteristics delay in the simulation, the pure delay inertia link of the hydraulic system is increased, and the time-lag of the hydraulic system is simulated to make the results more realistic and effective.

In the ADAMS/CAR modules, the topology of the components of each subsystem is defined. The subsystems of the four-wheel steering vehicle are built. Here, the rear-wheel steering subsystem is constructed as follows.

The rear-wheel steering subsystem is composed of the steering wheel, steering shaft, steering column, steering output shaft, steering rack and other components, and is connected with the rear suspension and body through the data communicator. In order to achieve a four-wheel steering control, setting input and output variables are needed for the rear-wheel steering subsystem in the model. Creating a state variable, and make it correspond to the control signals of the rear-wheel steering. The size of this control signal is matched to the steering angle signals of the rear-wheel by the gain link, and the state variable is set to be the input variable by using the Build-controls Toolkit in ADAMS. So, in the co-simulation, through the predefined input and output interface in the co-simulation, the steering-angle control signals calculated by the controller models of SIMULINK can be transferred to the vehicle models in ADAMS as the input values at each simulation step.

After completing the above steps and inputting the vehicle structural parameters, the following dynamic equations can be established:

$$
\begin{bmatrix} M & \Phi_q^{\,T} \\ \Phi_q & 0 \end{bmatrix} \begin{bmatrix} \ddot{q} \\ \lambda \end{bmatrix} = \begin{bmatrix} Q^A \\ \gamma \end{bmatrix}
\tag{5.33}
$$

where M is the generalized mass matrix, Q^A is the generalized force matrix, Φ_q is the Jacobian matrix where $\Phi_q = 0$, q is the system generalized coordinates, λ is the Lagrange multiplier array, and γ is the right term of the system acceleration equations.

The kinematic analysis of each subsystem can normally be operated by simulating the multi-body models. The major physical parameters of the vehicle are shown in Table 5.2, and the virtual prototype model is shown in Figure 5.7.

Table 5.2 Main structural parameters of the vehicle.

Parameter	Value
Body mass/kg	1247.5
Body roll moment of inertia/kg.m^2	300
Body pitch moment of inertia/kg.m^2	1067.2
Body yaw moment of inertia/kg.m^2	1181.8
Front mass/Rear mass/kg	37.6/43
Wheelbase/mm	2800
Tread/mm	1540 (front)/1540 (rear)
Body centroid and rear axle distance/mm	1320
Body centroid height/mm	450
Caster angle/(°)	3
Kingpin inclination/(°)	13
Camber/(°)	0.5
Toe angle/ (°)	0.1

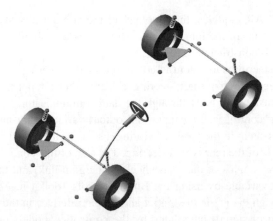

Figure 5.7   Vehicle virtual prototype model.

5.4.4   *Simulation Results and Analysis*

To validate the correctness and validity of the vehicle model and control strategies, co-simulation and analysis are carried out in this section based on ADAMS and MATLAB. The simulation results are shown in Figure 5.8. In the simulation, the speed is 80 km/h, and the road's surface adhesive coefficient is 0.8.

The simulation results show that, compared with the linear control of the front-wheel steering and four-wheel steering systems, the yaw rate, sideslip angle, and other parameters of the hybrid control system are markedly improved through the rear-wheel angle control of the four-wheel steering system. The handling and stability of the vehicle are effectively improved. The lateral displacement responses show that, when the four-wheel steering system has the same steering-wheel input, the decrease of the lateral displacement and steering sensitivity may affect the vehicle driving performance. Theory analysis and simulation results show that the constructed hybrid control system of the yaw rate feedback and neural network adaptive control of the four-wheel steering vehicle can effectively improve handling, stability, and security. This proves the good effects of the hybrid control algorithm. In addition, the control effects of the four-wheel steering vehicle in different road adhesive coefficients have been studied, and the system performance is improved effectively.

5.5   Electric Power Steering System and Control Strategy

The control strategy of a traditional electric power steering system (EPS) is mainly based on a linear vehicle model of two degrees of freedom. An improved steering portability and aligning performance can be obtained through the design of the control strategy. On a low adhesive road, the linear region and amplitude of the tyre self-aligning torque are obviously decreased. When steering on a low adhesive road at higher speeds, the self-aligning torque generated by the road and tyres will be greatly reduced. The driver tends to turn the steering wheel, and this operation will lead to slip trend and go into a dangerous state. When returning,

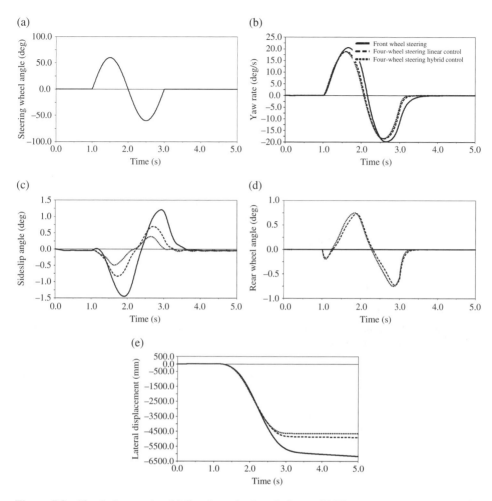

Figure 5.8 Simulation results. (a) Steering-wheel angle input. (b) The yaw rate response. (c) The sideslip angle response. (d) The rear wheel angle response. (e) The lateral displacement response.

as the self-aligning torque is not enough to overcome the internal friction of the steer system, the aligning performance is insufficient. This period involves a delay in adjusting the steering which makes the vehicle more difficult to handle, with the possibility of accidents.

The EPS dynamics model is established in this section. The torque values are obtained from the motor current and the torque sensor. Eliminating other factors, the self-aligning torque on the present road is obtained by a mathematical fitting. Meanwhile, the additional self-aligning torque on the ideal road is calculated through the steering wheel angle signals. Combining with the road estimation algorithm, a new EPS control strategy has been designed which includes a power-assisted control by adjusting the current and the aligning control by a time-varying sliding mode.

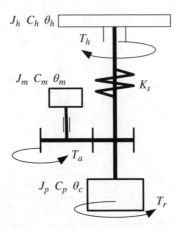

Figure 5.9 EPS dynamics model.

5.5.1 EPS Model

The EPS model is shown in Figure 5.9.

For convenience, the front wheel and steering mechanism are simplified as the steering shaft, and the dynamics equations are as follows[4,5]:

$$\theta_m = N_1 \theta_c \tag{5.34}$$

$$\theta_c = N_2 \delta_f \tag{5.35}$$

$$T_h - T_s = J_h \ddot{\theta}_h + C_h \dot{\theta}_h \tag{5.36}$$

$$T_s = K_s \left(\theta_h - \theta_c \right) \tag{5.37}$$

$$N_1 T_a + T_s = T_r + J_p \ddot{\theta}_c + C_p \dot{\theta}_c \tag{5.38}$$

where θ_m is the assisting-motor angle; θ_c is the steering shaft angle; N_1 is the transmission ratio of the motor to the steering shaft; N_2 is the transmission ratio of the steering shaft to the front wheels; δ_f is the front-wheel angle; T_h is the steering wheel torque; T_s is the value of torque sensor measurement; K_s is the torsion stiffness coefficient; θ_h is the steering wheel angle; J_h is the steering wheel moment of inertia; C_h is the steering wheel damping coefficient; T_a is the assisting-motor torque; T_r is the steering-torque acting on the steering pinion; J_p is the equivalent moment of inertia of the steering shaft; and C_p is the equivalent damping coefficient of the steering shaft.

The direct current motor is adopted by the system, then:

$$U = L_m \dot{I} + RI + K_b \dot{\theta}_m \tag{5.39}$$

$$T_a = K_a I - J_m \ddot{\theta}_m - C_m \dot{\theta}_m \tag{5.40}$$

where U, I are the voltage and current of the motor, respectively; L_m, R are the inductance and resistance of the motor, respectively; K_b is the counter-electromotive force coefficient of the motor; K_a is the torque coefficient of the motor; J_m is the moment of inertia of the motor; and C_m is the damping coefficient of the motor.

5.5.2 Steering Torque Model of the Steering Pinion

To use the steering torque and assisting-current of the motor, which could be measured to estimate the tyre self-aligning torque and the road adhesive coefficient, it is necessary to make sure that the steering torque T_r of the steering pinion follows the model[6,7]:

$$T_r = T_{align} + T_{f\_rp} sgn\left(\dot{\theta}_c\right) + \sigma\left(t\right)$$
$$T_{align} = \left(M_z + M_{sz}\right)/N_2$$

(5.41)

where T_{align} is the aligning torque equivalent to the steering pinion; M_z is the self-aligning torque of the front-wheel; M_{sz} is the aligning torque caused by gravity; T_{f_rp} is the friction torque of the steering system; and $\sigma(t)$ is the interference function related to the road.

In order to ensure the accuracy of the estimation algorithm of the road, the following will analyze the mathematical expression of each torque in equation (5.41).

5.5.2.1 Tyre Self-aligning Torque Model

Ignoring the influence of the vehicle roll factor, the nonlinear vehicle model (front-wheel steering) is used as shown in Figure 5.10, in which $i = 1, 2, 3, 4$ represent the left front, right front, left rear and right rear wheel, respectively.

The dynamic equations are:

$$m\left(\dot{u} - v\omega_r\right) = \left(F_{x1} + F_{x2} + F_{x3} + F_{x4}\right)$$

(5.42)

$$m\left(\dot{v} + u\omega_r\right) = \left(F_{y1} + F_{y2} + F_{y3} + F_{y4}\right)$$

(5.43)

$$\omega_r = \left[d\left(F_{x1} - F_{x2}\right)/2 + d\left(F_{x3} - F_{x4}\right)/2 + aF_{y1} + aF_{y2} - bF_{y3} - bF_{y4}\right]/I_z$$

(5.44)

where m, I_z are the vehicle mass and moment of inertia with respect to z-axis, respectively; u, v are the longitudinal and lateral speeds, respectively; F_x, F_y are the longitudinal and lateral forces, respectively; ω_r is the yaw rate; a, b are the distance of front, rear axle to the centroid, respectively; and d is the track width.

The decomposition of the contact forces of the tyres and road in the body heading Cartesian coordinate system is shown in Figure 5.11.

Assuming the steer angles of the left and right wheels are the same; thus, $\delta_1 = \delta_2 = \delta_f$, $\delta_3 = \delta_4 = 0$. From the analysis of Figure 5.11, the formula is expressed as follows:

$$\begin{Bmatrix} F_{xi} \\ F_{yi} \end{Bmatrix} = \begin{bmatrix} cos\delta_i & -sin\delta_i \\ sin\delta_i & cos\delta_i \end{bmatrix} \begin{Bmatrix} F_{xwi} \\ F_{ywi} \end{Bmatrix} \quad (i = 1,2,3,4)$$

(5.45)

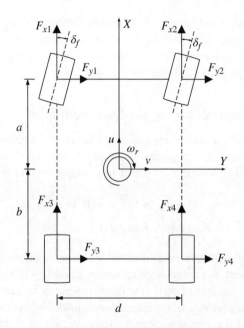

Figure 5.10   Vehicle model.

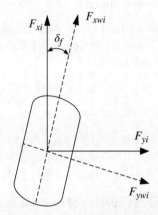

Figure 5.11   Tyre and ground forces.

where F_{xw}, F_{yw} are the tyre longitudinal and lateral forces, respectively.

Ignoring the effect of the moment of the inertia and air lifting force, then:

$$W_{1,2} = mg\frac{b}{2L} - ma_x\frac{h}{L} \pm ma_y\frac{hb}{dL}$$

$$W_{3,4} = mg\frac{a}{2L} + ma_x\frac{h}{L} \pm ma_y\frac{ha}{dL}$$

(5.46)

$$\alpha_{1,2} = \delta_f - \arctan\left[\left(v + a\omega_r\right)/\left(u \pm \frac{d}{2}\omega_r\right)\right]$$

(5.47)

$$\alpha_{3,4} = -\arctan\left[\left(v - b\omega_r\right)/\left(u \pm \frac{d}{2}\omega_r\right)\right]$$

where W_i is the vertical load of the tyres; L, h are the wheelbase and vehicle centroid height, respectively; α_i is the tyre sideslip angle; g is the acceleration of gravity; and a_x, a_y are the longitudinal and lateral accelerations of vehicle, respectively.

The nonlinear Dugoff tyre model is used in the simulation. Assuming the cornering stiffness of the left and right tyres is the same and the longitudinal stiffness of the tyres is also the same, the equations can be derived as follows[8]:

$$F_{xwi} = C_{si}\lambda_i\psi\left(\zeta_i\right)/\left(1 + \lambda_i\right)$$
$$F_{ywi} = K_{\alpha i}\tan\left(\alpha_i\right)\psi\left(\zeta_i\right)/\left(1 + \lambda_i\right)$$
$$\zeta_i = \frac{\mu W_i\left(1 + \lambda_i\right)\left(1 - \tau u\sqrt{\lambda_i^2 + \alpha_i^2}\right)}{2\sqrt{\left(C_{si}\lambda_i\right)^2 + \left(K_{\alpha i}\tan\left(\alpha_i\right)\right)^2}}$$

(5.48)

$$\psi\left(\zeta_i\right) = \begin{cases} \left(2 - \zeta_i\right)\zeta_i & \zeta_i < 1 \\ 1 & \zeta_i \geq 1 \end{cases} \left(i = 1, 2, 3, 4\right)$$

(5.49)

where F_{xw}, F_{yw} are the tyre longitudinal force and lateral force, respectively; λ is the wheel longitudinal slip ratio; τ is the speed impact factor; K_α, C_s are the tyre cornering stiffness and longitudinal stiffness, respectively; μ is the adhesion coefficient between the tyre and the road; ζ is the tyre dynamic parameter; and $\psi\left(\zeta\right)$ is the related function of ζ.

Tyre self-aligning torque results from the tyre lateral force and the offset distance of the tyres. The offset distance of the tyres is the sum of the mechanical offset caused by the kingpin caster and the pneumatic tyre offset. Mechanical offset t_m is treated as a certain value, and the pneumatic tyre offset t_p is different which is affected by the tyre cornering stiffness, road adhesive coefficient, slip angle, and other factors. Supposing t_{p0} is the initial value, the expression is given as follows:

$$t_p = t_{p0} - \text{sgn}\left(\alpha\right)\frac{t_{p0}K_\alpha}{3\mu W}\tan\left(\alpha\right)$$

(5.50)

$$M_z = \sum_{i=1}^{2}\left(t_m + t_{pi}\right)F_{ywi}$$

(5.51)

Under the condition of the road adhesive coefficient $\mu = 0.6$, the characteristics of the tyre cornering force F_y and self-aligning torque M_z changing with the tyre slip angle are as shown in Figure 5.12. The figure shows that, with the increase of the tyre slip angle, the self-aligning torque reaches saturation earlier than the cornering force. After reaching the

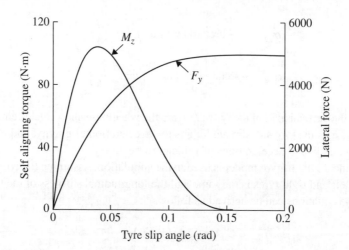

Figure 5.12 The characteristics of tyre lateral force and self-aligning torque.

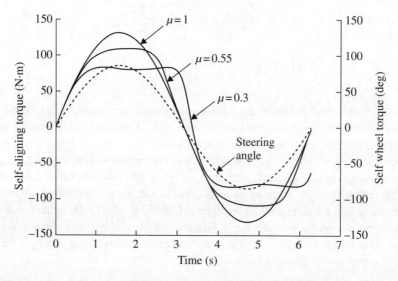

Figure 5.13 The tyre self-aligning torque under different adhesive road conditions.

peak, the tyre self-aligning torque decreases dramatically. Generally, the region of the self-aligning torque before the peak is called the linear region. With the decrease of the road adhesive coefficient, the linear region of the tyre self-aligning torque and the amplitude will decrease.

At the same time, at certain speeds, such as at 36 km/h, with the same steering-wheel angle control, if the road adhesive coefficient is low, the linear region of the tyre self-aligning torque and the amplitude will decrease. The simulation results are shown in Figure 5.13.

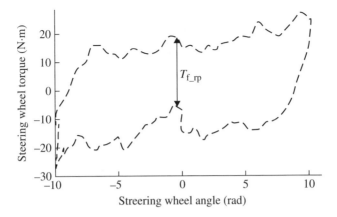

Figure 5.14   The internal friction torque of the steering system.

5.5.2.2   Model of Aligning Torque Caused by Gravity

The other part of the aligning torque is generated by the load of the front axle of the vehicle and the geometry of the front suspension. It can be described as[9]:

$$M_{sz} = W_f D_n \varphi \delta_f \qquad (5.52)$$

where D_n, φ are the kingpin offset and inclination angle, respectively; and W_f is the front axle load.

5.5.2.3   The Internal Friction of the Steering System and Road Disturbance

Under steering in in-situ conditions, the driver releases the steering-wheel, then it returns to the center position a little but cannot reach the center perfectly. The values of torque sensor will slowly recover to the initial values. The reason is due to internal friction torque T_{f_rp} of the steering system. The results measured by the experiments are shown in Figure 5.14.

In addition, the friction torque of the road and tyres can be regarded as the road related interference function $\sigma(t)$ in the process of vehicle driving.

5.5.3   *The Estimation Algorithm of the Road Adhesion Coefficient*

At present, the error generated by a variety of estimation algorithms of the road adhesive coefficient is relatively large, and the cost of accurately identifying the road's information is very high. With the commercial application of a torque/angle integrating sensor, it can provide the position and torque information of the steering wheel better than traditional torque sensors, and is convenient for the development of the aligning control algorithm. Combined with the estimation algorithm of the road adhesive

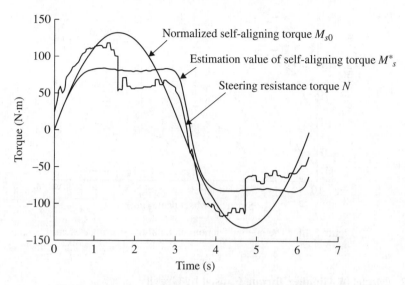

Figure 5.15   The estimation results of the self-aligning torque.

coefficient, the class of the identified adhesive coefficients can be divided into high, medium, and low, namely, μ_h, μ_m, and μ_l, as follows:

$$\mu^* = \begin{cases} \mu_h & \mu > 0.6 \\ \mu_m & 0.35 \leq \mu \leq 0.6 \\ \mu_l & \mu < 0.35 \end{cases} \tag{5.53}$$

where μ^* is the estimation value of the road adhesive coefficient.

Assuming the adhesive coefficient $\mu = 1$ and the simulation speed is 36km/h, when the vehicle is driving in steady state the maximum self-aligning torque of the front wheel is the additional self-aligning torque M_{s0}.

The motor's current is measured by the current sensor and combining equations (5.39)–(5.40), where the assisting-motor torque T_a is gained. By adding the steering torque and eliminating the aligning torque caused by gravity, the system's friction and other disturbances, such as the estimation values of the self-aligning torque M_s^*, can be obtained as shown in Figure 5.15. Then, through calculating M_s^*/M_{s0} and estimating the adhesive coefficient combined with Figure 5.15, the estimation values μ^* can be calculated. The estimation algorithm flow is shown in Figure 5.16.

5.5.4   Design of the Control Strategy

In the literature[4–5], there is a representative method of the traditional EPS control strategy. Generally, the motor reference current I_{ref} is determined by the power-assisted curves or look-up table method combined with the speed and torque sensor signals, and

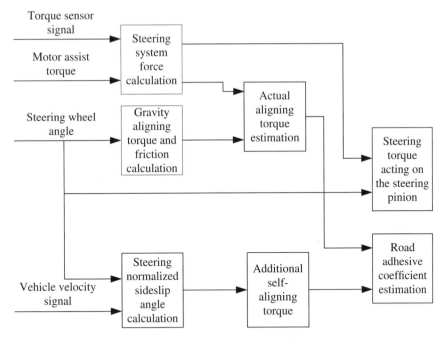

Figure 5.16 The estimation algorithm of the road adhesive coefficient.

then the power-assisted or aligning control will be achieved. But the slippery road is not considered; that is, the influence on the vehicle handling performance on the low adhesive road is not considered. In the actual steering, the driver continues to adjust the steering and aligning movement, so the power-assisted and aligning control strategies are designed respectively.

In the steering process, the motor current is reduced appropriately; thus, the power-assisted torque is reduced and the steering torque of the driver is increased, so the power-assisted current I_{assist} is designed as:

$$
\begin{aligned}
I_{assist} &= I_{ref} - I_{rev} \\
I_{rev} &= K_\mu \left(1 - \mu^*\right) \quad K_\mu > 0
\end{aligned}
\tag{5.54}
$$

where I_{rev} is the correction current, and K_μ is the current correction factor.

Under low adhesive conditions, when in the aligning process the self-aligning torque is not enough to overcome the internal friction of the steering system where the aligning ability is inadequate. The sliding mode controller can overcome the aligning torque changes in different adhesive coefficients and reduce the effect caused by friction and other uncertain factors on the system. Its application to the aligning control can obtain good results. However, the discontinuous switching of the traditional sliding mode controller will cause chattering, which will give a worse hand feel. Therefore, the control method can effectively eliminate the chattering by reducing the switching gain.

The time-varying sliding mode control method is able to make the system state in the sliding mode surface within the setting time. Therefore, the robustness of the approaching mode is strengthened, and the approaching mode of the traditional sliding mode control is eliminated.

Let,

$$e = \theta_c - \theta_d; \dot{e} = \dot{\theta}_c - \dot{\theta}_d \tag{5.55}$$

where θ_d is the desired steering-wheel angle, and e is the error.

From the simulation of three kinds of time-varying sliding surfaces in the literature[10], the method of variable slope sliding surface is better than the other two methods, and can be expressed as:

$$s = \begin{cases} \dot{e} + (At + B)e & t \le t_f \\ \dot{e} + \lambda e & t > t_f \end{cases}$$

$$A = \left[\lambda + \frac{\dot{e}_0}{e_0} \right] / t_f \tag{5.56}$$

$$B = -\frac{\dot{e}_0}{e_0}$$

where A, B, λ, t_f are constants and $\lambda > 0$, $t_f > 0$; e_0, $\dot{e}_0$ are the values of e and $\dot{e}$ at $t = 0$, respectively.

The estimation of $\hat{T}_r$ of the steering torque T_r, due to the moment of inertia J_m and the damping coefficient C_m of the motor are small, and can be neglected here. Therefore, equation (5.38) becomes:

$$\hat{T}_r = N_1 K_a I + T_s - J_p \ddot{\theta}_c - C_p \dot{\theta}_c \tag{5.57}$$

and,

$$\begin{aligned} T_r &= \hat{T}_r + \Delta T_r, \quad |\Delta T_r| \le \Delta T_{max} \\ C_P &= C_0 + \Delta C, \quad |\Delta C| \le \Delta C_{max} \\ J_P &= J_0 \Delta J, \quad \Delta J_{min} \le \Delta J \le \Delta J_{max} \end{aligned} \tag{5.58}$$

where ΔT_r is the uncertain value of T_r; ΔC, ΔJ are the uncertain values of C_P and J_P; C_0, J_0 are the certain values of C_P and J_P; ΔT_{max}, ΔC_{max} are the upper limit values of ΔT_r and ΔC; and ΔJ_{max}, ΔJ_{min} are the upper and lower limit values of ΔJ.

To solve the derivative of variable s in equation (5.56), we get:

$$\dot{s} = \begin{cases} \ddot{e} + (At + B)\dot{e} + Ae & t \le t_f \\ \ddot{e} + \lambda \dot{e} & t > t_f \end{cases} \tag{5.59}$$

From (5.59), the equivalent control law is as follows:

$$I_{eq} = \begin{cases} \left[\hat{T}_r - T_s + J_0\ddot{\theta}_d + C_0\dot{\theta}_c - J_0(At+B)\dot{e} - J_0Ae \right] / N_1K_a, & t \leq t_f \\ \left[\hat{T}_r - T_s + J_0\ddot{\theta}_d + C_0\dot{\theta}_c - J_0\lambda\dot{e} \right] / N_1K_a, & t > t_f \end{cases}$$ (5.60)

When aligning, the motor current I_{return} is designed as:

$$I_{return} = I_{eq} - \frac{J_0\kappa}{N_1K_a} sgn(s), \quad \kappa > 0$$ (5.61)

The Lyapunov function is defined as:

$$V = \frac{1}{2}s^2$$ (5.62)

By the Lyapunov stability theorem, when $\dot{V} = s\dot{s} \leq -\eta|s|(\eta > 0)$, the system is stable, thus,

$$\dot{V} = -s\frac{\kappa}{\Delta J} sgn(s) + sf(\Delta J, \Delta T_r)$$ (5.63)

where:

$$f(\Delta J, \Delta T_r) = \begin{cases} \left(1 - \frac{1}{\Delta J}\right)\left[(At+B)\dot{e} + Ae - \ddot{\theta}_d\right] - \frac{1}{J_0\Delta J}(\Delta T_r + \Delta C\dot{\theta}_c), & t \leq t_f \\ \left(1 - \frac{1}{\Delta J}\right)(\lambda\dot{e} - \ddot{\theta}_d) - \frac{1}{J_0\Delta J}(\Delta T_r + \Delta C\dot{\theta}_c), & t > t_f \end{cases}$$

Obviously, $f(\Delta J, \Delta T_r)$ is bounded. When $\Delta J = \gamma_{max}$, $\Delta f_{max} = \|f(\Delta J, \Delta T_r)\|_\infty$, $\kappa = \gamma_{max}$, $(\Delta f_{max} + \eta)$, the Lyapunov stability criterion is satisfied.

In order to reduce the system chattering, the following saturation function $sat\left(\dfrac{s}{\varepsilon}\right)$ is used to instead of the symbol function $sgn(s)$ in equation (5.61). And thus:

$$sat\left(\frac{s}{\varepsilon}\right) = \begin{cases} 1 & s > \varepsilon \\ s/\varepsilon & -\varepsilon \leq s \leq \varepsilon \\ -1 & s < -\varepsilon \end{cases}$$ (5.64)

where ε is the thickness of the boundary layer.

Figure 5.17 illustrates the overall system control strategy.

5.5.5 Simulation and Analysis

The structure parameters of a vehicle in simulation are shown in reference[9]. With the road adhesive coefficients $\mu = 0.5$ and 0.3, at the speed of 36 km/h, using the same control method from the EPS described above, the steering torque and aligning performance are

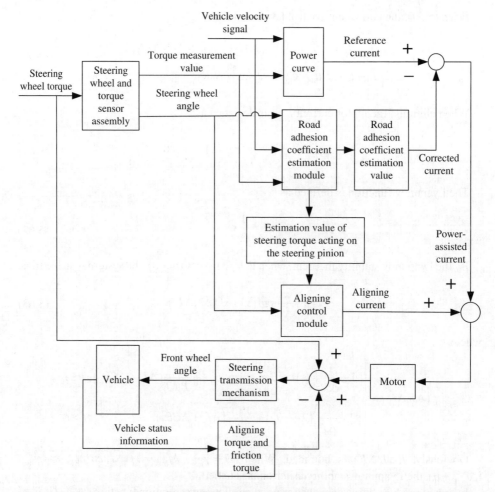

Figure 5.17 EPS control strategy.

simulated. When the adhesive coefficient $\mu = 1$, the performance of the designed controller in this section is similar to the performance of the traditional controller, so the latter will not be discussed.

5.5.5.1 Steering Torque

Figure 5.18(a) shows the simulation results when μ is 0.5 (in the figure, the traditional control strategy is presented in references[4] and[5]). In the steering process, the new control strategy presented in this section does not increase the steering torque. When the steering wheel returns, the steering torque is increased slightly (about 0.2 Nm). The motor provides a smaller aligning torque to improve the aligning performance.

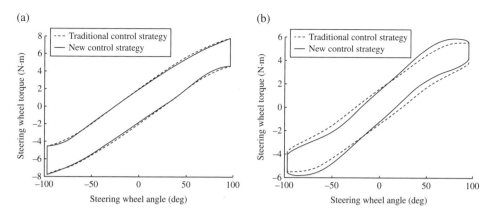

Figure 5.18 The simulation results of the steering torque. (a) Steering torque ($\mu = 0.5$). (b) Steering torque ($\mu = 0.3$).

Figure 5.18(b) shows the simulation results when μ is 0.3. In the steering process, the steering torque (about0.5 Nm) is slightly increased by the new control strategy. After the steering angles increase to a certain level, the steering torque is obviously reduced, so it can alert the driver that it is on a low adhesive road for suitable hand-feeling. Meanwhile, when the steering wheel returns, the steering torque is greatly increased (about 1.2 Nm), and a larger aligning torque is provided by the motor to improve the aligning performance.

5.5.5.2 Aligning Performance

Figure 5.19(a) shows the simulation results when μ is 0.5. In the aligning process, a large aligning torque can be provided by the motor to improve the aligning performance. This can be seen from the residual steering wheel angle shown in the simulation graph. Figure 5.19 shows that the residual steering wheel angle is close to 5° when the traditional control strategy in reference[5] is used, and the aligning process needs a longer period of time. However, a shorter time is expended for the aligning process by using the new control strategy, remaining with a steering wheel angle of about 4°. Therefore, the aligning performance is improved.

Figure 5.19(b) shows the simulation results when μ is 0.3. The figure shows that the residual steering wheel angle is close to 20° and the aligning process needs a longer period of time by using the traditional control strategy. However, using the new control strategy, a shorter time is expended for the aligning process, and the steering angle is slightly reduced.

5.5.6 Experimental Study

To verify the effectiveness of the proposed control strategy, a hardware-in-loop experimental study is carried out using LabVIEW software and a self-developed EPS test system. The experimental configuration is shown in Figure 5.20.

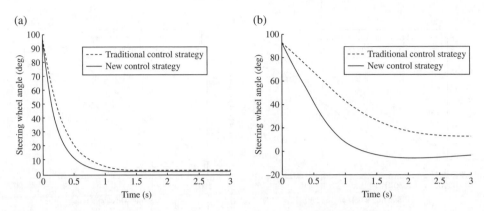

Figure 5.19 The simulation results of the aligning performance. (a) Aligning performance ($\mu = 0.5$). (b) Aligning performance ($\mu = 0.3$).

Figure 5.20 Experimental configuration of the developed EPS hardware-in-loop system.

Under the condition of a traditional low adhesive coefficient $\mu = 0.3$, the test data are taken for analysis. The steering torque test results are shown in Figure 5.21(a). The figure shows that, compared with the simulation results in Figure 5.18(b), although the amplitudes between the test and simulation results have little difference (about 1 Nm), the variation trend of the steering torque with the steering wheel angle is consistent.

Figure 5.21(b) shows the aligning test results: compared with the simulation results in Figure 5.19(b), the test results are basically consistent with the simulation results. Using the traditional control strategy, the residual steering wheel angle is more than 25°, and the aligning needs a longer period of time. However, using the new control strategy, a shorter time is expended for the aligning control, and the steering wheel slowly reaches the middle position. From a practical point of view, this is in a rational range.

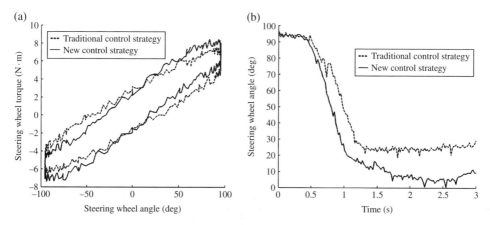

Figure 5.21 Experimental results. (a) Steering torque ($\mu = 0.3$). (b) Aligning performance ($\mu = 0.3$).

5.6 Automatic Lane Keeping System[11]

Lane keeping systems based on visual navigation include two key technologies: vehicle lateral control and lane detection. In recent years, in-depth research has been conducted by many scholars in this field. In the lateral control area, the nearest point between the vehicle and path is treated as the tracking point by some scholars, and the adaptive PID control and fuzzy preview control methods are designed. The nearest point to the vehicle is chosen as the tracking target, so there is no predictability about the road ahead. Furthermore, the fuzzy rules are difficult to establish, and the control effect is not good enough. In addition, the optimal controller of the autonomous navigation has been designed by some scholars based on the kinematics models, and a better control effect is achieved at low speeds. With the increase of speed, the gap between the kinematics models and the actual vehicle models grows. Therefore, this method is inapplicable to road tracking control of high-speed driving vehicles.

Based on the "preview following theory" for directional control of a vehicle, better path tracking can be achieved. Vehicle lateral control is affected greatly by the selection of the preview points. Under different road curvatures and longitudinal speeds, control effects of the same preview distance have great differences. In order to reduce the effect of the road curvature and preview distance on the lateral control and improve the control precision, a lateral control strategy is presented. This is used to plan a dynamic virtual path between the vehicle and preview points in real time, and produce a desired yaw rate based on the virtual path. Simulation and experimental results show that the proposed lateral control algorithm is valid.

5.6.1 Control System Design

The architecture of the lateral control system is shown in Figure 5.22. In the figure, according to the relative position of the vehicle preview points and vehicle state information from the coordinate transformation, the desired yaw rate as the input of the tracking

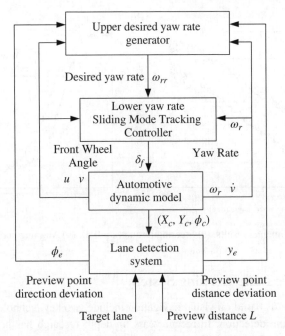

Figure 5.22   The architecture of the control system.

controller can be produced by the desired yaw rate generator. Based on the nonlinear vehicle dynamics models and the desired yaw rate, the desired vehicle state can be tracked by the yaw rate tracking controller, and the target path is achieved smoothly.

5.6.2   Desired Yaw Rate Generation

At a certain moment, X_c, Y_c are the position coordinates of the vehicle centroid in the global geodetic coordinate system (Figure 5.23), and ϕ_c is the angle between the vehicle longitudinal axis and the horizontal axis. The vehicle kinematics equations are described as follows:

$$\dot{X}_c = v_1 \cos(\phi_c + \beta) \tag{5.65}$$

$$\dot{Y}_c = v_1 \sin(\phi_c + \beta) \tag{5.66}$$

$$\dot{\phi}_c = \omega_r \tag{5.67}$$

where v_1 is the velocity, and β is the centroid sideslip angle.

Here, the sideslip angle in the left front is defined as positive, and the sideslip angle in the right front is defined as negative. By equations (5.65), (5.66), (5.67), we see that the motion position of a vehicle is determined by the yaw rate, centroid velocity, and sideslip angle. Because the centroid velocity is a vector, and it already contains the

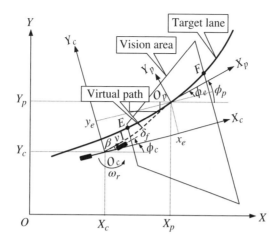

Figure 5.23 Coordinate system conversion of the vehicle.

sideslip angle information, the motion position of a vehicle is determined by the changes of the yaw rate and centroid velocity. Thus, according to the changes of the centroid velocity, the motion position of a vehicle can be determined by controlling the yaw rate. First, based on the local coordinate system, the path planning is carried out between the vehicle and the preview points. Second, based on the path planning, the yaw rate generator is designed, and the desired yaw rate is treated as the target parameters of the lateral control.

5.6.2.1 Coordinate Transformation

The coordinate transformation of a vehicle is shown in Figure 5.23. In the geodetic coordinate system XOY, X_p and Y_p are the coordinates of a point O_p located on the road in front of the vehicle preview, and ϕ_p is the angle between the tangent direction and transverse coordinate. In the global coordinate system, the relative position between the vehicle and preview point is (X_p, Y_p, ϕ_p), and now it can be transformed into the relative position (x_e, y_e, ϕ_e) in the local coordinate system of the vehicle.

According to the geometric relation in Figure 5.23, the relative position of the vehicle and preview point O_p in the local coordinate system (X_c, O_c, Y_c), can be expressed as:

$$P_e = \begin{bmatrix} x_e \\ y_e \\ \phi_e \end{bmatrix} = \begin{bmatrix} \cos\phi_c & \sin\phi_c & 0 \\ -\sin\phi_c & \cos\phi_c & 0 \\ 0 & 0 & 1 \end{bmatrix} \begin{bmatrix} X_p - X_c \\ Y_p - Y_c \\ \phi_p - \phi c \end{bmatrix} \tag{5.68}$$

where x_e is the preview distance; y_e is the lateral deviation between the preview point and the vehicle in the vehicle coordinate system; and ϕ_e is the directional deviation between the preview point and the vehicle in the same coordinate system.

5.6.2.2   Path Planning

In the local coordinate system (X_c, O_c, Y_c), a virtual path is constructed between the vehicle centroid and preview points in real time, as shown in Figure 5.23. Considering that the sideslip angle is relatively small, the direction of the actual speed at the centroid is supposed to be consistent with the vehicle longitudinal direction.

Assuming the equation of path planning is described as follows:

$$y(x) = A + Bx + Cx^2 + Dx^3 \tag{5.69}$$

The known conditions of virtual path are constructed:

$$y(0) = 0 \tag{5.70}$$

$$y(x_e) = y_e \tag{5.71}$$

$$\dot{y}(0) = 0 \tag{5.72}$$

$$\frac{\ddot{y}}{\left(1 + \dot{y}^2\right)^{(3/2)}}\bigg|_{x=0} = \rho \tag{5.73}$$

where ρ is the curvature of driving path; $A, B\ C, D$ are the coefficients of the planning path equation.

Combining equations (5.69)~(5.73), the planning curve equation is obtained as:

$$y(x) = \left[\omega_r / (2v_1)\right] x^2 + \frac{y_e - \omega_r / (2v_1) x_e^2}{x_e^3} x^3 \tag{5.74}$$

Obviously, when a vehicle is in the center of the lane and the driving direction is consistent with the tangent direction of the current path, the planning virtual path is the road equation of fitting the target lane central line. If several discrete preview points or a planning curve based on the curvature information at the preview points are selected, the curve in the preview points is more approximate to the actual road. Simulation and experiment results show that practical requirements can be met completely by using one preview point to plan a path. Here, one preview point is adopted to plan a virtual path.

5.6.2.3   Design of a Desired Yaw Rate Generator

It is assumed that a vehicle tracks a path smoothly and without deviation. The path equation $y(x)$ is known, the position of the vehicle centroid at the curve is (x, y), and the velocity is v_1. The direction of the velocity and driving curvature are consistent with the point which denotes the position of vehicle at the current curve. When the trajectory of the vehicle is a curve equation, the definition of the corresponding yaw rate is ω_d. The changing rate with time of the corresponding yaw rate, $\dot{\omega}_d$, is studied.

Curvature ρ is:

$$\rho = \frac{\ddot{y}}{\left(1+\dot{y}^2\right)^{(3/2)}}\,|_{x=0} \tag{5.75}$$

Changing rate of curvature ρ with time is:

$$\dot{\rho} = v_1 \left(\frac{d\rho}{dx}\right) / \left(\frac{d(S_d)}{dx}\right) \tag{5.76}$$

where S_d is the vehicle driving distance. Because driving curvature and road curvature are consistent, then the corresponding yaw rate ω_d is:

$$\omega_d = v_1 \cdot \rho \tag{5.77}$$

Combining equations (5.75)~(5.77), the changing rate of ω_d with the time is obtained:

$$\dot{\omega}_d = \dot{v}_1 \cdot \rho + \frac{v_1^2 \left[\ddot{y}\left(1+\dot{y}^2\right)-3\dot{y}\ddot{y}^2\right]}{\left(1+\dot{y}^2\right)^3} \tag{5.78}$$

In the local coordinate system (X_c, O_c, Y_c), the virtual path of the real-time planning equation (5.74) is substituted into equation (5.78), and the values at the origin position O_c are taken in the local coordinate system. At the current moment, when the vehicle approaches the target path along the virtual path, the corresponding changing rate of yaw rate ω_d is:

$$\dot{\omega}_d = \dot{v}_1 \omega_r / v_1 + 6v_1^2 \frac{y_e - \omega_r / (2v_1) x_e^2}{x_e^3} \tag{5.79}$$

The changing rate of yaw rate shown above represents the changing trend of the yaw rate when a vehicle is approaching the target path with the current speed at this time. It is relevant with the current speed, acceleration, yaw rate, and position preview point. The desired yaw rate can be expressed by the following equation:

$$\omega_{rr} = \omega_r + \varsigma\dot{\omega}_d \tag{5.80}$$

where ς is the scale factor associated with the control interval time, ω_{rr} is the desired yaw rate, and ω_r is the current yaw rate.

5.6.3 Desired Yaw Rate Tracking Control

In the simulation, consider that the front wheel of a rear-wheel drive vehicle has small longitudinal forces. For brevity, the longitudinal forces of the front wheel can be ignored, and the longitudinal forces of both sides of the rear wheels are equal. Therefore, the yaw moment acting on the vehicle is offered by the wheels' lateral forces. Supposing the yaw moment is M_z, and according to the nonlinear vehicle dynamics models in[11], it can be shown that:

$$M_z = a\left(F_{yfl}\cos\delta_f + F_{yfr}\cos\delta_f\right) - b\left(F_{yrl} + F_{yrr}\right) + \left(d/2\right)\left(F_{yfl}\sin\delta_f - F_{yfr}\sin\delta_f\right) \quad (5.81)$$

Generally, the front wheel angle is small, and thus $\cos\delta_f \approx 1$, $\sin\delta_f \approx 0$ is set, thereby equation (5.81) can be rewritten as:

$$M_z = a\left(F_{yfl} + F_{yfr}\right) - b\left(F_{yrl} + F_{yrr}\right) + \Delta M \quad (5.82)$$

In equation (5.82), ΔM is the additional torque which is the un-modeled part as the front wheel longitudinal forces are ignored, and the vehicle models are simplified; thus, they can be viewed as the system interference terms.

Taking the tyre model into the above equation, that is:

$$M_z = -\delta_f a\left(f_{fl}K_{yf} + f_{fr}K_{yfr}\right) - b\left(F_{yrl} + F_{yrr}\right) + \Delta M$$

$$+ a\left(f_{fl}K_{yfl}\arctan\left(\frac{v + \omega_r a}{u - \omega_r d/2}\right)\right) + a\left(f_{fr}K_{yfr}\arctan\left(\frac{v + \omega_r a}{u + \omega_r d/2}\right)\right) \quad (5.83)$$

$$= -\delta_f a\left(f_{fl}K_{yfl} + f_{fr}K_{yfr}\right) + T$$

where f_{fl} and f_{fr} are the coefficients associated with the maximum road adhesion[11]. T is a simple expression of the rest of the items. To reduce the effect of the system with the unmodeled parts and the parameter uncertainty and improve the system's robustness, a sliding mode controller is designed to track the desired yaw rate. The sliding switching surface is defined as:

$$s = \omega_r - \omega_{rr} \quad (5.84)$$

where ω_{rr} is the desired yaw rate, and ω_r is the current yaw rate. The derivation of the sliding surface is:

$$\dot{s} = \dot{\omega}_r - \dot{\omega}_{rr} = M_Z / I_Z - \dot{\omega}_{rr} \quad (5.85)$$

Let the sliding surface close to zero follow an exponential rate, thus, the output of the sliding mode controller is:

$$\delta_f = \delta_{fequ} + \lambda\,\text{sgn}\left(s\right) \quad (5.86)$$

The first term δ_{fequ} is an equivalent output of the sliding mode control. The second term ensures that, when the system is not in the sliding surface, the system is close to the ideal sliding surface, that is, $s\dot{s} \leq 0$. The Lyapunov function is described as:

$$J = \frac{1}{2}s^2 \quad (5.87)$$

Then, $\dot{J} = s\dot{s}$, $\dot{J} \le -\eta|s|$, $\eta > 0$. It can be proven that:

If
$$\lambda \ge \frac{\eta I_z}{a(f_{\mathrm{fl}}K_{y\mathrm{fl}} + f_{\mathrm{fr}}K_{y\mathrm{fr}})}, \quad t \to \infty, \text{ then } s \to 0.$$

So, the Lyapunov stability criterion is satisfied. Thus, the controller output is:

$$\delta_{\mathrm{f}} = \frac{T - I_z\dot{\omega}_{\mathrm{rr}} + \eta I_z \, \mathrm{sgn}(s)}{a\left(f_{\mathrm{fl}}K_{y\mathrm{fl}} + f_{\mathrm{fr}}K_{y\mathrm{fr}}\right)} \tag{5.88}$$

To weaken the chattering occurring around the sliding surface, the following saturation function $\mathrm{sat}\left(\dfrac{s}{\varepsilon}\right)$ is used instead of sign function $\mathrm{sgn}(s)$.

$$\mathrm{sat}\left(\frac{s}{\varepsilon}\right) = \begin{cases} \dfrac{s}{\varepsilon} & , \ |s| \le \varepsilon \\[2mm] \mathrm{sgn}\left(\dfrac{s}{\varepsilon}\right), & |s| > \varepsilon \end{cases} \tag{5.89}$$

where ε is the thickness of the boundary layer.

5.6.4 Simulation and Analysis

In the MATLAB/Simmulink simulation environment, in order to verify the effectiveness of the proposed algorithm, vehicle models are established and the control algorithm is simulated. Some vehicle parameters are shown in Table 5.3. Supposing that the road is formed by five curves as the simulation path, the simulation results of the lane keeping are compared with the position error feedback control method (shown in reference[12]) and the method proposed in this section. The longitudinal speed is 25 m/s, sampling time t and scale factor ς in equation (5.80) are 0.01.

The simulation results are shown in Figure 5.24.

Table 5.3 Some system parameters.

Parameter	Value
Vehicle mass m/kg	1704
Vehicle moment of inertia about z-axis I_z/kgm^2	3048
Distance between centroid and front, rear axis (a, b)/m	1.015, 1.675
Front and rear wheel track d/m	1.535
Front and rear cornering stiffness K_{yij}/(N/rad)	−105850, −79030
Front and rear longitudinal stiffness K_{xij}/(kN/m)	650,600
Centroid height h/m	0.542
Tyre rolling radius r_e/m	0.25
Preview distance x_e/m	16

From the tracking error curves, compared with the deviation feedback control, the tracking desired yaw rate control has a high accuracy for lane keeping. Also, it can be seen from the tracking error curves that when a vehicle tracks a large curvature, the tracking error can't get close to zero using the position deviation feedback control. This is because even if the deviation between the vehicle and the preview point is small, the deviation between the vehicle and the nearest point is still large. The control method of the tracking desired yaw rate can track the target path smoothly on any road and has better adaptability.

(a)

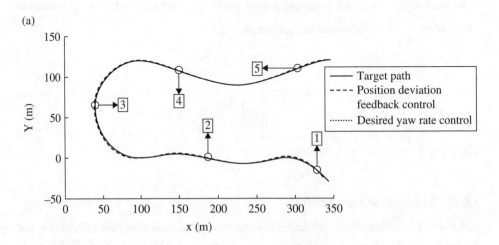

(b)

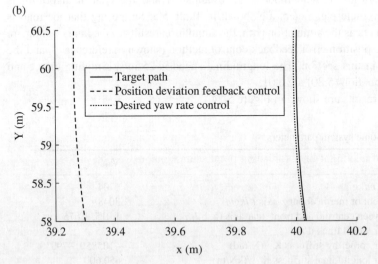

Figure 5.24 Simulation results. (a) Comparison of tracking trajectory. (b) Comparison of local enlarged trajectory. (c) Comparison of yaw rate, direction deviation, and distance deviation.

(c)

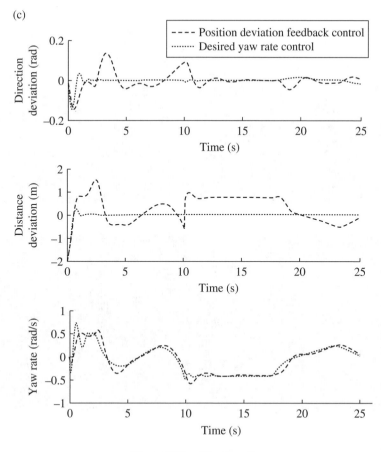

Figure 5.24   (Continued)

5.6.5   *Experimental Verification*

5.6.5.1   Lane Recognition

Lane recognition includes five parts, i.e., Gaussian smoothing filter, lane edge detection, Hough detection of a straight line, vanishing point search, and extraction of the left–right lane lines.

The characteristics of the gray-scale jumping are formed by the lane edge points and other points near the same row of the lane line. The points are generally not less than the width of the lane line, i.e., not less than or substantially equal to the number of pixels of the lane line. Therefore, if the edge points satisfy the width of the lane line, then they can be treated as edge points. The results of the fuzzy edge detection are shown in Figure 5.25(d). It can be seen that, compared with the common edge detection algorithm, the noise can be effectively suppressed by this algorithm.

(a)   (b)

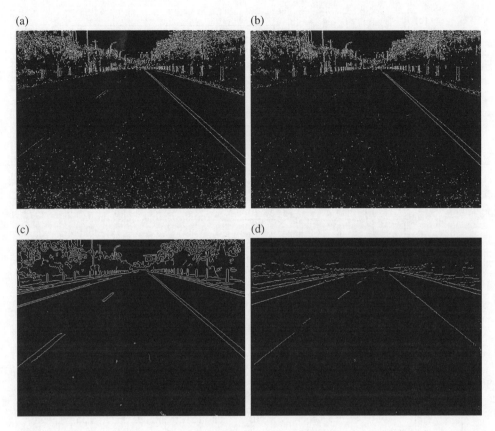

(c)   (d)

Figure 5.25 Comparison of fuzzy edge detection. (a) Sobel operator. (b) Prewitt operator. (c) Canny operator. (d) Edge detection.

The direction angles of the edge points are calculated, and the edge points which do not accord with the feature of the direction angles on the left and right lane lines are filtered. The numbers of edge points which are gathered at the vanishing points in different directions are ranked. Then, the parameter information of the straight lines focused on the vanishing points in different directions is available. According to the relative distance between lane lines, the detected lines are classified and extracted. Then the specific distribution locations of the lane lines are obtained. Meanwhile, optimizing the parameters of the lane lines, further information about the deviation parameters of a vehicle and road, and the curvature, is obtained[13]. Figure 5.26 shows some results of lane detection under typical roads.

5.6.5.2   Experimental System

As shown in Figure 5.27, the original vehicle steering system is modified and it has an automatic steering function. The experimental system includes the road image collection, vehicle state collection, host and slave computer control, and monitoring systems.

(a)   (b)

(c)   (d)

Figure 5.26   Detection results on different roads. (a) Character interference. (b) Sign interference. (c) Heavy rain. (d) Road of large curvature.

As a host computer, LabVIEW PXI8196 is responsible for collecting the signals of the vehicle-mounted sensors, such as the steering wheel angle, yaw rate, lateral acceleration, longitudinal velocity, and road images. The image acquisition card PXI-1411 is responsible for collecting road images from the CCD. Meanwhile, lane recognition and lateral control algorithms are performed in the host computer. The lane recognition module will calculate the relative positions of the vehicle preview points as an input of the lateral control module. As the slave computer, DSP2812 receives control instructions, and converts the instructions into a PWM pulse to control the steering motor to achieve the front-wheel steering. A PC monitors the signals about the vehicle-mounted sensors and the relative positions of the vehicle and the road.

Figure 5.28 shows the interface of the experimental system. Through the real-time display of the lane detection results and motion status signals of a vehicle in the front-end interface, the results of the vehicle lateral motion in different stages are tracked and detected. The lane detection and vehicle control algorithms can be updated and improved in real-time.

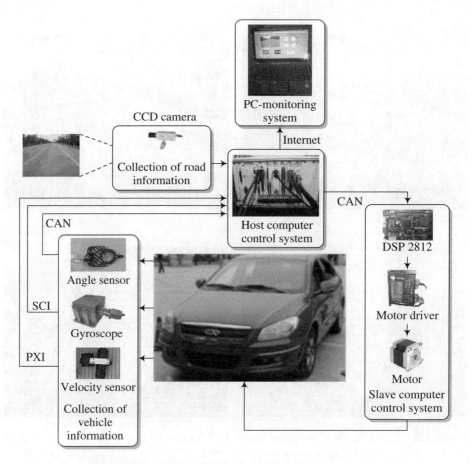

Figure 5.27 Schematic layouts of the experimental system.

5.6.5.3 Test Results and Analysis

Two methods of position deviation feedback and tracking desired yaw rate are used for the road tests. A large curved lane and a straight lane are selected as the test roads in order to compare the results.

Experimental results of two control methods are shown in Tables 5.4 and 5.5. It is clear from the tables that, when using the method of position deviation feedback to track the curved lane, the deviation is significantly increased than with the straight lane. When using the method of tracking desired yaw rate to track different lanes, the deviation is small. This indicates that the latter is less affected by the change of the road curvature.

When tracking the curved lane at low speeds by using the tracking desired yaw rate, the steady-state deviation is very small, about 0.03 m. When the position deviation feedback is used, there is a large lateral steady-state deviation, about 0.08 m. When the road curvature is large, even if the deviation between a vehicle and a preview point is in a smaller range, the deviation of the vehicle and the nearest point in the road center is

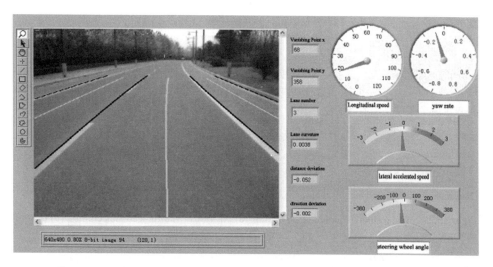

Figure 5.28   Test platform of the driver lateral assistance system based on vision.

Table 5.4   Distance deviation (straight lane).

Longitudinal speed u/(m/s)	Mean μ/m		Deviation σ^2/m^2	
	Position deviation feedback method	Desired yaw rate method	Position deviation feedback method	Desired yaw rate method
2	0.0757	0.0273	0.0338	0.0061
6	0.1993	0.0329	0.0728	0.0124
10	0.4713	0.0466	0.1375	0.0327

Table 5.5   Distance deviation (curved lane).

Longitudinal speed u/(m/s)	Mean μ/m		Deviation σ^2/m^2	
	Position deviation feedback method	Desired yaw rate method	Position deviation feedback method	Desired yaw rate method
2	0.0934	0.0382	0.0364	0.0082
6	0.2965	0.0590	0.0843	0.0216
10	0.6578	0.0806	0.1995	0.0508

still large. As shown in Figure 5.29, the results are consistent with the conclusions of the previous simulations.

It also can be seen that the deviation of two control methods increases when the speed increases because, even if the lane recognition can basically meet the real-time requirements, the response lag of the steering motor still exists. Even so, the lateral deviation of the lane-keeping system by the proposed method is always less than the deviation of the

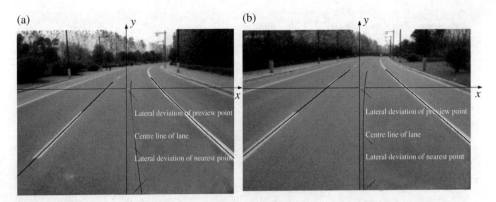

Figure 5.29 Test results of curved lane keeping. (a) Position deviation feedback control. (b) Desired yaw rate control.

Table 5.6 Percentage of deviation mean with speed increasing (%)

Type	Position deviation feedback method	Desired yaw rate method
Straight lane	40.82	5.13
Curved lane	54.36	13.61

feedback control. When the longitudinal speed is 10 m/s, the maximum lateral deviation is only 0.08 m. Table 5.6 shows that by using the method of tracking desired yaw rate, the percentage of deviation mean with the speed increasing is less than the position deviation feedback method. This indicates that the desired yaw rate method is less affected by the speed and has higher robustness.

References

[1] Yu Z S. Automobile Theory. Beijing: China Machine Press, 2009.
[2] Yu F, Lin Y. Vehicle System Dynamics. Beijing: China Machine Press, 2005.
[3] Song Y. Study on the control of vehicle stability system and four-wheel steering system and integrated system. Ph.D. Dissertation, Hefei, Hefei University of Technology, 2012.
[4] Wang Q D, Yang X J, Chen W W, et al. Modeling and simulation of electric power steering system. Transactions of the Chinese Society of Agricultural Machinery, 2004, 35(5): 1–4.
[5] Zhao L F, Chen W W, Qin M H, et al. Electric power steering application based on aligning and steering performance, Journal of Mechanical Engineering, 2009, 45(6): 181–187.
[6] Man H L, Seung K H, Ju Y C. Improvement of the steering feel of an electric power steering system by torque map modification. Journal of Mechanical Science and Technology, 2005, 19(3): 792–801.
[7] Yasui Y, Tanaka W, Muragishi Y, et al. Estimation of Lateral Grip Margin Based on Self-aligning Torque for Vehicle Dynamics Enhancement. SAE Paper 2004-01-1070, 2004.
[8] Dugoff H, Fancher P S. An Analysis of Tyre Traction Properties and Their Influence on Vehicle Dynamic Performance, SAE Paper 700377, 1970.

[9] Zhao L F, Chen W W, Liu G. Modeling and verifying of EPS at all operating conditions. Transactions of The Chinese Society of Agricultural Machinery, 2009, 40(10): 1–7.

[10] Jin Y Q, Liu X D, Qiu W, et al. Time-varying sliding mode controls in rigid spacecraft attitude tracking. Chinese Journal of Aeronautics, 2008, 21: 352–360.

[11] Wang J E, Chen W W. Vision guided intelligent vehicle lateral control based on desired yaw rate. Journal of Mechanical Engineering, 2012, 48(4): 108–115.

[12] Wang J M, Steiber J, et al. Autonomous ground vehicle control system for high-speed and safe operation. American Control Conference, 2008: 218–223.

[13] Zhou Y. Several Key Problem Research of the Intelligent Vehicle. Ph.D. Dissertation, Shanghai Jiao Tong University, 2007.

6

System Coupling Mechanism and Vehicle Dynamic Model

6.1 Overview of Vehicle Dynamic Model

With the development of automobile technology, the demands for vehicle handling, comfort, safety and other properties have become higher and higher. The corresponding control technology has emerged, and can be seen in the literature[1–6], which relates to vehicle suspension, steering, braking, and other subsystems. Since Segel made a comprehensive summary of vehicle dynamics on ImechE held in 1993[7], entitled "Vehicle ride and handling stability", vehicle dynamic model has been developed rapidly. With the requirements of improving the overall vehicle performance, the integrated control method has been proposed to achieve this goal.

Many scholars have carried out much research on the modeling of integrated systems. However, the established vehicle dynamics models are mostly a combination of various subsystems. The derived dynamic equations don't fully reflect the nonlinear coupled relationships and the interrelated effect among the vehicle longitudinal, lateral, and vertical movements. In the actual movements of a moving car, the longitudinal, lateral, and vertical motions are usually coupled tightly, and it is difficult to strictly separate them. In addition, the suspension, braking, and steering control input does not directly control the vehicle longitudinal, lateral, and vertical movements, nor the roll, pitch, and yaw movements although this is done indirectly through the impact of the tyre forces. Therefore, key isues are: the analysis of the braking, steering, suspension, and other chassis subsystem coupling mechanisms; the study of the nonlinear interaction between the tyre and road characteristics; and the establishment of the nonlinearly coupled dynamic models.

Integrated Vehicle Dynamics and Control, First Edition. Wuwei Chen, Hansong Xiao,
Qidong Wang, Linfeng Zhao and Maofei Zhu.
© 2016 John Wiley & Sons Singapore Pte. Ltd. Published 2016 by John Wiley & Sons, Ltd.

6.2   Analysis of the Chassis Coupling Mechanisms

The vehicle chassis is a complex system which includes brakes, steering, suspension and other subsystems. The suspension is a bridge between the body and wheel. The role of the suspension is to transfer the vertical, lateral, and longitudinal forces to the vehicle body. This will have a direct impact on the vehicle ride comfort. The steering system controls the direction of the vehicle and has a direct impact on the steering sensitivity, portability, and handling stability. The role of the braking system is to slow or stop the vehicle motion. The braking safety of a vehicle is determined by its braking performance and directional stability[8].

The suspension, steering, and braking systems determine the vehicle ride comfort, steering portability, handling stability, and driving safety. On the one hand, the movement of individual subsystems has an impact on the performance of the vehicle. However, there is a relationship between the motions of these subsystems. If a subsystem is improved based on individual performance (the design of mechanical structure or the optimization of control parameters), then the simple performance superposition of several subsystems cannot achieve an ideal comprehensive performance of a full vehicle. Therefore, in order to improve vehicle dynamic performance, the coupling relationship between the chassis subsystems must be studied in depth.

There is a strong coupling relationship between the longitudinal dynamics control systems (ABS, TCS, etc.), the lateral dynamics control systems (EPS, AFS and 4WS, etc.), and the vertical dynamics control systems (ASS, SASS, etc.)[9]. The coupling relationship is shown in Figure 6.1.

6.2.1   Coupling of Tyre Forces

A car tyre model describes the motion of the tyre, its 6-component forces, and the relationship between its inputs and outputs under specific working conditions. The longitudinal tyre slip ratio, slip angle, radial deformation, camber angle, rotary speed of wheels, and

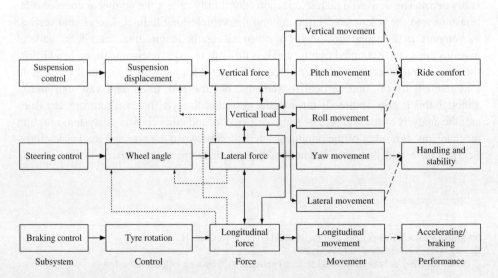

Figure 6.1   Coupling effects of typical chassis subsystems.

yaw angle determine the output of the models, such as tyre longitudinal force, lateral force, vertical force, roll torque, rolling resistance moment, and aligning torque. There is not only a nonlinear relationship between the inputs and outputs of the tyre model, but also a coupling relationship between the tyres and their 6-component forces.

1. The longitudinal force of a tyre is a nonlinear function of the vertical force and longitudinal slip rate. If the longitudinal slip rate increases to a certain value, the longitudinal force decreases slightly. In addition, if the longitudinal slip rate stays unchanged, there is a linear relationship between the longitudinal and vertical forces.
2. The lateral force of the tyre is a nonlinear function of the side slip angle and vertical force. If the vertical force is a constant value, there is a linear relationship between the lateral force and slip angle.
3. In certain cases of the side slip angle and the vertical force, the tyre lateral force and the longitudinal force influence each other. First, with the increase of the longitudinal slip rate, the tyre longitudinal force increases; if the slip rate reaches a certain value, the longitudinal force gradually decreases and is stabilized[10]. However, the tyre lateral force gradually decreases when the longitudinal slip rate increases; and when the longitudinal slip rate approaches 1, the tyre lateral force is almost 0.

6.2.2   Coupling of the Dynamic Load Distribution

The comprehensive performance of a full vehicle is determined by the external forces. For example, the yaw rate is related to the tyre lateral force, the vertical vibration acceleration of the vehicle body is determined by the vertical force of the sprung mass, and the braking distance is affected by the tyre longitudinal force. Thus, the tyre longitudinal force, lateral force, and sprung mass vertical force are directly or indirectly influenced by the tyre vertical loads. The tyre vertical loads consist of the dynamic and static load. The static load is determined by the vehicle parameters, and the calculation of the dynamic load is affected by the vehicle movements and inertial forces. The coupling of the dynamic load distribution is described as follows.

1. When a car is turning, the lateral force is changed as a result of the change of the steering-angle of the front wheel. In addition, the vertical load of the left tyre is not equal to the vertical load of the right tyre, which can affect the steady-state response and even make the car go from an understeer into an oversteer state.
2. When a car is steering and braking, the suspension movement will make the front and rear vertical load change, thereby affecting the front and rear longitudinal forces. Furthermore, when the tyre is operating in a saturation state, the increase of the lateral force results in the decrease of the longitudinal force, then the braking distance increases, which reduces the braking safety of the car.

6.2.3   Coupling of Movement Relationship

The vehicle movements are mainly controlled by the accelerator pedal, brake pedal, and steering wheel, and are made up of the longitudinal movement, vertical movement, lateral movement, roll movement, pitch movement, and yaw movement. These movements have a direct or indirect effect on braking, steering and suspension subsystems.

1. While a car is steering (or braking), the roll (or pitch) torque affects the sprung mass acceleration, which affects the vehicle ride comfort.
2. While a car is steering and braking, the change in the longitudinal and lateral acceleration results in the variation of the longitudinal inertia force and lateral inertia force. Then, a redistribution of the vertical load acting on the wheels occurs, and the lateral force and yaw rate changes, which directly affects the stability and safety of the car.

6.2.4   Coupling of Structure Parameters and Control Parameters

The chassis is a complex system made up of mechanical structures and control systems. There are two methods for designing a traditional chassis system[11].

1. The mechanical structure parameters and design objectives are known, the key issue is to choose the appropriate controller and control parameters.
2. The controller parameters are known, the design variables are the structure parameters.

The above two methods can be used to obtain a local optimum performance but not a global optimum performance. The reason is the coupling relationship between the mechanical structure system and control system.

In an integrated chassis control system, there is a strong coupling relationship between the control parameters and structural parameters. For example, when designing an active suspension system, first, the mechanical structure parameters (suspension stiffness, damping, sprung mass, etc.), are designed by optimization methods, and then a better static performance is obtained. Second, in order to improve the dynamic performance of the suspension, a controller is designed. However, it is difficult to reach the desired performance requirement, mainly because the internal coupling of the active and passive components is ignored.

6.3   Dynamic Model of the Nonlinear Coupling for the Integrated Controls of a Vehicle[12]

In order to build a nonlinear dynamic model which can reflect each subsystem coupled relationship, a 14-DOF nonlinear coupling dynamic model of a vehicle is built using the Newton-Euler method. Thus, the longitudinal, lateral, vertical, yaw, roll, and pitch movement and the rotation of the wheels are fully reflected[13].

In order to facilitate the derivation of the mathematical models, the following two parallel vehicle coordinate systems are defined.

1. The full vehicle coordinate system (x_c, y_c, z_c). The coordinate origin is the centroid position of the vehicle, the x_c and y_c axes are parallel to the ground, and the z_c axis is perpendicular to the ground.

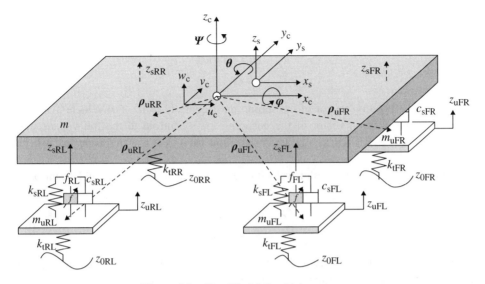

Figure 6.2   Simplified full vehicle system.

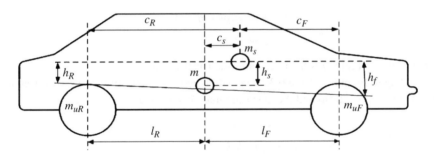

Figure 6.3   Relationship between the vehicle centroid and sprung mass centroid.

2. The sprung mass coordinate system(x_s, y_s, z_s). The coordinate origin is the center of the sprung mass.

In order to build the mathematical model of the full vehicle, the vehicle system is divided into the sprung mass and unsprung mass systems. The simplified model of the vehicle system is shown in Figure 6.2. The relationship between the centers of the vehicle mass and sprung mass is shown in Figure 6.3.

The position vector of the full vehicle centroid is R_c, the position vector of the sprung mass centroid is r_s, and the position vector of the unsprung mass centroid is r_{ui}.

The centroid motion of the full vehicle is:

$$\dot{R}_c = u_c \boldsymbol{i} + v_c \boldsymbol{j} + w_c \boldsymbol{k} \tag{6.1}$$

where u_c is the longitudinal velocity of the full vehicle centroid, v_c is the lateral velocity of the full vehicle centroid, and w_c is the vertical velocity of the full vehicle centroid.

According to the transformation relationship of the coordinate system, the position vector of the sprung mass centroid and the unsprung mass centroid are:

$$\begin{cases} \boldsymbol{r}_s = \boldsymbol{R}_c + (\varphi \boldsymbol{i} + \theta \boldsymbol{j} + \psi \boldsymbol{k}) \times \rho_s \\ \boldsymbol{r}_{ui} = \boldsymbol{R}_c + \psi \boldsymbol{k} \times \rho_{ui} \end{cases} \tag{6.2}$$

where ρ_s is the position vector of the sprung mass centroid to the full vehicle mass centroid, ρ_{ui} is the position vector of the unsprung mass centroid to the full vehicle mass centroid, $i = FL, FR, RL, RR$ are the left front, right front, left rear and right rear wheels, respectively.

From equation (6.2), the absolute velocities of the sprung mass centroid and the unsprung mass centroid are:

$$\begin{cases} \dot{\boldsymbol{r}}_s = \boldsymbol{R}_c + \bar{\omega}_s \times \rho_s \\ \dot{\boldsymbol{r}}_{ui} = \dot{\boldsymbol{R}}_c + \bar{\omega}_{ui} \times \rho_{ui} \end{cases} \tag{6.3}$$

where $\bar{\omega}_s$ is the angular velocity of the sprung mass around the centroid axis, and $\bar{\omega}_{ui}$ is the angular velocity of the unsprung mass around the centroid axis.

$\bar{\omega}_s$ and $\bar{\omega}_{ui}$ can be expressed as follows:

$$\begin{cases} \bar{\omega}_s = p\boldsymbol{i} + q\boldsymbol{j} + r\boldsymbol{k} \\ \bar{\omega}_{ui} = r\boldsymbol{k} \end{cases} \tag{6.4}$$

where p is the roll rate, q is the pitch rate, and r is the yaw rate.

According to Figures 6.2 and 6.3, the position vectors of the sprung mass centroid to the full vehicle mass centroid and the unsprung mass centroid are:

$$\begin{cases} \rho_s = c_s \boldsymbol{i} + h_s \boldsymbol{k}, \rho_{uF} = l_F \boldsymbol{i} - h_F \boldsymbol{k}, \rho_{uR} = -l_R \boldsymbol{i} - h_R \boldsymbol{k} \\ \rho_{uFL} = l_F \boldsymbol{i} - d_F \boldsymbol{j} - h_F \boldsymbol{k} \\ \rho_{uFR} = l_F \boldsymbol{i} + d_F \boldsymbol{j} - h_F \boldsymbol{k} \\ \rho_{uRL} = -l_R \boldsymbol{i} - d_R \boldsymbol{j} - h_R \boldsymbol{k} \\ \rho_{uRR} = -l_R \boldsymbol{i} + d_R \boldsymbol{j} - h_R \boldsymbol{k} \end{cases} \tag{6.5}$$

where l_F, l_R are the distances between the vehicle centroid and the front and rear axles, respectively; h_F, h_R are the vertical distances from the front and rear sprung mass centroids to the roll axis, respectively; d_F, d_R are the half front wheel track and the half rear wheel track, respectively; c_s is the longitudinal distance from the full vehicle centroid to the sprung mass centroid; and h_s is the vertical distance from the full vehicle centroid to the sprung mass centroid.

The absolute velocity vectors of the sprung mass and the unsprung mass are:

$$\begin{cases} \dot{\boldsymbol{r}}_s = (u_c + h_s q)\boldsymbol{i} + (v_c + c_s r - h_s p)\boldsymbol{j} + (w_c - c_s q)\boldsymbol{k} \\ \dot{\boldsymbol{r}}_{uFL} = (u_c + d_F r)\boldsymbol{i} + (v_c + l_F r)\boldsymbol{j} + w_c \boldsymbol{k} \\ \dot{\boldsymbol{r}}_{uFR} = (u_c - d_F r)\boldsymbol{i} + (v_c + l_F r)\boldsymbol{j} + w_c \boldsymbol{k} \\ \dot{\boldsymbol{r}}_{uRL} = (u_c + d_R r)\boldsymbol{i} + (v_c - l_R r)\boldsymbol{j} + w_c \boldsymbol{k} \\ \dot{\boldsymbol{r}}_{uRR} = (u_c - d_R r)\boldsymbol{i} + (v_c - l_R r)\boldsymbol{j} + w_c \boldsymbol{k} \end{cases} \tag{6.6}$$

The acceleration vectors of the sprung mass and the unsprung mass are:

$$
\begin{cases}
\ddot{r}_s = \left(\dot{u}_c + h_s \dot{q} - v_c r + w_c q + h_s r p + c_s r^2 \right) i \\
\quad + \left(\dot{v}_c + c_s \dot{r} - h_s \dot{p} + u_c r - w_c p + h_s q r \right) j \\
\quad + \left[\dot{w}_c - u_c q + v_c p + c_s r p - h_s \left(p^2 + q^2 \right) \right] k \\
\ddot{r}_{uFL} = \left(\dot{u}_c + d_F \dot{r} \right) i + \left(\dot{v}_c + l_F \dot{r} \right) j + \dot{w}_c k \\
\ddot{r}_{uFR} = \left(\dot{u}_c - d_F \dot{r} \right) i + \left(\dot{v}_c + l_F \dot{r} \right) j + \dot{w}_c k \\
\ddot{r}_{uRL} = \left(\dot{u}_c + d_R \dot{r} \right) i + \left(\dot{v}_c - l_R \dot{r} \right) j + \dot{w}_c k \\
\ddot{r}_{uRR} = \left(\dot{u}_c - d_R \dot{r} \right) i + \left(\dot{v}_c - l_R \dot{r} \right) j + \dot{w}_c k
\end{cases}
\tag{6.7}
$$

In order to simplify the calculations, we assume the vehicle longitudinal velocity is constant. According to equation (6.6) and the Newton-Euler method, the differential equations of a vehicle longitudinal, lateral, vertical, roll, yaw, and pitch motion are obtained.

The longitudinal movement of a vehicle is:

$$
\begin{aligned}
& m \left(\ddot{x}_c - \dot{y}_c r \right) + m_s \left(h_s \ddot{\theta} + \dot{z}_s \theta + h_s r \varphi + c_s r^2 \right) \\
& = - \left(F_{xFL} + F_{xFR} \right) \cos \delta_F - \left(F_{yFL} + F_{yFR} \right) \sin \delta_F - F_{xRL} - F_{xRR}
\end{aligned}
\tag{6.8}
$$

where m is the full vehicle mass, m_s is the sprung mass, F_{xi} is each longitudinal force of the four wheels, F_{yi} is each lateral force of the four wheels, and δ_F is the steering-wheel angle.

The lateral movement of a vehicle is:

$$
\begin{aligned}
& m \left(\ddot{y}_c + u_c r \right) + m_s \left(-h_s \dot{p} + h_s q r - w_c p \right) \\
& = \left(F_{xFL} + F_{xFR} \right) \sin \delta_F + \left(F_{yFL} + F_{yFR} \right) \cos \delta_F + F_{yRL} + F_{yRR}
\end{aligned}
\tag{6.9}
$$

The vertical movement of a vehicle is:

$$
\begin{aligned}
& m_s \left(\ddot{z}_c - u_c q + v_c p \right) - m_s h_s \left(q^2 + p^2 \right) + m_s c_s r p + k_{sFL} \left(z_{sFL} - z_{uFL} \right) \\
& + k_{sFR} \left(z_{sFR} - z_{uFR} \right) + k_{sRL} \left(z_{sRL} - z_{uRL} \right) + k_{sRR} \left(z_{sRR} - z_{uRR} \right) + c_{sFL} \left(\dot{z}_{sFL} - \dot{z}_{uFL} \right) \\
& + c_{sFR} \left(\dot{z}_{sFR} - \dot{z}_{uFR} \right) + c_{sRL} \left(\dot{z}_{sRL} - \dot{z}_{uRL} \right) + c_{sRR} \left(\dot{z}_{sRR} - \dot{z}_{uRR} \right) = 0
\end{aligned}
\tag{6.10}
$$

where k_{si} is the spring stiffness of the four suspensions; c_{si} is the shock absorber damping of the four suspensions; z_{si} is the vertical displacement of the four sprung masses; and z_{ui} is the vertical displacement of the four unsprung masses.

The roll movement of a vehicle is:

$$
\begin{aligned}
& I_{xu} \ddot{\varphi} - I_{xzu} \ddot{\psi} - \left(I_{zs} - I_{ys} - m_s h_s h \right) q r - I_{zxs} p q + m_s h w_c p - m_s \left(\dot{v}_c + u_c r \right) h = m_s g h \varphi \\
& + k_{sFL} \left(z_{sFL} - z_{uFL} \right) d_F - k_{sFR} \left(z_{sFR} - z_{uFR} \right) d_F + k_{sRL} \left(z_{sRL} - z_{uRL} \right) d_R - k_{sRR} \left(z_{sRR} - z_{uRR} \right) d_R \\
& + c_{sFL} \left(\dot{z}_{sFL} - \dot{z}_{uFL} \right) d_F - c_{sFR} \left(\dot{z}_{sFR} - \dot{z}_{uFR} \right) d_F - c_{sRL} \left(\dot{z}_{sRL} - \dot{z}_{uRL} \right) d_R - c_{sRR} \left(\dot{z}_{sRR} - \dot{z}_{uRR} \right) d_R
\end{aligned}
\tag{6.11}
$$

where I_{xu} is the rotary inertia of the sprung mass around the x_c axis; I_{xzu} is the product of inertia of the sprung mass around the x_c and z_c axes; I_{ys} is the rotary inertia of the sprung mass around the y_s axis; I_{zs} is the rotary inertia of the sprung mass around the z_s axis; g is the acceleration of gravity; and h is the height of the full vehicle centroid.

The yaw movement of a vehicle is:

$$I_z \ddot{\psi} - I_{zxu} \ddot{\varphi} + I_{zxu} qr + \left(I_{ys} - I_{xs} \right) pq - m_s c_s w_c p = d_F \left[\left(F_{xFL} - F_{xFR} \right) \cos \delta_F + \left(F_{yFL} - F_{yFR} \right) \sin \delta_F \right]$$

$$+ d_R \left(F_{xRL} - F_{xRR} \right) + l_F \left[\left(F_{xFL} + F_{xFR} \right) \sin \delta_F + \left(F_{yFL} + F_{yFR} \right) \cos \delta_F \right] - l_R \left(F_{yRL} + F_{yRR} \right)$$

(6.12)

where I_z is the rotary inertia of the vehicle around the z_c axis, and I_{xs} is the rotary inertia of the sprung mass around the x_s axis.

The pitch movement of a vehicle is:

$$I_{ys} \ddot{\theta} + \left(I_{xs} - I_{zs} \right) pr - I_{zxs} \left(r^2 - p^2 \right) = k_{sFL} \left(z_{sFL} - z_{uFL} \right) c_F + k_{sFR} \left(z_{sFR} - z_{uFR} \right) c_F$$

$$- k_{sRL} \left(z_{sRL} - z_{uRL} \right) c_R - k_{sRR} \left(z_{sRR} - z_{uRR} \right) c_R + c_{sFL} \left(\dot{z}_{sFL} - \dot{z}_{uFL} \right) c_F + c_{sFR} \left(\dot{z}_{sFR} - \dot{z}_{uFR} \right) c_F \quad (6.13)$$

$$- c_{sRL} \left(\dot{z}_{sRL} - \dot{z}_{uRL} \right) c_R - c_{sRR} \left(\dot{z}_{sRR} - \dot{z}_{uRR} \right) c_R$$

The vertical movements of the unsprung mass are:

$$\begin{cases} m_{uFL} \ddot{z}_{uFL} + c_{sFL} \left(\dot{z}_{uFL} - \dot{z}_{sFL} \right) + k_{sFL} \left(z_{uFL} - z_{sFL} \right) - k_{tFL} \left(z_{0FL} - z_{uFL} \right) = 0 \\ m_{uFR} \ddot{z}_{uFR} + c_{sFR} \left(\dot{z}_{uFR} - \dot{z}_{sFR} \right) + k_{sFR} \left(z_{uFR} - z_{sFR} \right) - k_{tFR} \left(z_{0FR} - z_{uFR} \right) = 0 \\ m_{uRL} \ddot{z}_{uRL} + c_{sRL} \left(\dot{z}_{uRL} - \dot{z}_{sRL} \right) + k_{sRL} \left(z_{uRL} - z_{sRL} \right) - k_{tRL} \left(z_{0RL} - z_{uRL} \right) = 0 \\ m_{uRR} \ddot{z}_{uRR} + c_{sRR} \left(\dot{z}_{uRR} - \dot{z}_{sRR} \right) + k_{sRR} \left(z_{uRR} - z_{sRR} \right) - k_{tRR} \left(z_{0RR} - z_{uRR} \right) = 0 \end{cases} \quad (6.14)$$

where m_{ui} is the unsprung mass, k_{ti} is the equivalent spring stiffness of the tyres.

The rotational movements of tyres are:

$$\begin{cases} I_{uFL} \dot{\omega}_{FL} = F_{xFL} R_{FL} - T_{bFL} \\ I_{uFR} \dot{\omega}_{FR} = F_{xFR} R_{FR} - T_{bFR} \\ I_{uRL} \dot{\omega}_{RL} = F_{xRL} R_{RL} - T_{bRL} \\ I_{uRR} \dot{\omega}_{RR} = F_{xRR} R_{RR} - T_{bRR} \end{cases} \quad (6.15)$$

where I_{ui} is the rotary inertia of the four wheels; ω_i is the angular velocity of the four wheels; R_i is the radius of the four wheels; and T_{bi} is the braking torque of the four wheels, respectively.

The vertical loads of tyres are calculated as[14]:

$$\begin{cases} F_{zFL} = \dfrac{1}{L}\left[l_R g + \dot{u}_c h_0 + \dot{v}_c \dfrac{h_0\left(Ll_R - h_0\right)}{LR_0}\right]m/2 + \dot{v}_c k_{\phi F}\left(m - m_s\right) \\[4mm] F_{zFR} = \dfrac{1}{L}\left[l_R g + \dot{u}_c h_0 + \dot{v}_c \dfrac{h_0\left(Ll_R - h_0\right)}{LR_0}\right]m/2 - \dot{v}_c k_{\phi F}\left(m - m_s\right) \\[4mm] F_{zRL} = \dfrac{1}{L}\left[l_F g - \dot{u}_c h_0 - \dot{v}_c \dfrac{h_0\left(Ll_R - h_0\right)}{LR_0}\right]m/2 + \dot{v}_c k_{\phi R}\left(m - m_s\right) \\[4mm] F_{zRR} = \dfrac{1}{L}\left[l_F g - \dot{u}_c h_0 - \dot{v}_c \dfrac{h_0\left(Ll_R - h_0\right)}{LR_0}\right]m/2 - \dot{v}_c k_{\phi R}\left(m - m_s\right) \end{cases} \tag{6.16}$$

where F_{zi} is the vertical load of the four wheels; L is the wheelbase; R_0 is the turning radius; h_0 is the distance from the vehicle centroid to the roll axis; $k_{\phi F}$ is the equivalent roll stiffness of the front axle; and $k_{\phi R}$ is the equivalent roll stiffness of the rear axle.

The slip angles of the tyres are obtained:

$$\begin{cases} \alpha_{FL} = -\arctan\dfrac{v_c + l_F r}{u_c + d_F r} + \delta_F, \alpha_{FR} = -\arctan\dfrac{v_c + l_F r}{u_c - d_F r} + \delta_F \\[4mm] \alpha_{RL} = -\arctan\dfrac{v_c - l_R r}{u_c + d_R r}, \alpha_{RR} = -\arctan\dfrac{v_c - l_R r}{u_c - d_R r} \end{cases} \tag{6.17}$$

6.4 Simulation Analysis

6.4.1 Simulation

In order to verify the results of the analysis of the vehicle dynamics coupling mechanism, according to the nonlinear Magic Formula tyre model described in Chapter 2, the nonlinear coupling dynamic simulation models are built, as shown in Figure 6.4.

The simulation models contain three translational movements along the axes (longitudinal, lateral, and vertical), three rotational movements around the axes (roll, pitch, and yaw), the tyre vertical load calculations, the tyre slip angle calculations, the nonlinear Magic Formula tyre model, and the road input model. The road input model uses a filtered white noise, which is defined as follows[15]:

$$\dot{z}_0(t) = -2\pi f_0 z_0(t) + 2\pi\sqrt{G_0 u_c}\, w(t) \tag{6.18}$$

where G_0 is the road roughness coefficient, f_0 is the lower cut-off frequency, and $w(t)$ is the Gaussian white noise.

The transfer function between the left and right road input is $G_{RL}(s)$, and the transfer function between the front and rear road input is $G_{FR}(s)$[12].

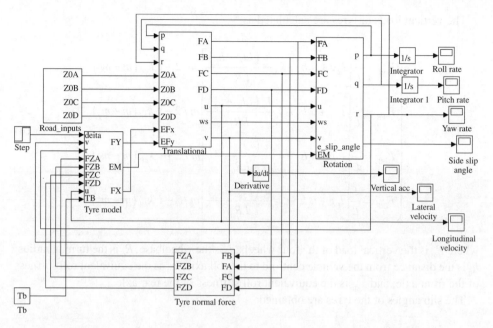

Figure 6.4   The simulation models of a nonlinear coupled dynamic system.

$$G_{RL}(s) = \frac{3.1815 + 0.2063s + 0.0108s^2}{3.223 + 0.59s + 0.0327s^2} \tag{6.19}$$

$$G_{FR}(s) = \frac{1 - \dfrac{t_d}{2}s + \dfrac{t_d^2}{12}s^2}{1 - \dfrac{t_d}{2}s + \dfrac{t_d^2}{12}s^2} \tag{6.20}$$

where $t_d = L / u_c$.

The simulation parameters are shown in Table 6.1.

6.4.2   Results Analysis

6.4.2.1   Simulation for Vehicle Ride Comfort

1. Setting the initial braking speed to 70km/h, the simulation condition studied is emergency braking. The tyre vertical load, the pitch angle of the vehicle body, and the vertical acceleration of vehicle centroid are shown in Figures 6.5–6.7.

   In Figure 6.5, because the vertical loads of the tyres are changed, the front wheel vertical loads increase, and the rear wheel vertical loads are reduced. In Figure 6.6,

Table 6.1　The vehicle model parameters.

Parameter	Value	Unit
m/m_s	1375/1055	kg
L	2.4	m
l_F/l_R	1.19/1.21	m
c_s	0.15	m
c_F/c_R	1.08/1.32	m
h/h_s	0.94/0.1	m
d_F/d_R	0.7/0.6	m
$k_{si}\,(i=FL,FR,RL,RR)$	40/40/35/35	kN/m
$c_{si}\,(i=FL,FR,RL,RR)$	1.4/1.4/1.2/1.2	kN.s /m
$I_{xs}/I_{ys}/I_{zs}$	1100/3000/4285	kg.m^2
I_{xu}	1996	kg.m^2
I_{xzu}	377.8	kg.m^2
I_z/I_{zx}	5428/47.5	kg.m^2
$k_{ti}\,(i=FL,FR,RL,RR)$	220/220/220/220	kN.m^{-1}
$I_{ui}\,(i=FL,FR,RL,RR)$	12/12/12/12	kN.m^{-1}
$R_i\,(i=FL,FR,RL,RR)$	0.25/0.25/0.25/0.25	m
$k_{\phi F}/k_{\phi R}$	7989/5096	N/rad
f_0	0.1	Hz
G_0	5×10^{-6}	m^3/cycle

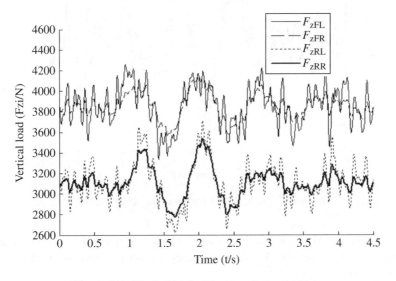

Figure 6.5　Vertical load of the tyres during braking.

because of the emergency braking, the pitch angle of the vehicle body increases greatly. In Figure 6.7, the emergency braking not only causes the front and rear axle loads to transfer and the body posture to change, but also causes the vehicle centroid vertical acceleration to increase greatly, which also affects the ride comfort.

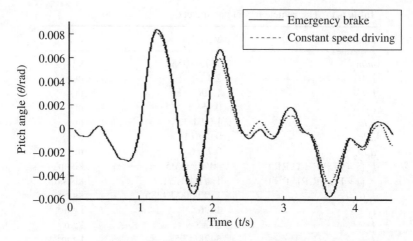

Figure 6.6 Pitch angle of the vehicle body.

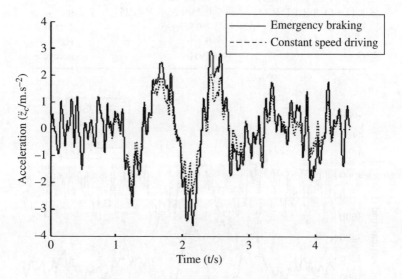

Figure 6.7 Vertical acceleration of the vehicle centroid.

The vehicle braking distance curves are shown in Figure 6.8. Because the suspension damper consumes the vehicle vibration energy, the braking distance is reduced in the coupling model.

2. Setting the simulation time to 5s, the initial speed of the vehicle to 50km/h, and a step input applied to the steering wheel. The results of the vertical tyre load, lateral acceleration, and roll angle are shown in Figures 6.9–6.11.

In Figure 6.9, the vertical loads cause a lateral shift. The features seen are that the vertical loads of the inside of the tyres are reduced, while the vertical loads of the outside of the tyres are increased, but the sum of the longitudinal force is constant.

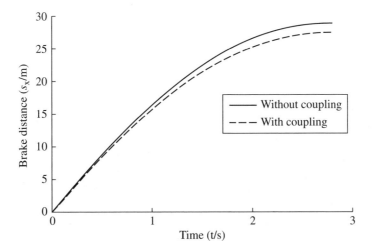

Figure 6.8   Brake distance curves.

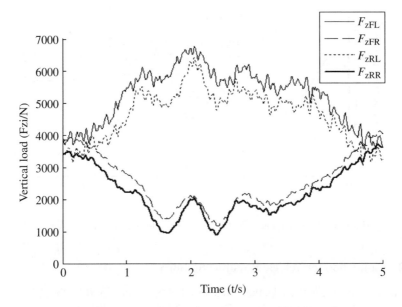

Figure 6.9   Vertical loads of the tyres during steering.

However, due to the tyre longitudinal force and lateral force being coupled to each other, the braking distance increases under steering and braking conditions. In Figure 6.10, with the steering peak angle increasing, the body roll angle becomes larger and larger, and the ride comfort deteriorates. In Figure 6.11, with the steering angle increasing, the lateral acceleration becomes larger and larger, and even results in the tyre skidding, so the driving safety is degraded.

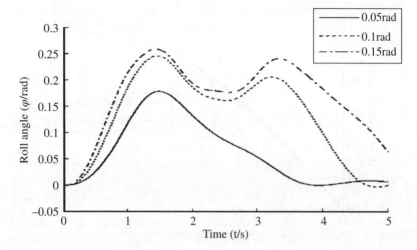

Figure 6.10   Roll angles during steering.

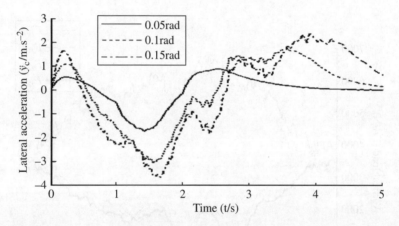

Figure 6.11   Lateral acceleration of the vehicle centroid during steering.

6.4.2.2   Simulation for Vehicle Handling Stability

The road surface level is B grade, the initial speed of the vehicle is 70km/h, and a sinusoidal curve is applied to the steering wheel (Figure 6.12). The road adhesion coefficients are 0.9 and 0.5 respectively, and the simulation results are shown in Figures.6.13–6.15.

In Figures 6.13 and 6.14, with the steering angle increasing, the yaw rate and the sideslip angle increase linearly in the linear model. However, the tyre lateral force becomes saturated with the sideslip angle increasing in the nonlinear model, and the deviation between the linear model and the nonlinear model becomes larger and larger.

In Figures 6.15 and 6.16, the yaw rate and sideslip angle of the linear model stay in a linear increasing state, even if the road adhesion coefficient is very low. However, the lateral force of the tyre reaches saturation, an increase of tyre sideslip angle does

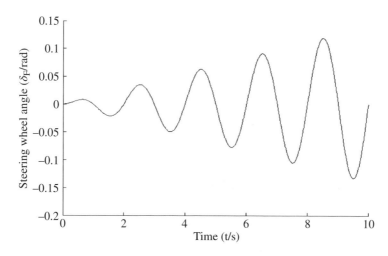

Figure 6.12   The curve of the steering wheel angle.

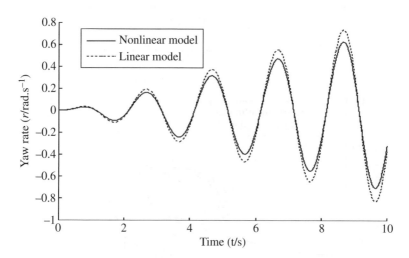

Figure 6.13   Yaw rate curves (adhesion coefficient is 0.9).

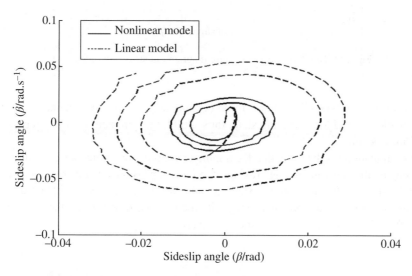

Figure 6.14   Sideslip angle phase diagram (adhesion coefficient is 0.9).

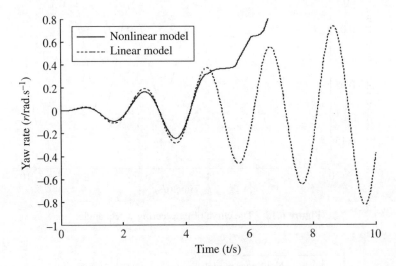

Figure 6.15   Yaw rate (adhesion coefficient is 0.5).

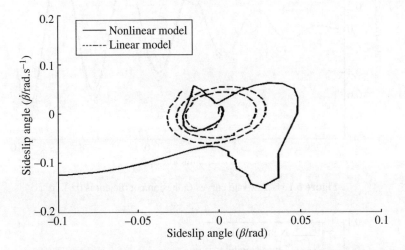

Figure 6.16   Sideslip angle phase diagram (adhesion coefficient is 0.5).

not make the corresponding lateral force become large, so a driver cannot even control the vehicle in a sudden cornering state; thus, skidding, spin, and other dangerous conditions may occur.

The simulation results show that the nonlinear coupling dynamic models can reflect the coupling relationship among the braking, steering, suspension, and other subsystems. On a low adhesion coefficient road, the yaw rate and sideslip angle obtained from the vehicle dynamic models cannot keep increasing with the increase of the front wheel angle, which can reflect the nonlinear characteristics of the tyres.

References

[1] Sato Y, Ejiri A, Iida Y, et al. Micro-G emulation system using constant tension suspension for a space manipulator. Proceedings of the IEEE International Conference on Robotics and Automation Sacramento. California, 1991.

[2] Yao Y S, Mei T. Dynamic modeling and simulation on suspension module. Chinese Journal of Mechanical Engineering, 2006, 42(7): 30–34.

[3] Liao Y G, Du H I. Modeling and analysis of electric power steering system and its effect on vehicle dynamic behavior. International Journal of Vehicle Autonomous Systems, 2003, 1(3): 351–362.

[4] Ikenaga S, Lewis F L, Campos J, et al. Active suspension control of ground vehicle based on a full vehicle model. Proceedings of the American Control Conference Chicago, Illinois June, 2000.

[5] Lee B R, Sin K H. Slip-ration control of ABS using sliding mode control. School of Mechanical and Automotive Engineering University of Ulsan, 1993, 1(2):72–77.

[6] Segawa M, Nakano S, Nishibara O, et al. Vehicle stability control strategy for steer by wire system. JSAE Review, 2001, 22(4): 383–388.

[7] Segel L. An overview of developments in road vehicle dynamics: Past, present and future. Proceedings of Imech E Conference on Vehicle Ride and Handling, 1993.

[8] Chen J R. Automobile Structure. Beijing: Machinery Industry Press, 2000.

[9] Lu S B. Study on Vehicle Chassis Key Subsystems and its Integrated Control Strategy. PhD Thesis, Chongqing: Chongqing University, 2009.

[10] Wang Q D, Wang X, Chen W W. Coordination control of active front wheel steering and ABS. Agricultural Machinery Journal, 2008, 39(3): 1–4.

[11] Wang Q D, Jiang W H, et al. Simultaneous optimization of mechanical and control parameters for integrated control system of active suspension and electric power steering. Chinese Journal of Mechanical Engineering, 2008, 44(8): 67–72.

[12] Zhu M F. Research on Decoupling Control Method of Vehicle Chassis and Time-delay Control of Chassis Key Subsystems. PhD Thesis, Hefei University of Technology, Hefei, 2011.

[13] Cui S M, Xiao L S, et al. A research on computer simulation for vehicle handling dynamics with 18 DOFs. Automotive Engineering, 1998, 20(4): 212–219.

[14] Zhu H. Vehicle Chassis Integrated Control Based on Magneto-rheological Semi-active Suspension. PhD Thesis, Hefei: Hefei University of Technology, 2009.

[15] Yu F, Lin Y. Automotive System Dynamics. Beijing: Machinery Industry Press, 2008.

7

Integrated Vehicle Dynamics Control: Centralized Control Architecture

7.1 Principles of Integrated Vehicle Dynamics Control

Current and future motor vehicles are incorporating increasingly sophisticated chassis control systems to improve vehicle handling, stability, and comfort. These chassis control systems include vehicle stability control (VSC), active suspension system (ASS), electrical power steering (EPS), and active four-wheel steering control (4WS), etc. These control systems are generally designed by different suppliers with different technologies and components to accomplish certain control objectives or functionalities. Especially when equipped into vehicles, control systems often operate independently and thus result in a parallel vehicle control architecture. In such a parallel vehicle control architecture, inevitably there occur interaction and performance conflict among the control systems occur ineviably because the vehicle motions in the vertical, lateral, and longitudinal directions are coupled together in nature. To address the problem, an approach of using an integrated vehicle control system was proposed around the 1990s[1]. An integrated vehicle control system is an advanced system that coordinates all the chassis control systems and components to improve the overall vehicle performance including handling stability, ride comfort, and safety, through creating synergies in the use of sensor information, hardware, and control strategies of different control systems[1,2]. As a result, the application of integrated vehicle control systems brings a number of advantages, including: (1) coordinating the interactions among the different subsystems; (2) further exploiting the potentials of each subsystem through integrating the function of the different subsystems with different work domains; (3) reducing the number of sensors and actuators by sharing and integrating the related ones. As shown in Figure 7.1, a better pareto-optimal solution of the vehicle overall performance is achieved through creating synergies amongst the different subsystems.

Integrated Vehicle Dynamics and Control, First Edition. Wuwei Chen, Hansong Xiao, Qidong Wang, Linfeng Zhao and Maofei Zhu.
© 2016 John Wiley & Sons Singapore Pte. Ltd. Published 2016 by John Wiley & Sons, Ltd.

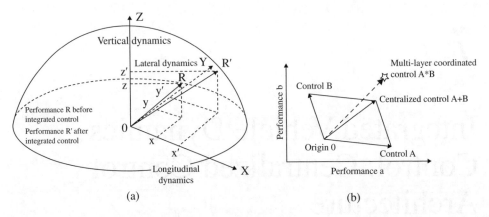

Figure 7.1 Principle of an integrated vehicle control system.

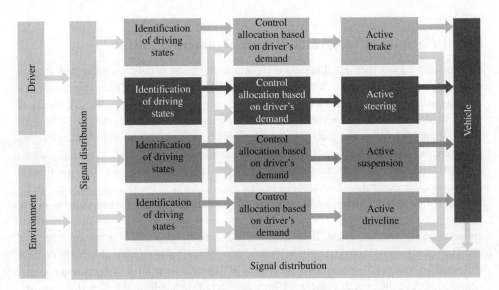

Figure 7.2 Decentralized (or Parallel) architecture.

A number of control techniques have been designed to achieve the goal of functional integration of the chassis control systems. These control techniques can be classified into three categories according to the extent of function integration of the subsystems, as suggested by Gordon et al.[2] and Yu et al.[3]: (1) decentralized or parallel control; (2) centralized control; and (3) multilayer control. In the decentralized control architecture shown in Figure 7.2, the subsystems of the vehicle are relatively independent and communicate with each other through the onboard network (CAN or LIN) to achieve their local control targets conveniently and flexibly. However, due to the lack of a global control target for the decentralized control architecture, the control architecture can only serve as a combined control structure of the vehicle subsystems at most. Compared to the parallel structure with

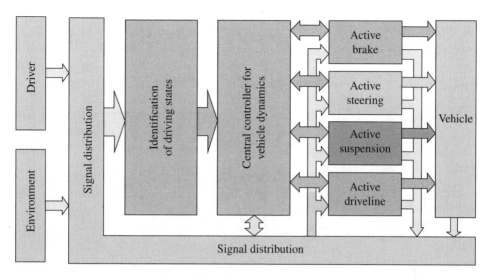

Figure 7.3 Centralized architecture.

standalone subsystems, the decentralized control architecture is superior through taking advantage of integrating and sharing the information of sensors and actuators.

Most of the control techniques used in the previous studies in recent years fall into the second category. Examples include nonlinear predictive control[4], random sub-optimal control[5], robust H_∞[6], sliding mode[7], and artificial neural networks[8]. In the centralized architecture shown in Figure 7.3, a single central controller collects all the vehicle operation information, including information from the sensors and the state estimators, and then generates control commands to the subsystem actuators by applying a global multi-objective optimization algorithm. Therefore, both the advantages and disadvantages are obvious. The centralized architecture has the advantages of controlling and observing all the subsystems in an integrated manner. However, the disadvantages cannot be ignored: the curse of dimensionality caused by the increasing number of subsystems results in tremendous design difficulties. Moreover, the failure of the centralized controller inevitably leads to a total failure of the whole chassis control system. Finally, when the centralized architecture needs to include more required subsystems, the entire centralized architecture has to be redesigned since the architecture lacks flexibility.

In contrast, multilayer control has not yet been applied extensively to integrated vehicle control. It is indicated by a relatively small volume of research publications [2,9–14]. The multilayer control architecture shown in Figure 7.4 consists of two layers. The upper layer controller monitors the driver's intentions and the current vehicle state. Based on these input signals, the upper layer controller is designed to coordinate the interactions amongst all the subsystem controllers in order to achieve the desired vehicle state. Thereafter, the control commands are generated by the upper layer controller and distributed to the corresponding individual lower layer controllers. Finally, the individual lower layer controllers execute respectively their local control objectives to control the vehicle dynamics.

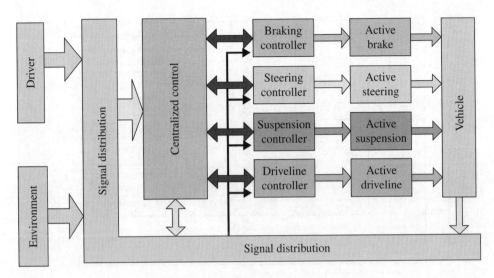

Figure 7.4   Multilayer control architecture.

In this chapter, the applications of the centralized control architecture are introduced by using various control methods to fulfill the integrated control goal for different subsystems.

7.2   Integrated Control of Vehicle Stability Control Systems (VSC)

A vehicle stability control system (VSC) is an integrated control system through the function integration of the anti-lock brake system (ABS) and traction control system (TCS) with the active yaw moment control system (AYC). VSC maintains the lateral stability of the vehicle by controlling the longitudinal forces between the tyres and road. As discussed in Section 3.5, the widely-used direct yaw moment control (DYC) method was briefly introduced to achieve the aims of the VSC. To fully explore the work principles of VSC, a control strategy for the sideslip angle of the vehicle center of gravity (CG) is proposed by using dynamic limits of the road surfaces in order to examine the effects on the sideslip angle for different road surfaces. Furthermore, a method for estimating the road adhesion coefficient is proposed by applying both the extended Kalman filter and neural network since estimation of the road adhesion coefficient is an important topic in the area of VSC and is also the basis of designing the control strategy of a VSC[15].

7.2.1   Sideslip Angle Control

As mentioned in Section 3.5 above, the two crucial states to determine the vehicle stability include the yaw rate and sideslip angle. The yaw rate measures the vehicle angular velocity around its vertical inertia axis, and the sideslip angle reflects the deviation of the vehicle on its current driving direction. Therefore, both states must be taken as control targets when designing the VSC.

Moreover, the effects of the sideslip angle resulting from different road surfaces must be taken into consideration. There are two main reasons. First, the stability limit that the vehicle is able to achieve is different for different road surfaces. For example, the stability limit for the road surface with a higher adhesion coefficient is larger than that with a lower adhesion coefficient. Second, the control of the sideslip angle is fulfilled through adjusting the longitudinal forces between the tyres and the road, and the longitudinal forces are directly related to the adhesion coefficient. Therefore, the control strategy for the sideslip angle is proposed by using dynamic limits of road surfaces in order to examine the effects on the sideslip angle for different road surfaces.

7.2.1.1   Development of the Sideslip Angle Control Strategy

7.2.1.1.1   *Dynamic Characteristics of the Sideslip Angle*
We first investigate the dynamic characteristics of the sideslip angle through performing a simulation study of a 7-DOF vehicle dynamic model. The vehicle is assumed to drive on a road with the adhesion coefficient of 0.3, and the double lane change maneuver is performed. The relationship between the sideslip angle and the sideslip angular velocity is shown in Figure 7.5. In the figure, when the absolute value of the sideslip angle is less than 0.02 rad, the vehicle stays stable; when it is larger than 0.02 rad, the absolute value of the rate of the sideslip angle increases drastically. This phenomenon shows that the vehicle tends to become unstable. Since vehicle stability is directly related to the sideslip motion of the vehicle, this motion must be bounded in order to keep the vehicle stable. Thus, the aim of the sideslip angle controller is to bind the sideslip angle within a suitable region in

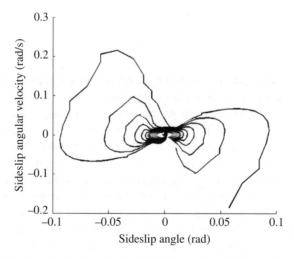

Figure 7.5   Simulation results for the relationship between the sideslip angle and sideslip angular velocity.

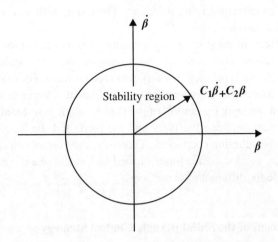

Figure 7.6   Stability region in the phase plane of the sideslip motion.

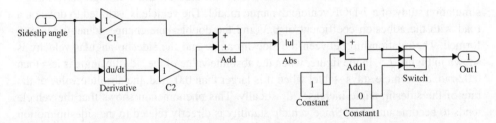

Figure 7.7   Block diagram of the proposed sideslip angle controller.

which the vehicle stays stable. As shown in Figure 7.6, the suitable stability region is defined in the phase plane of the sideslip motion:

$$\left| C_1\dot{\beta} + C_2\beta \right| < 1 \tag{7.1}$$

The suitable stability region is achieved by selecting suitable values of the parameters C_1 and C_2. Thus, the sideslip angle controller is proposed in Figure 7.7.

To demonstrate the effectiveness of the proposed sideslip angle controller, simulation investigations are performed for different driving conditions. First, the driving condition is set as follows: the vehicle is assumed to drive at a constant speed of 120 km/h on a road with a high adhesion coefficient of 0.9, and a double lane change maneuver is performed. As shown in Figures 7.8 and 7.9, the simulation results demonstrate that the sideslip angle is bounded at a relatively small value, and the sideslip motion is stable. In addition, the other driving condition is also performed: in this case, the vehicle speed is set to 60 km/h on a road with a low adhesion coefficient of 0.4, and the double lane change maneuver is also performed. As shown in Figures 7.10 and 7.11, the simulation results demonstrate that the VSC is able to restrain the sideslip at a relatively small value, and hence the sideslip motion stays stable.

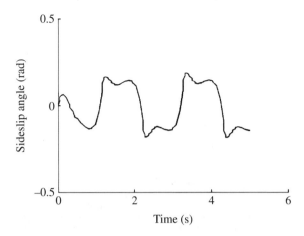

Figure 7.8 Response of the sideslip angle.

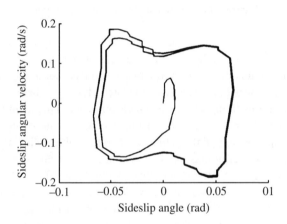

Figure 7.9 Phase plane of the sideslip motion.

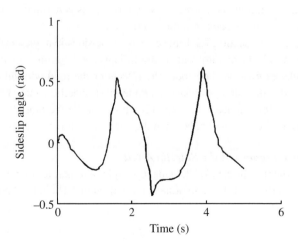

Figure 7.10 Response of the sideslip angle.

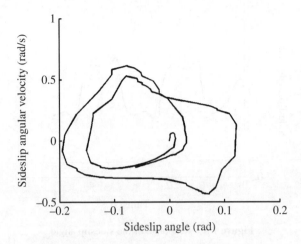

Figure 7.11   Phase plane of the sideslip motion.

However, as shown in Figure 7.10, the peak value of the sideslip angle is quite large and the phenomenon contradicts reality since the vehicle cannot stay stable with such a large sideslip angle. The simulation results show that it is inappropriate to define directly the handling limit as the control objectives since the lateral tyre force has already been close or even beyond the saturation point when the vehicle approaches the handling limit.

Therefore, an effective control method must determine the control objectives to generate the corrective yaw moment to pull the vehicle back to the stable region before it approaches the handling limit. The definitions of the reference region and the control region for designing the sideslip angle controller are illustrated in Figure 7.12. There are two boundaries, the inner boundary and outer boundary, which define the reference region and the control region, respectively. When the vehicle state lies inside the reference region, the vehicle is considered to be stable and no control action is required. When the vehicle state reaches the control region, which is bounded by the inner boundary and the outer boundary, the VSC is actuated and thus the corrective yaw moment is generated by the sideslip angle controller to pull the vehicle back into the reference region.

As discussed earlier in Section 7.2.1, the effects of the sideslip angle resulting from different road surfaces must be taken into consideration. Thus the determination of the above-mentioned two boundaries must also consider the effects of the road adhesion coefficients. As shown in Figure 7.13, the outer boundary is defined as a specific value of the sideslip angle when the lateral tyre force reaches the saturation point, while the inner boundary is defined as a specific value of the sideslip angle when the lateral tyre force reaches the linear limit.

7.2.1.1.2   *Outer Boundary of the Sideslip Angle*
To determine the outer boundary of the sideslip angle, a dynamic boundary is constructed by considering the effects of the road adhesion coefficients. The lateral acceleration of the C.G. is given as:

$$a_y = v_x \dot{r} + \dot{v}_y \tag{7.2}$$

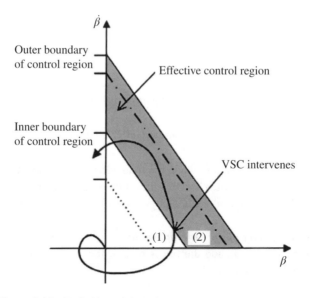

Figure 7.12   Definition of the reference region and control region.

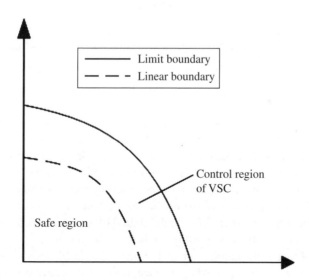

Figure 7.13   Definition of the two boundaries for designing the sideslip angle controller.

Considering the sideslip angle as relatively small, we have $v_y = v_x \tan \beta$. Therefore, the above equation can be rewritten as:

$$a_y = v_x \dot{r} + \dot{v}_x \tan \beta + \frac{v_x \dot{\beta}}{\sqrt{1 + \tan^2 \beta}} \qquad (7.3)$$

Since $a_y \leq \mu g$, and the latter two terms in equation (7.3) are relatively small compared to the first term, the upper limit of the yaw rate r is selected as:

$$r_{max} = 0.85\mu g/v_x \qquad (7.4)$$

Accordingly, the upper limit of the sideslip angle is chosen as:

$$\beta_{max} = \arctan(0.02\mu g) \qquad (7.5)$$

According to the above equation, when the road adhesion coefficient $\mu = 0.9$, the sideslip angle equals to 0.17 rad; when $\mu = 0.3$, the value is 0.08 rad. The above equation can be adjusted according to different vehicle physical parameters.

7.2.1.1.3 Inner Boundary of the Sideslip Angle
When the vehicle state lies inside the linear region, the yaw rate r is derived from the 2-DOF linear vehicle dynamic model:

$$r = \frac{v_x/L}{1 + Kv_x^2}\delta_f \qquad (7.6)$$

where $K = \dfrac{m}{L^2}\left(\dfrac{a}{k_r} - \dfrac{b}{k_f}\right)$. When $K = 0$, the above equation is given as follows:

$$r = v_x\delta_f/L \qquad (7.7)$$

The above equation shows that the steady state gain of the yaw rate is linear with the steering angle of the front wheel when the vehicle lies inside the linear region. Therefore, it is possible to determine whether the vehicle lies inside the linear region by examining whether the above linear relationship exists. A simulation study is performed to demonstrate the relationship of the two variables. As shown in Figures 7.14 and 7.15, the simulation results illustrate that the yaw rate r is linear with the steering angle of the front wheel δ_f when δ_f is smaller than 0.05 rad. However, with the increase on the steering angle δ_f, the relationship tends to be nonlinear.

As a matter of fact, only an approximately linear relationship exists for the yaw rate r and the steering angle of the front wheel δ_f since most vehicles have understeer characteristics. The relationships for the cases of understeer and neutral steer are illustrated in Figure 7.16. Therefore, a weighting function is constructed as follows to compensate for the nonlinear relationship

$$k = \begin{cases} 1 & v_x < 9m/s \\ c \cdot e^{v_x/7} & v_x \geq 9m/s \end{cases} \qquad (7.8)$$

The weighting function is illustrated in Figure 7.17.

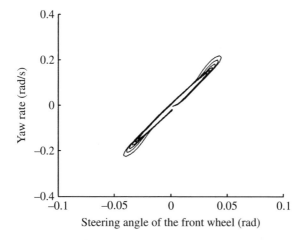

Figure 7.14 Relationship between the steady state gain of the yaw rate and the steering angle of the front wheel.

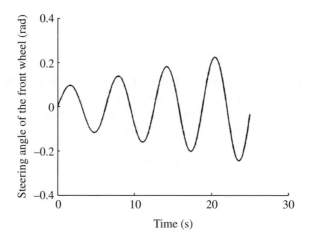

Figure 7.15 Steering angle of the front wheel.

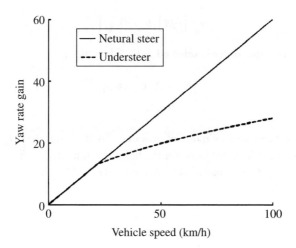

Figure 7.16 Relationships between r and δ_f for the cases of understeer and neutral steer.

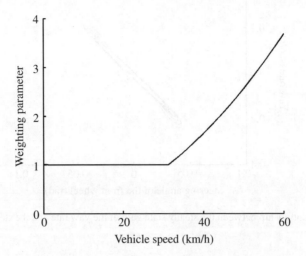

Figure 7.17   Weighting parameter for the yaw rate.

7.2.1.2   Sideslip Angle Controller Design

The nonlinear sliding mode control method is applied to the design of the sideslip angle controller since the controller is actuated mainly in the nonlinear region[16]. The state space equation of the 2-DOF vehicle dynamic model is derived as follows, with the assumptions of a constant forward speed and a small sideslip angle:

$$\dot{x} = Ax + Bu + E\delta_f \tag{7.9}$$

where

$$A = \begin{pmatrix} -\left(k_f + k_r\right)/mv_x^2 & \left(k_f l_f + k_r l_r\right)/mv_x^2 - 1 \\ \left(k_r l_r - k_f l_f\right)/I_z v_x & -\left(k_f l_f^2 + k_r l_r^2\right)/I_z v_x \end{pmatrix}, \ E = \left[k_f/mv_x \quad k_f l_f/I_z\right]^T, \ x = \left[\beta \quad r\right]^T,$$

$$u = \left[\Delta M\right], B = \left[0 \quad 1\right]^T.$$

A system with the same order is selected as the ideal model:

$$\begin{cases} \dot{x}_m = A_m x_m + B_m r_m \\ y = C_m x_m \end{cases} \tag{7.10}$$

where ΔM is the corrective yaw moment generated by the controller; $x_m \in R^n$ is the state of the ideal model; $r_m \in R^n$ is the input for the bounded model; $y_m \in R^n$ is the output of the model; (A, B) and (A_m, B_m) is controllable, respectively; and (A_m, C_m) is observable. Let the sliding hyper plane be:

$$\sigma = Sx_m \tag{7.11}$$

Decomposing the input matrix B as:

$$B = \begin{bmatrix} B_1 \\ B_2 \end{bmatrix}$$

(7.12)

and det $B_2 \neq 0$, we obtain:

$$x_m = Tx$$

(7.13)

where $T = \begin{bmatrix} I_{n-m} & -B_1 B_2^{-1} \\ 0 & I_m \end{bmatrix}$, and det $T \neq 0$. Thus,

$$x = T^{-1} \begin{bmatrix} x_{m1} \\ x_{m2} \end{bmatrix}$$

(7.14)

where $x_{m1} \in R^{n-m}$ and $x_{m2} \in R^m$. Substituting equation (7.13) into equations (7.10) and (7.11), we have:

$$\begin{cases} \dot{x} = T^{-1}ATx + T^{-1}Bu \\ \sigma = STx \end{cases}$$

(7.15)

where,

$$T^{-1}AT = \begin{bmatrix} A_{11} & A_{12} \\ A_{21} & A_{22} \end{bmatrix}, T^{-1}B = \begin{bmatrix} 0 \\ B_2 \end{bmatrix}, ST = \begin{bmatrix} S_1 & S_2 \end{bmatrix}$$

The expression for the canonical system is derived as:

$$\begin{cases} \dot{x}_1 = A_{11}x_1 + A_{12}x_2 \\ \dot{x}_2 = A_{21}x_1 + A_{22}x_2 + B_2 u \\ \sigma = S_1 x_1 + S_2 x_2 \end{cases}$$

(7.16)

Through the following transformation of coordinates:

$$\begin{bmatrix} x_1 \\ x_2 \end{bmatrix} = \begin{bmatrix} x_1 \\ \sigma \end{bmatrix}$$

i.e.

$$\begin{cases} x_1 = x_1 \\ x_2 = S_2^{-1}\sigma - S_2^{-1}S_1 x \end{cases}$$

(7.17)

Equation (7.16) becomes:

$$\begin{cases} \dot{x}_1 = \left(A_{11} - A_{12}S_2^{-1}S_1\right)x_1 A_{12}S_2^{-1}\sigma \\ \dot{\sigma} = \left[\left(S_1A_{11} + S_2A_{21}\right) - \left(S_1A_{12} + S_2A_{22}\right)S_2^{-1}S_1\right]x_1 \\ \qquad + \left(S_1A_{12} + S_2A_{22}\right)S_2^{-1}\sigma + S_2B_2u \end{cases} \tag{7.18}$$

When the system reaches the switch plane, we obtain:

$$\begin{cases} \sigma = S_1x_1 + S_2x_2 = 0 \\ \dot{\sigma} = S_1\dot{x}_1 + S_2\dot{x}_2 = 0 \end{cases} \tag{7.19}$$

Substituting equation (7.19) into equation (7.18),

$$\dot{x}_1 = \left(A_{11} - A_{12}K\right)x_1 \tag{7.20}$$

where $K = S_2^{-1}S_1$, and K can be determined by pole assignment. And hence, the hyper plane matrix of the system is expressed as:

$$S = \begin{bmatrix} S_1 & S_2 \end{bmatrix} = \begin{bmatrix} S_2K & S_2 \end{bmatrix} = S_2\begin{bmatrix} K & I_m \end{bmatrix} \tag{7.21}$$

Assuming $C_m = C$, the state error and its derivative are defined as:

$$e = x_m - x \tag{7.22}$$

$$\dot{e} = A_m e + \left(A_m - A\right)x + B_m r_m - Bu \tag{7.23}$$

The sliding mode function for the error space is defined as:

$$\sigma_e = Se \tag{7.24}$$

Its derivative is given as:

$$\dot{\sigma}_e = S\dot{e} = S\left[A_m e + \left(A_m - A\right)x + B_m r_m - Bu\right] \tag{7.25}$$

When the matrices S and B are invertible, the equivalent control law is given as:

$$u_{eq} = \left(SB\right)^{-1}S\left[A_m e + \left(A_m - A\right)x + B_m r_m\right] \tag{7.26}$$

Substituting equation (7.26) into equation (7.23),

$$\dot{e} = \left[I - B\left(SB\right)^{-1}\right]\left[A_m e + \left(A_m - A\right)x + B_m r_m\right] \tag{7.27}$$

Let $A_m - A = BK_1$, $B_m = BK_2$, and $E\delta = BK_3$. The above system error equation (7.27) can be rewritten as:

$$\dot{e} = \left[I - B\left(SB\right)^{-1}S\right]A_m e \tag{7.28}$$

And the error equation (7.23) can be expressed as:

$$\dot{e} = A_m e + B\left(K_1 x + K_2 r_m - u - K_3\right) \tag{7.29}$$

Defining the following transformation for the error e,

$$e' = Te \tag{7.30}$$

Let $e = x_m$ and $\sigma_e = \sigma$. Combining equation (7.29) and equation (7.30), the switch hyper plane of the error is given as:

$$\dot{e}' = \left(A'_{m11} - A'_{m12} S_2^{-1} S_1\right) e' \tag{7.31}$$

Defining the control input for the system as

$$u = \Delta M = Kx \tag{7.32}$$

To make the system converge on to the sliding surface, the following condition must be satisfied:

$$\sigma\dot{\sigma} = SxS\left(A - BK\right) = SxSB\left[\left(SB\right)^{-1} SA - K\right]x = SxSB\left(\alpha - K\right)x < 0 \tag{7.33}$$

where $\alpha = (SB)^{-1} SA = [\alpha_1, \alpha_2, \ldots, \alpha_n]$, only considering the continuous term of ΔM, we have:

$$\begin{cases} \Delta M = Kx, K = \left[k_1, k_2, \ldots, k_n\right] \\ k_i = \alpha_i + \mu_i \sigma SB x_i, \ \mu_i > 0 \ \left(i = 1, 2, \ldots, n\right) \end{cases} \tag{7.34}$$

Substituting the above equation into equation (7.33), the constraint condition is derived as follows:

$$\sigma\dot{\sigma} = \sigma SB\left(\alpha - K\right)x = -\sum_{i=1}^{n} \mu_i \left[\sigma_i(x)\right]^2 < 0 \tag{7.35}$$

7.2.1.3   Simulation Study

To demonstrate the effectiveness of the proposed sideslip angle controller, simulation investigations are performed for different driving conditions. First, the driving condition is set as follows: the vehicle is assumed to drive at a constant speed of 60km/h and 120km/h, respectively. The road adhesion coefficient is selected as 0.4 and 0.9, respectively. The double lane change maneuver is performed. For comparison, the commonly-used controller with a static boundary is also applied. The simulation results for the adhesion coefficient of 0.9 and 0.4 are illustrated in Figures 7.18–7.20, and Figures 7.21–7.23, respectively. As shown in Table 7.1, a quantitative analysis of the simulation results is also performed to better demonstrate the simulation results.

It can be observed from the simulation results that the proposed sideslip angle controller is able to bound the sideslip angle at a relatively small value, and hence the lateral stability

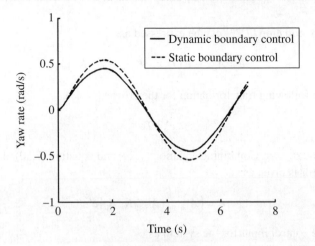

Figure 7.18   Yaw rate.

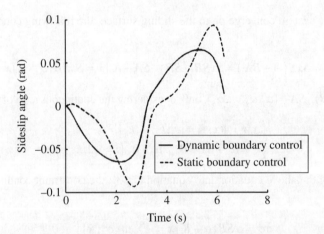

Figure 7.19   Sideslip angle.

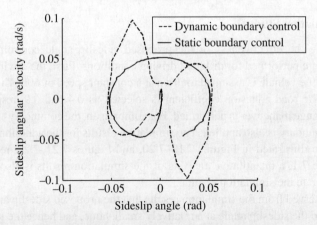

Figure 7.20   Phase plane of the sideslip motion.

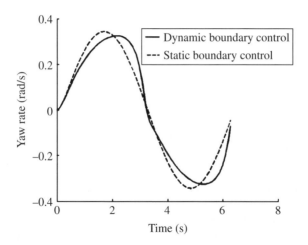

Figure 7.21 Yaw rate.

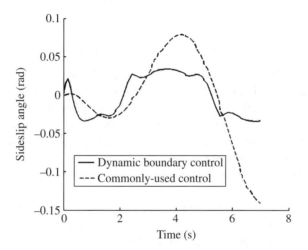

Figure 7.22 Sideslip angle.

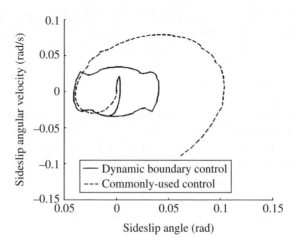

Figure 7.23 Phase plane of the sideslip motion.

Table 7.1 Comparison of the simulation results.

Adhesion coefficient	Control objective	Maximum		
		Controller with static boundary	Controller with dynamic boundary	Improvement
0.9	Yaw rate	0.52	0.43	17%
	Sideslip angle	0.095	0.061	36%
0.4	Yaw rate	0.34	0.32	6%
	Sideslip angle	0.079	0.046	42%

of the vehicle is achieved for both the roads with the high adhesion coefficient and low adhesion coefficient. However, the commonly-used controller with the static boundary only performs well on the road with a high adhesion coefficient. As shown in Figure 7.23, the vehicle cannot stay stable on the road with a low adhesion coefficient.

7.2.2   Estimation of the Road Adhesion Coefficient

Estimation of road adhesion coefficient is crucial in developing VSC since it is the basis for implementing the VSC. There are two main reasons: first, an effective control strategy of VSC must consider the effects of the road adhesion coefficients on the stability limits. Second, VSC must be able to precisely adjust the tyre force to execute the control commands. Adjusting the tyre force depends mainly on whether or not the road adhesion coefficient is able to be estimated precisely.

A large number of estimation methods have been developed through the brake driving condition. During the process of braking, the relationship between the road adhesion coefficient and brake efficiency factor is constructed, and thus the road adhesion coefficient is calculated[17, 18]. However, there is no severe braking when the VSC intervenes since the VSC works mainly under the steer driving condition. Therefore, it is necessary to develop methods to estimate the road adhesion coefficient for the VSC under the steer driving condition. As discussed earlier, with the increase of the sideslip angle, the lateral tyre force increases from the linear region to the nonlinear region and is close to or even beyond the saturation point. In addition, the inner boundaries of the sideslip angle are different with respect to the different road adhesion coefficients[19, 20]. Therefore, the method for estimating the road adhesion coefficient is developed through determining precisely the point that the vehicle reaches the nonlinear region, and thus calculating the corresponding sideslip angle.

However, if the lateral tyre force does not have a distinct transformation from the linear region to the nonlinear region when the change of the steering angle is quite small, it is necessary to take this case into account when developing the estimation methods. Figure 7.24 shows the simulation results of the sideslip angles for a road adhesion coefficient of 0.4 and 0.9, respectively, when the same yaw rate illustrated in Figure 7.25 is maintained. The simulation results demonstrate that the sideslip angle for the low road adhesion coefficient is larger than that for the high road adhesion coefficient when the yaw

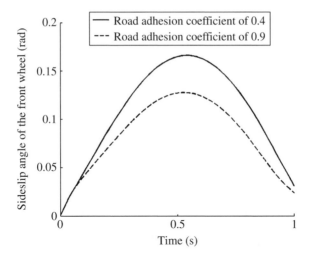

Figure 7.24   Sideslip angle for different road adhesion coefficient.

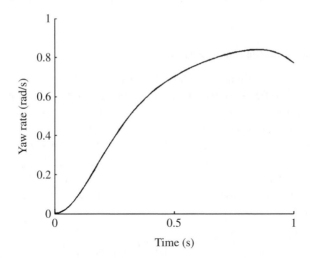

Figure 7.25   Yaw rate.

rate stays the same. The reason is that the sideslip angle must increase for the low road adhesion coefficient in order to provide the same lateral tyre force. Therefore, this characteristic is applied to design the estimation method when the change of the steering angle is quite small.

Moreover, as illustrated in Figure 7.26, it is observed that the vertical load has a great effect on the sideslip angle and the lateral tyre force. In this case, it may not be precise enough to estimate the road adhesion coefficient by using only one sideslip angle out of the four wheels. Therefore, the estimation method is proposed to use the sideslip angles of the two front wheels since the sideslip angles of the two rear wheels are relatively small.

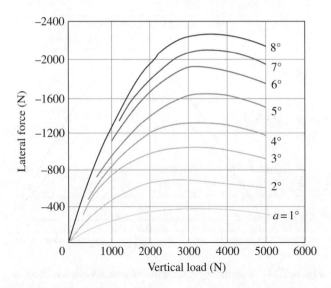

Figure 7.26   Effects of vertical load on the wheel sideslip characteristics.

7.2.2.1   Estimation of the Sideslip Angle

As discussed above, the estimation of the sideslip angle is important for the estimation of the road adhesion coefficient. thus, the accuracy of the estimation of the road adhesion coefficient is mainly determined by the accuracy of the sideslip angle. To achieve this aim, an extended Kalman filter is used to accurately estimate the vehicle velocity v_x and v_y, and then the sideslip angle is calculated according to the wheel model. For the Kalman filter developed in this section, the vehicle velocity, which is calculated from the 2-DOF vehicle dynamic model, is used as the estimated value, while the acceleration calculated from the 7-DOF vehicle model is used for calculating the measured value. Thus, the Kalman filter is derived as:

$$x_{k+1} = A_k x_k + B u_k + \omega \tag{7.36}$$

$$y_k = C x_k + \varepsilon \tag{7.37}$$

where $x_k = \begin{pmatrix} v_{xk} \\ v_{yk} \end{pmatrix}; u_k = \begin{pmatrix} a_{xk} \\ a_{yk} \end{pmatrix}; y_k = \begin{pmatrix} v_{xak} \\ v_{yak} \end{pmatrix}; A_k = \begin{pmatrix} 1 & T\varepsilon_k \\ -T\varepsilon_k & 1 \end{pmatrix}; B = \begin{pmatrix} T & 0 \\ 0 & T \end{pmatrix}; C = \begin{pmatrix} 1 & 0 \\ 0 & 1 \end{pmatrix};$

a_x and a_y are the longitudinal and lateral acceleration, respectively; T is the period of sampling cycle; v_{xa} and v_{ya} are the longitudinal and lateral vehicle velocity calculated from the 2-DOF vehicle dynamic model, respectively; k is the number of iterations; ε and ω are the measured error and prediction error of the system model. It is assumed that they are independent of each other and subject to the Gaussian distribution, and their covariances are denoted as R and Q. Therefore, the expanded Kalman filter proceeds in two steps. In the first step, the sampling value and error increment between the two samplings are calculated according to equation (7.38) and equation (7,39).

$$\hat{x}_{k|k-1} = A_k x_{k-1|k-1} + Bu_{k-1} \tag{7.38}$$

$$P_{k|k-1} = F_k P_{k-1|k-1} F^T + Q \tag{7.39}$$

where $\hat{x}_{k|k-1}$ is the prediction value; $P_{k|k-1}$ is the covariance of the prediction error; and $F_k = \dfrac{\partial f(x)}{\partial(x)}\bigg|_{x=\hat{x}_{k-1|k-1}}$ is the dynamic matrix obtained by the linearized system state equation when calculating $\hat{x}_{k-1|k-1}$. For the second step, the measured value is amended according to the system prediction value and prediction error during the sampling, as expressed in equations (7.40)–(7.42).

$$K_k = P_{k|k-1} H_k^T \left(H_k P_{k|k-1} H_k^T + R \right)^{-1} \tag{7.40}$$

$$\hat{x}_{k|k} = \hat{x}_{k-1|k-1} + K_k \left[y_k - C\hat{x}_{k-1|k-1} \right] \tag{7.41}$$

$$P_{k|k} = P_{k|k-1} - K_k H_k P_{k|k-1} \tag{7.42}$$

where $H_k = \dfrac{\partial f(x)}{\partial(x)}\bigg|_{x=\hat{x}_{k|k-1}}$ is the matrix obtained by the linearized system output equation in the prediction process.

7.2.2.2 Proposed Estimation Method of the Road Adhesion Coefficient

The block diagram of the estimation method of the road adhesion coefficient is shown in Figure 7.27. First, the vehicle longitudinal and lateral velocity is calculated by the expanded Kalman filter according to the outputs of the 7-DOF vehicle dynamic model and the 2-DOF vehicle dynamic model. Then, the parameters required for the road estimation method are calculated from the 7-DOF vehicle dynamic model, including the yaw rate gain, front steering angle, and yaw rate. Finally, the adhesion coefficient is estimated by the trained neural network. Obviously, the whole estimation process is an open loop system.

The linear boundary limitation shown in Section 7.2.1 introduces the approach to determine if the vehicle is under a nonlinear state according to the yaw rate gain. However, the yaw rate gain is a fixed value, and hence it is not accurate enough to determine the vehicle state according to the r/δ_f threshold. To overcome this difficulty, the error Back Propagation (BP) neural network[21–25] is adopted since it is effective in handling nonlinear problems because of the learning ability of the neural network algorithm. Moreover, the genetic algorithm optimization method is applied to the BP neural network. Therefore, the accuracy of determination of the vehicle state can be improved significantly through heavy learning on some typical test results.

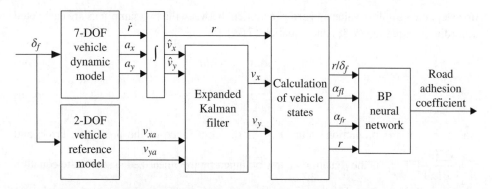

Figure 7.27   Block diagram of the road estimation method.

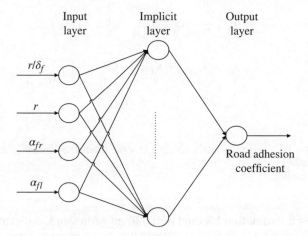

Figure 7.28   Three-layer BP neural network structure.

The BP neural network uses a three-layer feed forward structure as illustrated in Figure 7.28. There are four nodes on the input layer, including the yaw rate gain r/δ_f, side slip angles of the two front wheels α_{fl}, α_{fr}, and yaw rate r; one output node on the output layer, i.e., the road adhesion coefficient; the nodes on implicit layer are determined by the test results during the learning process.

A genetic algorithm is used to optimize the weighting parameters of each node in the neural network. E is defined as the overall training error of the network

$$\min E = f\{x_1, x_2, \cdots, x_s\} = f\{w_1, w_2, \cdots, w_M, \theta_1, \theta_2, \cdots, \theta_K\} \tag{7.43}$$

where $x_i\,(i = 1, 2, \cdots, s)$ is a set of chromosome; s is the summation of the number of the weighting parameters and the number of the thresholds of all the nodes; w_i is the i-th connection weighting parameter of the network; M is the total number of the connection

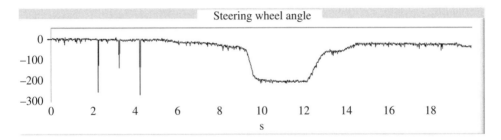

Figure 7.29   Steering angle of the front wheel under the maneuver of step steering.

weighting parameters; θ_K is the threshold of the K-th neuron; and K is the total number of neurons on the implicit and output layers. In addition, the following weighting parameters must be determined in the neural network: the connection weighting parameters w_{ik} between the nodes on the input and implicit layer, and the connection weighting parameters w_{kp} between the nodes on the implicit and output layer; θ_K is the threshold of the neuron on the implicit layer; and θ_p is the threshold of the neuron on the output layer. The following steps of the optimization process are performed:

1. Code the network connection weighting parameters by real numbers.
2. Generate randomly an initial population using the small cluster generation method.
3. Evaluate the performance of the individuals according to a fitness function. The fitness function $f(x)$ is defined as the reciprocal of the error, i.e., $f(x) = 1/E(x)$.
4. Obtain the initial network connection weighting parameters by decoding every individual. Then, the overall error is calculated by inputting the initial network connection weighting parameters and the samples.
5. Select, crossover, and mutate the parent population and produce the next generation of population.
6. Calculate the fitness value of each individual in the current generation and sort them in an ascending order.
7. Obtain the optimal initial weighting parameters of the BP network by decoding the optimal individual. Then calculate the overall error E after adjusting the weighting parameters.
8. If the overall error E is less than the assigned target value, the training is terminated. Otherwise, the weighting parameters obtained from the current optimization process is used as the initial weighting parameter of the next training, and step (5) is repeated.

The training sample of the proposed BP neural network is selected from the VSC test results performed on a test vehicle[15]. The adhesion coefficient of the test road is approximately 0.8. Figures 7.29–7.32 show the measured steering angle at the wheel and the yaw rate under the maneuver of step steering and double lane change. The side slip angles of the two front wheels can be calculated by the test results.

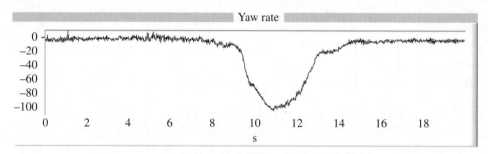

Figure 7.30   Yaw rate under the maneuver of step steering.

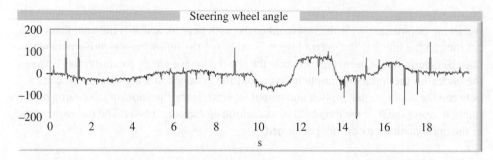

Figure 7.31   Steering angle of the front wheel under the maneuver of double lane change.

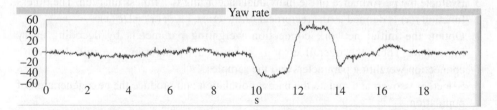

Figure 7.32   Yaw rate under the maneuver of double lane change.

7.2.2.4 Simulation Investigation

To demonstrate the performance of the proposed estimation method of the road adhesion coefficient a simulation investigation is performed by selecting the road adhesion coefficient as 0.9 and 0.4, respectively, and the vehicle speed is set as 60 km/h. A sinusoidal input is given as the steering angle. The simulation model is constructed in Simulink as shown in Figure 7.33, and the simulation results are illustrated in Figures 7.34 and 7.35.

It can be seen from Figures 7.34 and 7.35, and Table 7.2 that the proposed estimation method is able to estimate accurately the road adhesion coefficient for both high and low adhesion coefficients, with an acceptable error. In addition, it can be observed that there are small undulations in the simulation results since the proposed estimation method is open loop; hence, it lacks feedback and self-adjusting mechanisms to compensate for the estimation results.

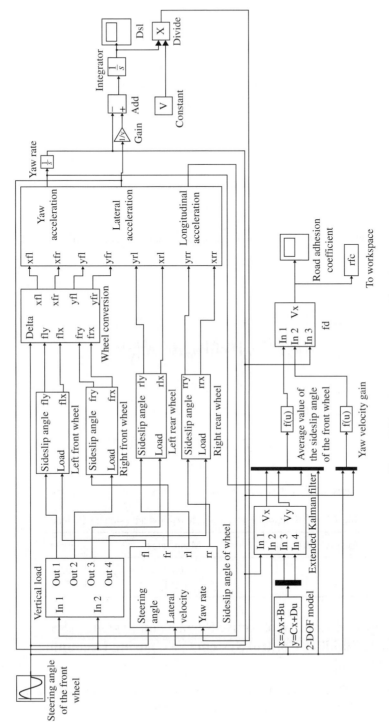

Figure 7.33 Simulation model in Simulink.

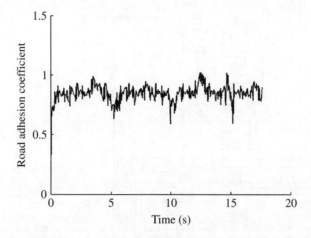

Figure 7.34 Estimation of a high road adhesion coefficient of 0.9.

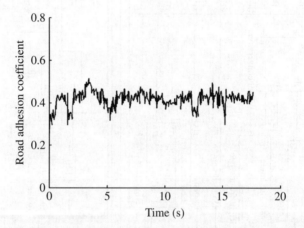

Figure 7.35 Estimation of a low road adhesion coefficient of 0.4.

Table 7.2 Estimation results for high and low adhesion coefficients.

Adhesion coefficient	Mean value	Error
0.9	0.87	3.3%
0.4	0.41	2.5%

7.3 Integrated Control of Active Suspension System (ASS) and Vehicle Stability Control System (VSC) using Decoupling Control Method

Vehicle Stability Control (VSC) system generates a proper yaw moment on the vehicle through the tyre braking or driving forces, and hence improve the vehicle performance in both the lateral and yaw motions. In addition, the active suspension system (ASS) is able

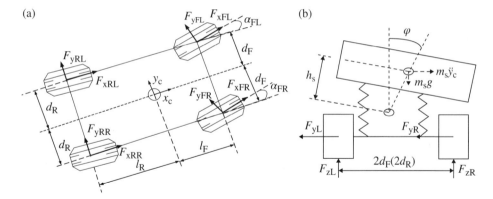

Figure 7.36   7-DOF Vehicle dynamic model. (a) Longitudinal and lateral motion. (b) Pitch motion.

to control the vehicle attitude and regulate the vehicle vertical load transfer during pitch and roll motions by adjusting the suspension stiffness and damping characteristics. Therefore, the main purpose of integrating the VSC with the ASS is to improve the overall vehicle performance, including the lateral stability and ride comfort, through the coordinated control of the VSC and ASS system, especially under critical driving conditions.

7.3.1   Vehicle Dynamic Model

To develop the integrated control of VSC and ASS, the 7-DOF dynamic model is established by considering the interactions between the VSC and ASS, which is analyzed in Chapter 6. The dynamic model shown in Figure 7.36 includes both the VSC and ASS, and the equations of motion can be derived as follows.

Lateral motion

$$m\dot{x}_c\left(\dot{\beta}+r\right)+m_s\left(-\dot{z}_c\dot{\varphi}-h_s\ddot{\varphi}+h_s\dot{\theta}r\right)=2\left(F_{yF}\cos\delta_f+F_{yR}+k_{\alpha F}E_F\varphi+k_{\alpha R}E_R\varphi\right) \quad (7.44)$$

Yaw motion

$$I_z\dot{r}-I_{zx}\ddot{\varphi}+I_{zx}\dot{\theta}r+\left(I_{ys}-I_{xs}\right)\dot{\varphi}\dot{\theta}-m_sl_s\dot{z}_s\dot{\varphi}$$
$$=2\left(F_{yF}\cos\delta_fl_F+F_{yR}l_R+l_Fk_{\alpha F}E_F\varphi-l_Rk_{\alpha R}E_R\varphi\right)+\Delta M \quad (7.45)$$

Vertical motion

$$m_s\left(\ddot{z}_s-\dot{x}_c\dot{\theta}+\dot{y}_c\dot{\varphi}\right)-m_sh_s\left(\dot{\theta}^2+\dot{\varphi}^2\right)+m_sl_sr\dot{\theta}=-\left(F_{FL}+F_{FR}+F_{RL}+F_{RR}\right) \quad (7.46)$$

$$\begin{aligned}
F_{FL} &= -k_{sFL}\left(z_{sFL} - z_{uFL}\right) - c_{sFL}\left(\dot{z}_{sFL} - \dot{z}_{uFL}\right) + f_{FL} \\
F_{FR} &= -k_{sFR}\left(z_{sFR} - z_{uFR}\right) - c_{sFR}\left(\dot{z}_{sFR} - \dot{z}_{uFR}\right) + f_{FR} \\
F_{RL} &= -k_{sRL}\left(z_{sRL} - z_{uRL}\right) - c_{sRL}\left(\dot{z}_{sRL} - \dot{z}_{uRL}\right) + f_{RL} \\
F_{RR} &= -k_{sRR}\left(z_{sRR} - z_{uRR}\right) - c_{sRR}\left(\dot{z}_{sRR} - \dot{z}_{uRR}\right) + f_{RR}
\end{aligned}$$

where

Roll motion

$$\begin{aligned}
&I_{xu}\ddot{\varphi} - I_{xzu}\dot{r} - \left(I_{zs} - I_{ys} - m_s h_s h\right)\dot{\theta}r - I_{zxs}\dot{\varphi}\dot{\theta} + m_s h\dot{z}_c\dot{\varphi} - m_s\left(\ddot{y}_c + \dot{x}_c r\right)h \\
&= \left(F_{FL} - F_{FR}\right)d_F + \left(F_{RL} - F_{RR}\right)d_R
\end{aligned} \tag{7.47}$$

Pitch motion

$$I_{ys}\ddot{\theta} + \left(I_{xs} - I_{zs}\right)\dot{\varphi}r - I_{zxs}\left(r^2 - \dot{\varphi}^2\right) - \left(F_{FL} + F_{FR}\right)\left(l_F - l_s\right) + \left(F_{RL} + F_{RR}\right)\left(l_F + l_s\right) = 0 \tag{7.48}$$

$$m_{ui}\ddot{z}_{ui} = F_i \qquad i = FL, FR, RL, RR \tag{7.49}$$

where m, m_s, and m_u are the vehicle mass, sprung mass, and unsprung mass, respectively; x_c, y_c, and z_c are the Cartesian coordinates of the vehicle center of gravity; x_s, y_s, and z_s are the Cartesian coordinates of the center of gravity of the sprung mass; u_c is the vehicle longitudinal speed; φ and θ are the pitch and roll angles, respectively ; r is the yaw rate of the vehicle; δ_f is the steering angle of the front wheel; F_{YF} and F_{YR} are the front and rear lateral tyre forces, respectively; E_F and E_R are the roll camber coefficients of the front and rear wheels, respectively; $k_{\alpha F}$ and $k_{\alpha R}$ are the cornering stiffness of the front and rear tyres, respectively; h is the height of the vehicle center of gravity; h_s is the vertical distance between the centers of gravity of both the vehicle and the sprung mass; I_z is the moment of inertia of the vehicle mass about axis z_c; I_{zx} is the product of inertia of the vehicle mass about axis x_c and z_c; I_{xu} is the moment of inertia of the sprung mass about axis x_c; I_{xs}, I_{ys}, and I_{zs} are the moments of inertia of the sprung mass about axis x_s, y_s, z_s, respectively; I_{zxu} is the product of inertia of the sprung mass about axis x_c and z_c; I_{zxs} is the product of inertia of the sprung mass about axis x_s and z_s; ΔM is the vehicle corrective yaw moment generated by VSC; l_s is the longitudinal distance between the centers of gravity of the vehicle mass and sprung mass; l_F and l_R are the longitudinal distances between the vehicle center of gravity and the front and rear axles, respectively; f_{FL}, f_{FR}, f_{RL}, and f_{RR} are the front-left, front-right, front-left, and rear-left, and rear-right control forces of the active suspension, respectively; z_{ui} is the vertical displacement of the i-th unsprung mass; c_{si} is the damping coefficient of the i-th damper; k_{si} is the suspension stiffness of the i-th suspension; d_F and d_R are the half of front and rear wheel track, respectively.

7.3.2   2-DOF Reference Model

The 2-DOF vehicle linear dynamic model is adopted as the vehicle reference model to generate the desired vehicle states in this study since the 2-DOF model reflects the desired relationship between the driver's steering input and the vehicle yaw rate.

The equations of motion are expressed as follows by assuming a small sideslip angle and a constant forward speed.

$$mu_c\left(\dot{\beta}+r\right)=-2k_{\alpha F}\left(\beta+l_F\frac{r}{u_c}-\delta_f\right)-2k_{\alpha R}\left(\beta-l_R\frac{r}{u_c}\right) \tag{7.50}$$

$$I_z\dot{r}=-2k_{\alpha F}\left(\beta+l_F\frac{r}{u_c}-\delta_f\right)l_F+2k_{\alpha R}\left(\beta-l_R\frac{r}{u_c}\right)l_R+\Delta M \tag{7.51}$$

7.3.3 Lateral Force Model

To simplify the design of the integrated control system, a small sideslip angle is assumed and hence the tyre displacement is linear. Therefore, the front and rear lateral forces are derived by considering the vehicle roll steering effect:

$$F_{yF}=k_{\alpha F}\left(\delta_f-\beta-l_F\frac{r}{u_c}+G_F\varphi\right) \tag{7.52}$$

$$F_{yR}=k_{\alpha R}\left(-\beta+l_R\frac{r}{u_c}+G_R\varphi\right) \tag{7.53}$$

where G_F and G_R are the roll steering coefficients of the front and rear axles, respectively.

7.3.4 Integrated System Control Model

The state variables are defined as follows for the integrated VSC and ASS control system, by combining equations (7.44)–(7.49), and equations (7.52) and (7.53).

$$x=\left(z_{uFL}\quad z_{uFR}\quad z_{uRL}\quad z_{uRR}\quad \dot{z}_{uFL}\quad \dot{z}_{uFR}\quad \dot{z}_{uRL}\quad \dot{z}_{uRR}\quad z_c\quad \dot{z}_c\quad \varphi\quad \dot{\varphi}\quad \theta\quad \dot{\theta}\quad r\quad \beta\right)^T \tag{7.54}$$

In addition, the variables of the external disturbance for the integrated control system are defined as:

$$w=\left(z_{0FL}\quad z_{0FR}\quad z_{0RL}\quad z_{0RR}\quad \delta_f\right)^T \tag{7.55}$$

where z_{0i} is the stochastic excitation of each tyre generated by the road unevenness; and δ_f is the steering angle of the front wheel generated by the driver. As mentioned earlier in this chapter, the VSC system generates an additional yaw moment to track the desired vehicle states, and the ASS adjusts the suspension stiffness and damping characteristics to improve the vehicle ride comfort, and also indirectly improves the handling stability through regulating the load transfer. Therefore, the control input variables for the integrated control system are defined as:

$$u=\left(\Delta M\quad f_{FL}\quad f_{FR}\quad f_{RL}\quad f_{RR}\right)^T \tag{7.56}$$

The goal of the integrated VSC and ASS control system is to improve the vehicle handling stability and ride comfort. Therefore, the output variables of the integrated control system include the vehicle yaw rate r, the sideslip angle β, the vertical acceleration at the vehicle center of gravity $\ddot{z}_c$, the suspension deflection f_d, and the vehicle roll angle φ, by considering the measurability of these signals,

$$y = \begin{pmatrix} r & \beta & \ddot{z}_c & f_d & \varphi \end{pmatrix}^{\mathrm{T}} \tag{7.57}$$

The state equation and the output equation are then obtained as:

$$\begin{cases} \dot{x} = Ax + B_1 u + B_2 w + f(x,t) \\ y = Cx + Du \end{cases} \tag{7.58}$$

where A, C are the 16×16 input matrix and the 5×16 output matrix, respectively; B_1, B_2 are the 16×1 input matrices; D is the 16×1 direct transfer matrix; $f(x,t)$ is the coupling term of the state variable with size of 16×1.

It is clear that the VSC/ASS integrated control system defined in equation (7.58) is a typical multivariable nonlinear system. Due to the correlations between the tyre longitudinal and vertical forces, and also the interactions among the roll, pitch, and lateral motions, the VSC and ASS are highly coupled. The coupling effects, i.e., a certain control input affecting multiple outputs, are caused by the coupling correlation term included in the state variable. Therefore, it is required to decouple the above-mentioned five control loops and hence achieve that a certain output is controlled solely by one control input in order to improve the overall vehicle performance.

7.3.5   Design of the Decoupling Control System

The decoupling method of nonlinear system is applied to the integrated VSC and ASS system established in equation (7.58) to derive the state feedback control law[26].

7.3.6   Calculation of the Relative Degree

According to the decoupling theory of nonlinear system, the calculation of the relative degree of the original integrated control system is required in order to apply the state feedback control and transform the original nonlinear coupled system into the independent decoupled subsystems. The calculation process is described as follows: first, the derivatives of the control output y are computed with respect to time. Various orders of the derivatives continue to be computed until the input variable u is included explicitly in the output derivative function. Thus, the corresponding derivative order is the system relative degree. In addition, the rank of the Jacobian matrix can be determined through the Interactor algorithm of nonlinear systems[27]. The detailed calculation of the relative degree is demonstrated below.

1. Perform the derivative of the control output variable y_1 with order $R_1 = 1$:

$$y_1 = r$$

$$\dot{y}_1 = \dot{r} = \frac{1}{I_z}\Big[I_{zx}\ddot{\varphi} - I_{zx}\dot{\theta}r - \left(I_{ys} - I_{xs}\right)\dot{\varphi}\dot{\theta} + m_s l_s \dot{z}_c \dot{\varphi} + 2F_{yF}\cos\delta_f l_F$$

$$+ 2F_{yR}l_R + 2l_F k_{\alpha F}E_F\varphi - 2l_R k_{\alpha R}E_R\varphi + \Delta M \Big]$$

Let $Y_1 = \dot{y}_1$, then $\partial Y_1/\partial u^T = \begin{pmatrix} 0 & 0 & 0 & 0 & \dfrac{1}{I_z} \end{pmatrix}$, and $t_1 = \text{rank}\left(\partial Y_1/\partial u^T\right) = 1$.

2. Perform the derivative of the system output variable y_2 with order $R_2 = 1$:

$$y_2 = \beta$$

$$\dot{y}_2 = \dot{\beta} = -r - \frac{1}{mu_c}\Big[m_s\left(-\dot{z}_c\dot{\varphi} - h_s\ddot{\varphi} + h_s\dot{\theta}r\right) + 2\left(F_{yF}\cos\delta_f + F_{yR} + k_{\alpha F}E_F\varphi + k_{\alpha R}E_R\varphi\right)\Big]$$

Since the control input u is not included explicitly in $\dot{y}_2$, the derivative of the system output y_2 with order $R_2 = 2$ is then computed:

$$\dot{y}_2 = \ddot{\beta} = -\frac{1}{mu_c}\Big[m_s\left(-\dot{z}_c\dot{\varphi} - h_s\ddot{\varphi} + h_s\dot{\theta}r\right) + 2\left(F_{yF}\cos\delta_f + F_{yR} + k_{\alpha F}E_F\varphi + k_{\alpha R}E_R\varphi\right)\Big] - \frac{1}{I_z}[I_{zx}\ddot{\varphi}$$

$$- I_{zx}\dot{\theta}r - \left(I_{ys} - I_{xs}\right)\dot{\varphi}\dot{\theta} + m_s l_s \dot{z}_c \dot{\varphi} + 2F_{yF}\cos\delta_f l_F + 2F_{yR}l_R + 2l_F k_{\alpha F}E_F\varphi - 2l_R k_{\alpha R}E_R\varphi + \Delta M]$$

Let $Y_2 = \begin{pmatrix} Y_1 & \ddot{y}_2 \end{pmatrix}^T$, $\partial Y_2/\partial u^T = \begin{pmatrix} 0 & 0 & 0 & 0 & \dfrac{1}{I_z} \\ \dfrac{-I_{zx}d_F}{I_z I_x} & \dfrac{I_{zx}d_F}{I_z I_x} & \dfrac{-I_{zx}d_R}{I_z I_x} & \dfrac{I_{zx}d_R}{I_z I_x} & \dfrac{-1}{I_z} \end{pmatrix}$, then

$t_2 = \text{rank}\left(\partial Y_2/\partial u^T\right) = 2$.

3. For the integrated system output variable $y_3 = \ddot{z}_c$, it can be seen that the control input variables $f_{FL}, f_{FR}, f_{RL},$ and f_{RR} are included in y_3, then $R_3 = 0$.

$$y_3 = \ddot{z}_c = -\frac{1}{m_s}\left(F_{FL} + F_{FR} + F_{RL} + F_{RR}\right) + u_c\dot{\theta} - \dot{y}_c\dot{\varphi} + h_s\left(\dot{\theta}^2 + \dot{\varphi}^2\right) - c_s r\dot{\theta}$$

Let $Y_3 = \begin{pmatrix} Y_1 & \ddot{y}_2 & y_3 \end{pmatrix}^T$, then $\partial Y_2/\partial u^T = \begin{pmatrix} 0 & 0 & 0 & 0 & \dfrac{1}{I_z} \\ \dfrac{-I_{zx}d_F}{I_z I_x} & \dfrac{I_{zx}d_F}{I_z I_x} & \dfrac{-I_{zx}d_R}{I_z I_x} & \dfrac{I_{zx}d_R}{I_z I_x} & \dfrac{-1}{I_z} \\ \dfrac{-1}{m_s} & \dfrac{-1}{m_s} & \dfrac{-1}{m_s} & \dfrac{-1}{m_s} & 0 \end{pmatrix}$, and

$t_3 = \text{rank}\left(\partial Y_3/\partial u^T\right) = 3$.

4. Similarly, the derivatives of the system outputs y_4 and y_5 are performed. We obtain $R_4 = 2$ and $R_5 = 3$; and the system Jacobian matrices $\partial Y_4 / \partial u^T$ and $\partial Y_5 / \partial u^T$ are full ranked,i.e., the ranks are 4 and 5, respectively.

Therefore, the relative degree of the original integrated control system is $R = \begin{pmatrix} 1 & 2 & 0 & 2 & 3 \end{pmatrix}^T$ according to the definition of the relative degree.

7.3.7 Design of the Input/Output Decoupling Controller

For the multivariable coupled integrated system, the purpose of the decoupling controller is to make a certain control input u_i $(i = 1,2,3,4,5)$ rely solely on the system state variable x and some other independent reference variables v_i $(i = 1,2,3,4,5)$ through developing the state feedback law. Thus, the system control input satisfies the following relationship:

$$u = \bar{\alpha}(x) + \bar{\beta}(x)v + \bar{r}(x)w \qquad (7.59)$$

When the state feedback law defined in equation (7.59) is applied on the coupled integrated system, the i-th component of the closed loop system output y_i is affected solely by the i-th reference variable v_i, and therefore the decoupling of the control channels of the close loop system is achieved.

According to the nonlinear decoupling control theory, the relative degree $R = \begin{pmatrix} 1 & 2 & 0 & 2 & 3 \end{pmatrix}^T$ of the integrated control system is obtained by computation, and also the system Falb-Wolovich matrix (i.e., decoupling matrix) $E(x)$ at the equilibrium point is given as:

$$E(x) = \begin{pmatrix} L_{g1}L_f^{r_1-1}y_1(x) & L_{g2}L_f^{r_1-1}y_1(x) & L_{g3}L_f^{r_1-1}y_1(x) & L_{g4}L_f^{r_1-1}y_1(x) & L_{g5}L_f^{r_1-1}y_1(x) \\ L_{g1}L_f^{r_2-1}y_2(x) & L_{g2}L_f^{r_2-1}y_2(x) & L_{g3}L_f^{r_2-1}y_2(x) & L_{g4}L_f^{r_2-1}y_2(x) & L_{g5}L_f^{r_2-1}y_2(x) \\ L_{g1}L_f^{r_3-1}y_3(x) & L_{g2}L_f^{r_3-1}y_3(x) & L_{g3}L_f^{r_3-1}y_3(x) & L_{g4}L_f^{r_3-1}y_3(x) & L_{g5}L_f^{r_3-1}y_3(x) \\ L_{g1}L_f^{r_4-1}y_4(x) & L_{g2}L_f^{r_4-1}y_4(x) & L_{g3}L_f^{r_4-1}y_4(x) & L_{g4}L_f^{r_4-1}y_4(x) & L_{g5}L_f^{r_4-1}y_4(x) \\ L_{g1}L_f^{r_5-1}y_5(x) & L_{g2}L_f^{r_5-1}y_5(x) & L_{g3}L_f^{r_5-1}y_5(x) & L_{g4}L_f^{r_5-1}y_5(x) & L_{g5}L_f^{r_5-1}y_5(x) \end{pmatrix}$$
$$(7.60)$$

And the system matrix $b(x)$ is obtained as:

$$b(x) = \begin{pmatrix} L_f^{r_1}y_1(x) & L_f^{r_2}y_2(x) & L_f^{r_3}y_3(x) & L_f^{r_4}y_4(x) & L_f^{r_5}y_5(x) \end{pmatrix}^T \qquad (7.61)$$

Therefore, the state feedback is defined as:

$$\begin{aligned} \bar{\alpha}(x) &= -E^{-1}(x)b(x) \\ \bar{\beta}(x) &= E^{-1}(x) \end{aligned} \qquad (7.62)$$

When the state feedback control law $u_1(x)$ is applied to the coupled integrated control system, the coupled integrated system is transformed into a decoupled system with independent control channels. The state feedback control law is represented as:

$$u_1(x) = \bar{\alpha}(x) + \bar{\beta}(x)v = -E^{-1}(x)b(x) + E^{-1}(x)v \qquad (7.63)$$

7.3.8 Design of the Disturbance Decoupling Controller

The purpose of the system disturbance decoupling is to fulfill the independence between the control output y in the close loop system and the external disturbance w through designing an appropriate state feedback law. For the integrated control system, the state feedback control law is constructed as follows by assuming that the system external disturbance is measurable.

$$u_2(x) = -B_1^{-1}B_2w \qquad (7.64)$$

Therefore the state feedback control law of the coupled integrated system is designed by combining the developed input/output decoupling controller given in equation (7.63) and the disturbance decoupling controller given in equation (7.64).

$$u = u_1(x) + u_2(x) = -E^{-1}(x)b(x) + E^{-1}(x)v - B_1^{-1}B_2w \qquad (7.65)$$

7.3.9 Design of the Closed Loop Controller

The decoupled integrated system not only eliminates the coupling effects between the control channels, but reduces the influence of the external disturbance on the system control output variable. However, the independent reference variable v in the proposed state feedback control law is unable to improve the control performance of the integrated system since a corrective action is not applied to the independent reference variable. To overcome the problem, a composite controller is proposed through integrating the close loop controller and the decoupling controller in order to improve the overall quality of the system response. As illustrated in Figure 7.37, the proposed integrated control system is decoupled into five independent single-variable systems, and then the closed loop controller is applied to effectively control the decoupled integrated control system.

7.3.10 Design of the ASS Controller

A PID controller is applied to improve the control performance of the closed loop ASS. By considering the ASS control target, the inputs of the PID controller are selected to include the differences e between the desired and the actual values of the vertical acceleration of the vehicle center of gravity, the suspension deflection, and the roll angle, which are given as:

$$e = \left(\ddot{z}_{cd} - \ddot{z}_c \quad f_{dd} - f_d \quad \varphi_d - \varphi \right)^{\mathrm{T}} \qquad (7.66)$$

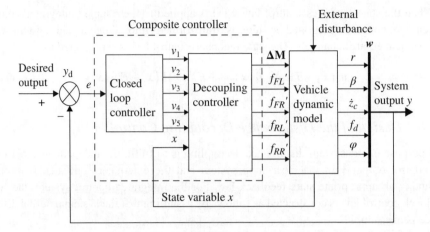

Figure 7.37   Block diagram of the integrated control system.

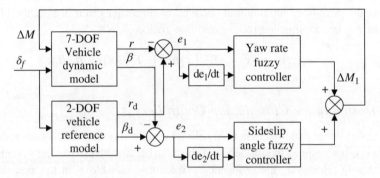

Figure 7.38   Block diagram of the fuzzy control system for the VSC.

Therefore, the PID control law is constructed as:

$$v = \begin{pmatrix} v_2 & v_3 & v_4 & v_5 \end{pmatrix} = K_P \left(e(t) + \frac{1}{T_I} \int_0^t e(t) \, dt + T_D \frac{de(t)}{dt} \right) \tag{7.67}$$

7.3.11   Design of the VSC Controller

A fuzzy control strategy is used for the design of the VSC system, and the block diagram of the proposed VSC control system is shown in Figure 7.38. In the VSC fuzzy control system, the yaw rate and the sideslip angle are selected as the control objectives. As shown in the figure, the VSC fuzzy control system has two input variables, the tracking errors e_1 and e_2, and the differences of the errors ec_1 and ec_2 for the yaw rate and the

Table 7.3 Fuzzy rule bases for yaw rate.

ΔM_1		ec_1						
		NB	NM	NS	ZE	PS	PM	PB
e_1	NB	PB	PB	PB	PB	PM	ZE	ZE
	NM	PB	PB	PB	PB	PM	ZE	ZE
	NS	PM	PM	PM	PM	ZE	NS	NS
	ZE	PM	PM	PS	ZE	NS	NM	NM
	PS	PS	PS	ZE	NM	NM	NM	NM
	PM	ZE	ZE	NM	NB	NB	NB	NB
	PB	ZE	ZE	NM	NM	NB	NB	NB

Table. 7.4 Fuzzy rule bases for the sideslip angle.

ΔM_2		ec_2						
		NB	NM	NS	ZE	PS	PM	PB
e_2	NB	PB	PB	PM	PM	PS	ZE	ZE
	NM	PB	PB	PM	PM	PS	ZE	ZE
	NS	PB	PB	PM	PM	PS	ZE	NM
	ZE	PB	PM	PM	ZE	NM	NM	NB
	PS	PM	PM	ZE	NS	NM	NM	NB
	PM	ZE	ZE	NS	NS	NM	NM	NB
	PB	ZE	ZE	NS	NM	NM	NM	NB

sideslip angle, respectively. The output variables are defined as the corrective yaw moments ΔM_1 and ΔM_2. Thus, the overall corrective yaw moment is defined as a linear combination of the two:

$$\Delta M = n\Delta M_1 + (1-n)\Delta M_2 \qquad (7.68)$$

where n is the weighting coefficient.

To determine the fuzzy controller output for the given error and its difference, the decision matrix of the linguistic control rules is designed and presented in Tables 7.3 and 7.4, respectively. In the tables, seven fuzzy sets are used to represent the states of the inputs and outputs, i.e., {PB,PM,PS,ZE,NS,NM,NB}. A trigonometric function is adopted as the basic membership function, and a trapezoidal function is used for the fuzzy boundary. In addition, the dividing density is relatively higher around the zero value (ZE) of the membership function of the fuzzy input, while it is relatively smaller at a distance from the ZE value, in order to improve the control sensitivity. These rules are determined based on expert knowledge and a large number of simulation results performed in the study. Finally, the outputs of the fuzzy controllers ΔM_1 and ΔM_2 are defuzzified by applying the centroid method to the fuzzy output.

7.3.12   Simulation Investigation

In order to evaluate the performance of the developed integrated control system, i.e., the centralized control system using decoupling control method, a simulation investigation is performed. The performance and dynamic characteristics of the integrated control system are analyzed using MATLAB/Simulink. The road excitation is set as the filtered white noise expressed in equation (7.58). After tuning the parameter setting for the integrated control system, we select $K_P = \text{diag}\{1.8, 1.8, 1.3\}$, $T_I = \text{diag}\{1.56, 3.75, 0.82\}$, and $T_D = \text{diag}\{0.77, 2.5, 0.4\}$ for the closed loop ASS controller; the weighting coefficient $n = 0.85$ in the VSC system fuzzy controller. The vehicle physical parameters are presented in Table 7.5. The centralized control using decoupling control method control and the decentralized control (i.e., the VSC and ASS subsystem controllers work independently) are compared to demonstrate the performance of the integrated control system. Three driving conditions are performed, including step steering input, single lane change, and double lane change. The following discussions are made by comparing the centralized control system with the decentralized control system on the corresponding performance indices.

Table 7.5   Vehicle physical parameters.

Symbol (unit)	Value
m(kg)	3018
m_s(kg)	2685
m_{ui}(kg)	333/4
r_0(m)	0.4
h(m)	0.938
h_s(m)	0.1
H(m)	0.838
d_F/d_R(m)	0.8/0.9
l_F/l_R(m)	1.84/1.88
l_s(m)	0.15
k_{ti}(i=1,2, 3,4)(N/m)	420000(1,2)/350000(3,4)
k_{si}(i=1, 2, 3,4)(N/m)	44444(1,2)/35000(3,4)
c_{si}(i=1,2,3,4) (N.s/m)	1200(1,2)/900(3,4)
$k_{\infty F}/k_{\infty R}$ (N/rad)	29890/50960
I_z(kg.m^2)	10437
I_{zx}(kg.m^2)	2030
I_{js}(j=x,y,z) (kg.m^2)	1744/3000/9285
I_{xu}(kg.m^2)	1996
I_{xzu}(kg.m^2)	377.8
G_f/G_r	0.114/0.1
E_f/E_r	0.8/0.6
J_p(kg.m^2)	0.06
k_s(N.m/rad)	90
B_s(N.m.s/rad)	0.3
d(m)	0.1
G(dimensionless)	20
G_0(m^3/cycle)	5.0×10^{-6}

(1) Step steering input maneuver

The simulation is conducted according to GB/T6323.2-94 controllability and stability test procedure for automobiles – steering transient response test (steering wheel angle step input). The step steering input to the wheel is set as 1.57 rad and the vehicle drives around a circle at a constant speed of 60 km/h. The road adhesion coefficient is selected as 0.6. The simulation results are shown in Figure 7.39.

It is clearly shown in Figure 7.39(a)–(c) that the peak value of the vehicle vertical acceleration for the centralized control is reduced by 30.6% from 2.48 m.s^{-2} to 1.72 m.s^{-2}, the peak value of the roll angle is reduced by 8.1% from 0.099 rad to 0.091 rad, and the peak value of the suspension deflection is reduced by 14.6% from 0.048 m to 0.041 m, compared with those for the decentralized control. The results indicate that the centralized integrated control system is able to decrease the influence from the external disturbance on the system control output through applying the disturbance decoupling controller since the road excitation has the major effect on the vehicle ride comfort.

It is observed that in Figure 7.39(d) and (e) that the overshoots of the yaw rate and side-slip angle for the centralized control are reduced by 13.1% and 7.2%, respectively, compared with those for the decentralized control. In addition, the settling time of the two performance indices are lessened by 37.5% and 26.4%, respectively. It is evident that the centralized control system using decoupling control method is able to improve effectively the transient characteristics of handling stability, and also suppress significantly the steady state responses of the yaw rate and sideslip angle.

(2) Single lane change maneuver

The simulation is performed according to the GB/T6323.1-94 controllability and stability test procedure for automobiles – Pylon course slalom test. For the maneuver of a single lane change, the amplitude of the front wheel steering angle is set as 0.08 rad and the frequency as 0.3 Hz. The road adhesion coefficient and the vehicle speed are assumed to be 0.6 and 60km/h, respectively.

It is clearly illustrated in Figure 7.40 that the peak value of the yaw rate and the sideslip angle for the centralized control are reduced greatly by 33.3% from 0.24 rad/s to 0.16 rad/s, and by 22.2% from 0.09 rad to 0.07 rad, respectively; and the corresponding settling time by 26.8% from 4.1s to 3s, and by 31.9% from 4.7s to 3.2s, respectively, compared with those for the decentralized control. Similarly, the peak value of the roll angle is decreased by 31.7% from 0.082 rad to 0.056 rad. The results indicate that the centralized control system is able to maintain effectively the vehicle trajectory and hence improve the vehicle handling stability, compared with the decentralized control system.

(3) Double lane change maneuver

In order to investigate the adaptability of the developed centralized control system with respect to the variations of the vehicle physical parameters, three vehicle parameters are manipulated with a variation of ±10% by applying a sinusoidal function, including the vehicle mass, wheel base, and height of the vehicle center of gravity. However, the design and parameter setting of the decoupling controller are kept the same. The simulation is performed according to the GB/T6323.1-94 test. For the double lane change maneuver, the amplitude of the front steering angle is set

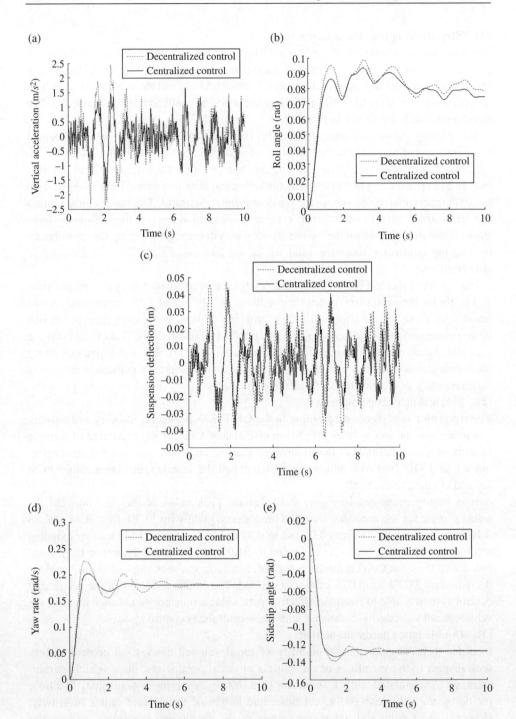

Figure 7.39 Comparison of the responses for the maneuver of step steering input. (a) Vertical acceleration. (b) Roll angle. (c) Suspension deflection. (d) Yaw rate. (e) Sideslip angle.

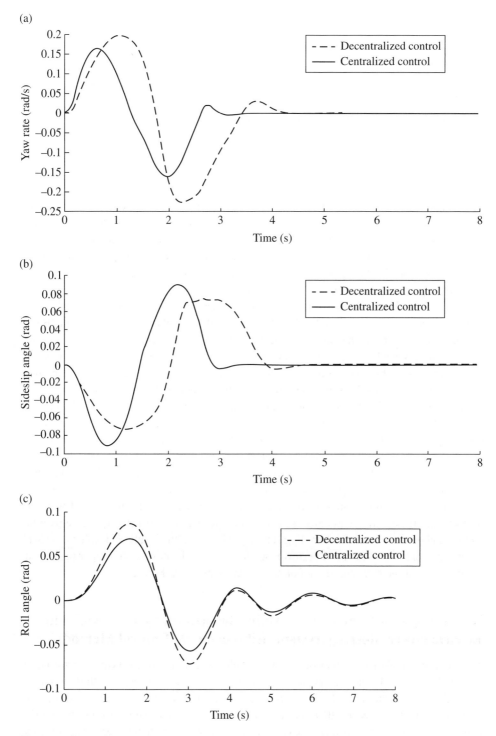

Figure 7.40   Comparison of responses for the single lane change maneuver. (a) Yaw rate. (b) Sideslip angle. (c) Roll angle.

as 0.06 rad, and frequency as 0.5 Hz, the road adhesion coefficient as 0.5, and the initial vehicle speed as 50 km/h.

It is observed in Figure 7.41 that the four performance indices for the centralized control are reduced slightly compared with those for the decentralized control. The results indicate that the adaptability of the centralized control system is insufficient to adapt the variation of the vehicle physical parameters since the accurate mathematical model and specific system physical parameters are required to develop the decoupling controller.

7.3.13   Experimental Study

To validate the effectiveness of the centralized integrated control system, a hardware-in-the-loop (HIL) experimental study is conducted based on LabVIEW PXI. As shown in Figure 7.42, the developed HIL system consists of a host computer, a client computer, an interface system, and the VSC and ASS actuators. The client computer (PXI-8196 manufactured by National Instruments Inc.) collects the signals measured by the sensors, which include the pressure of each brake wheel cylinder, the pressure of the brake master cylinder, and the vertical acceleration of the sprung mass at each suspension. These signals are in turn provided to the host computer (PC) through a LAN (local area network) cable. Based on these input signals, the host computer computes the vehicle states and the desired vehicle motions, such as the desired yaw rate. Thereafter, the host computer generates control commands to the client computer. Through the hardware interface circuits, the client computer in turn sends the control commands to the corresponding actuators.

Two driving conditions are performed, including the step steering input and double lane change, by assuming that the initial vehicle speed is 72km/h, and the road adhesion coefficient is 0.6. As illustrated in Figure 7.43 for the double lane change maneuver, the centralized control system using decoupling control method is able to track closely the desired yaw rate generated from the 2-DOF reference model with only a 10.3% amplitude difference. In addition, the peak value of the sideslip angle is restrained at a relatively small value of 0.1 rad, although there is a deviation from the desired sideslip angle. The results indicate that the centralized control system is able to maintain effectively the vehicle trajectory and hence improve its handling stability. Moreover, the small peak value of the roll angle represents a good control performance for the vehicle attitude. A similar pattern can be observed for the step steering input maneuver as shown in Figure 7.44.

7.4   Integrated Control of an Active Suspension System (ASS) and Electric Power Steering System (EPS) using H_∞ Control Method

Numerous external disturbances occur when a vehicle is being driven. Typical disturbances include lateral winds and stochastic excitations from the road surface. Both disturbances affect the vehicle lateral and vertical motions, respectively. On the other hand, the two motions interact with each other and have great effects on both vehicle stability and ride comfort. To suppress the disturbances and hence improve the vehicle overall performance,

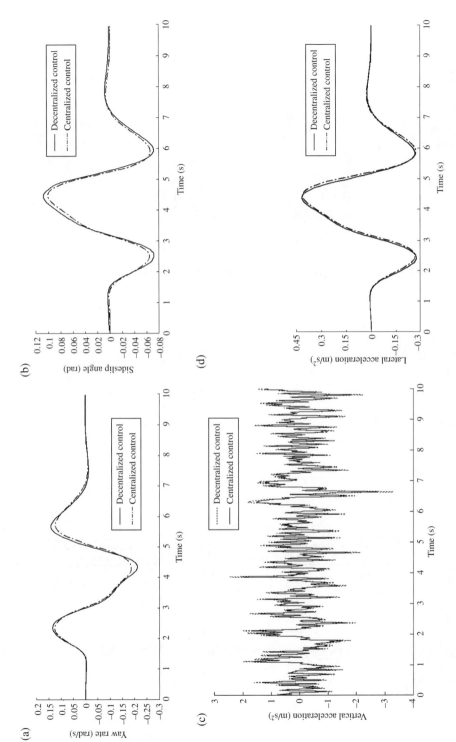

Figure 7.41 Comparison of responses for the double lane change maneuver. (a) Yaw rate. (b) Sideslip angle. (c) Vertical acceleration. (d) Lateral acceleration.

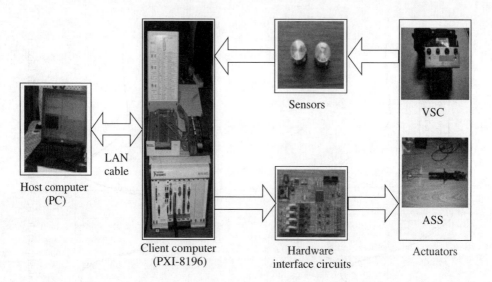

Figure 7.42 Experimental configuration of the developed integrated control system.

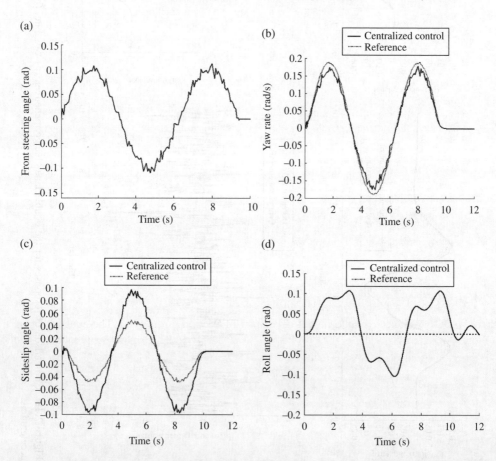

Figure 7.43 Comparison of responses for the double lane change maneuver. (a) Front steering angle. (b) Yaw rate. (c) Sideslip angle. (d) Roll angle.

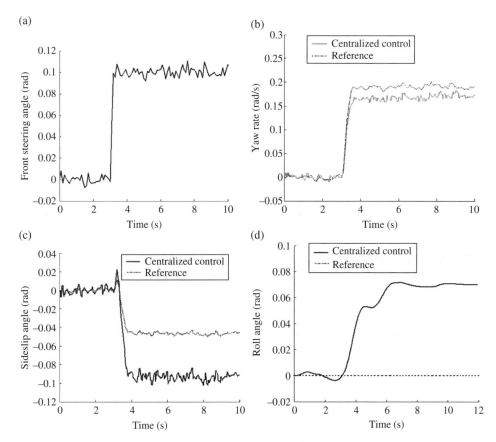

Figure 7.44 Comparison of responses for the step steering input maneuver. (a) Front steering angle. (b) Yaw rate. (c) Sideslip angle. (d) Roll angle.

an integrated control method is applied to achieve the function integration of both the steering and suspension systems through coordinating the interactions between the vehicle lateral and vertical motions[28].

7.4.1   Vehicle Dynamic Model

The 7-DOF vehicle dynamic model developed in Section 7.3 is used. The equations of motion are the same as equations (7.44)–(7.49), except that the corrective yaw moment is omitted in equation (7.45).

7.4.2   EPS Model

The following governing equations can be obtained by applying a force analysis on the steering gear of the EPS system:

$$T_m + k_n \left(\delta_h - \delta_1 \right) = T_r + J_p \ddot{\delta}_1 + B_p \dot{\delta}_1 \tag{7.69}$$

where T_m is the assist torque applied on the steering column; T_c is the hand torque applied on the steering wheel and $T_c = k_n \left(\delta_h - \delta_1 \right)$; k_n is the torsional stiffness of the torque sensor; δ_h is the rotation angle of the steering wheel; δ_1 is the rotation angle of the pinion, and hence the steering angle of the front wheel δ_f can be calculated as $\delta_f = \delta_1 / G$, and G is the speed reduction ratio of the rack-pinion mechanism; J_p is the equivalent moment of inertia of multiple parts reflected on the pinion axis, including the motor, the gear assist mechanism, and the pinion; B_p is the equivalent damping coefficient reflected on the pinion axis; and T_r is the aligning torque transferred from the tyres to the pinion, $T_r = \dfrac{2}{G} d k_{\alpha F} \left(\dfrac{\delta_1}{G} - \dfrac{l_f}{u_c} r - \beta \right)$, where d is the pneumatic trail of the front tyre. The state variable is defined as:

$$x = \left(z_{uFL} \quad z_{uFR} \quad z_{uRL} \quad z_{uRR} \quad \dot{z}_{uFL} \quad \dot{z}_{uFR} \quad \dot{z}_{uRL} \quad \dot{z}_{uRR} \quad z_c \quad \dot{z}_c \quad \varphi \quad \dot{\varphi} \quad \theta \quad \dot{\theta} \quad r \quad \dot{y}_c \quad \delta_1 \quad \dot{\delta}_1 \right)^{\mathrm{T}}$$

(7.70)

The external disturbances are defined as the stochastic excitation of the road unevenness to each wheel z_{0i}, and the lateral wind disturbance, which is given as:

$$w = \left(z_{0FL} \quad z_{0FR} \quad z_{0RL} \quad z_{0RR} \quad z_w \right)^{\mathrm{T}}$$

(7.71)

The control input U is defined as the four active suspension forces f_i, and the assist torque T_m:

$$U = \left[f_{FL} \quad f_{FR} \quad f_{RL} \quad f_{RR} \quad T_m \right]^T$$

(7.72)

The system state equation is constructed as:

$$\dot{X} = A(X) + B \begin{bmatrix} S_w w \\ U \end{bmatrix}$$

(7.73)

where $A(X)$ is the polynomial column vector of the state variable; $S_w = diag \left(S_{w1}, S_{w2}, S_{w3}, S_{w4}, S_{w5} \right)$ corresponds to the weighting coefficients of the road excitation and lateral wind perturbation, respectively.

The multiple performance indices are selected by considering the vehicle handling stability, ride comfort, and energy consumption of the ASS. They include the yaw rate r, and sideslip angle β, roll angle φ, vehicle vertical acceleration $\ddot{z}_c$, pitch angle θ, assist torque T_m, and control forces f_i of the ASS. Therefore, the system penalty function is proposed as:

$$Z = S_Z \left[r \quad \ddot{z}_c \quad \phi \quad \theta \quad \beta \quad f_{FL} \quad f_{FR} \quad f_{RL} \quad f_{RR} \quad T_m \right]^T$$

(7.74)

where $S_Z = diag \left(S_1, S_2, S_3, S_4, S_5, S_6, S_7, S_8, S_9, S_{10} \right)$ is the weighting coefficient matrix.

The system output is defined as follows by considering the measurability of the signals:

$$Y = \left[\ddot{z}_c \quad r \quad \dot{\phi} \quad \dot{\theta} \quad \dot{y}_c \quad f_d \right]^T$$

(7.75)

Therefore, the state equation and output equation of the nonlinear vehicle dynamic system is obtained as:

$$\begin{cases} \dot{X} = A(X) + B_1 w + B_2 U \\ Z = C_1 X + D_{12} U \\ Y = C_2 X \end{cases} \tag{7.76}$$

where B_1 and B_2 are the input 18×5 matrices; C_1 and C_2 are the output matrices with size of 10×18 and 6×18, respectively; D_{12} is the matrix of size 10×5.

7.4.3 Design of Integrated Control System

As discussed earlier in the chapter, the integrated control of the EPS and ASS is a complex nonlinear control problem since there are uncertainties on the structure and parameters, along with some unmodeled dynamics, etc. In addition, the complexity of the system is further increased by the external disturbances. To overcome the problem, the H_∞ control method is applied to design the complex integrated control system since it has advantages in simultaneously achieving the robust stabilization and performance of the control system. Although H_∞ control is applied to linear systems in general, the same methodology can be used for nonlinear systems. Then, H_∞ control for nonlinear systems becomes a so-called L_2 gain constrained control. Moreover, H_∞ techniques can be used to minimize the closed loop impact of the disturbances. The structure of the proposed integrated control system is shown in Figure 7.45.

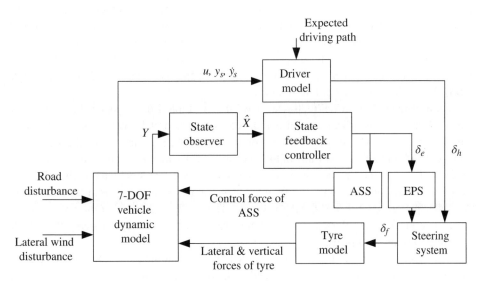

Figure 7.45 Block diagram of the integrated control system of EPS and ASS.

As a matter of fact, not all the signals of the integrated control system can be obtained and, even if they can, the cost of the controller is increased significantly. Therefore, a H_∞ state observer is required to realize the feedback control. The state observer is constructed as follows:

$$
\begin{cases}
\dot{\hat{X}} = A\left(\hat{X}\right) + B_1 w + B_2 U + Q\left(\hat{X}\right)\left(Y - \hat{Y}\right) \\
\hat{Y} = C_2 \hat{X}
\end{cases}
\tag{7.77}
$$

where $\hat{X}$ is the state vector of the observer; $\hat{Y}$ is the observer output; and $Q\left(\hat{X}\right)$ is the output gain. The aim to solve for the observer is to find the output gain $Q\left(\hat{X}\right)$ and the detailed solution of the output gain is provided in the reference[15].

7.4.4 Simulation Investigation

To demonstrate the effectiveness of the developed integrated control system, a simulation investigation is performed. The vehicle physical parameters are given in Table 7.5 of Section 7.3.12. After tuning, the matrices of the weighting coefficients are selected as:

$$
S_w = \mathrm{diag}\left(0.03, \quad 0.03, 0.03, \quad 0.03, 0.002\right)
$$

$$
S_Z = \mathrm{diag}\left(0.5, 0.1, 0.8, 0.8, 0.9, 0.005, 0.005, 0.005, 0.005, 0.01\right)
$$

The vehicle speed is set as $u_c = 25\,m/s$ and the expected input trajectory of the vehicle is illustrated in Figure 7.46. It is observed in the figure that the vehicle travels straight forward first and then around a circle. The stochastic road excitation is applied all the way through, while the lateral wind disturbance is exerted after the vehicle turns and then reaches a steady state condition. In this chapter, it is assumed that the vehicle encounters an abrupt (step) lateral wind disturbance F_w with an amplitude of $1500\,N$ at time $t_2 = 3s$, and disappears at time $t_3 = 4s$. The proposed integrated control system is compared with the two other systems: only with an EPS (named single EPS), only with an ASS (named single ASS). The following observations are made.

As illustrated in Figure 7.47(a, b; see page 248) and Table 7.6, the peak value of the sideslip angle for the integrated control is reduced by 6.06% and 13.89% respectively, compared with that for the single EPS control and single ASS control, after the steering is applied. In addition, the peak value of the sideslip angle for the integrated control is reduced by 28.99% and 14.04% respectively, after the vehicle encounters the lateral wind. The settling time of the sideslip angle for the integrated control is also decreased for both cases. A similar pattern is observed for the yaw rate. The results indicate that the impact of the abrupt lateral wind disturbance on the vehicle is restrained effectively and hence the vehicle handling stability is improved.

It is observed clearly in Figure 7.47(c) and Table 7.7 that the peak value of the steering torque for the integrated control is reduced by 8.78% and 16.18% respectively, compared with that for the single ASS control and single EPS control. In addition, the steady state

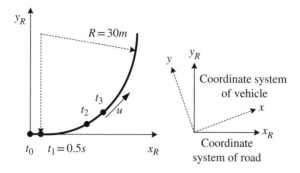

Figure 7.46 Expected vehicle trajectory input.

Table 7.6 Responses for handling stability.

Performance index	Control method	Peak value		Response Time(s)	
		Steering	Lateral Wind	Steering angle	Lateral Wind
Sideslip angle	Single EPS control	−0.033	−0.057	2.022	4.985
β (rad)	Single ASS control	−0.036	−0.069	2.125	5.434
	Integrated system control	−0.031	−0.049	2.018	4.751
Yaw rate r	Single EPS control	0.215	0.342	1.851	4.895
(rad.s^{-1})	Single ASS control	0.238	0.716	1.878	5.284
	Integrated system control	0.212	0.289	1.845	3.826

value of the steering torque for the integrated control is reduced by 13.45% and 14.96% respectively, and the settling time is decreased by 8.22% and 10.32% respectively. The results demonstrate that the integrated control system is able to maintain both steering agility and good road feel, and at the same time effectively restrain the impact of the abrupt lateral wind disturbance on the vehicle.

Figure 7.47(d–h) and Table 7.8 illustrate that these performance indices on ride comfort, including the vehicle vertical acceleration, roll angle, pitch angle and suspension deflection, are reduced for the integrated control, compared with that for the single ASS control and single EPS control. For brevity, the vehicle vertical acceleration is selected to show the improvement of the integrated control system over the other two control systems. As shown in Figure 7.47(d), the PSD (power spectrum density) value of the vehicle vertical acceleration for the integrated control is decreased significantly compared with the single EPS control in the human body-sensitive frequency region of 1–12 Hz. Moreover, it is observed that the PSD value for the integrated control is decreased greatly in the frequency region of 8–12 Hz compared with the single ASS control, although there is no big difference between the two in the frequency region of 1–4 Hz, which is the resonant frequency region of the sprung mass. Therefore, the results indicate that the vehicle ride comfort is improved significantly by the integrated control system compared with both the single EPS and single ASS control systems.

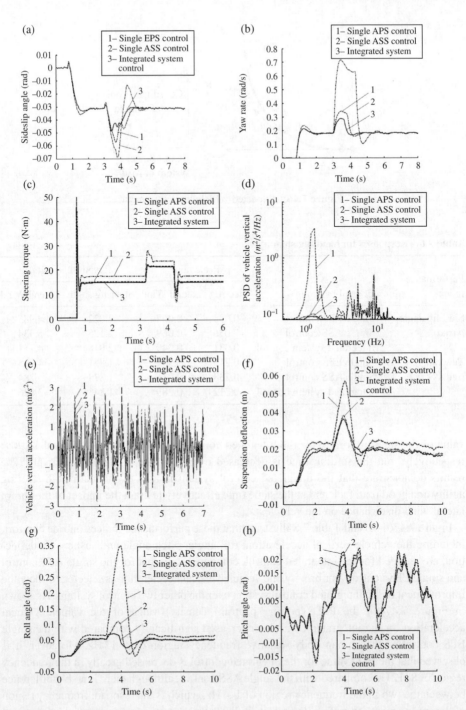

Figure 7.47 Comparisons of the responses for the three control systems. (a) Sideslip angle. (b) Yaw rate. (c) Steering torque. (d) PSD of vehicle vertical acceleration. (e) Vehicle vertical acceleration. (f) Suspension deflection. (g) Roll angle. (h) Pitch angle.

Table 7.7   Steering torque.

Control Method	Maximum (Nm)		Response Time (s)		Steady State Value (Nm)	
	Steering	Lateral Wind	Steering	Lateral Wind	Steering	Lateral Wind
Single EPS control	45.85	22.17	1.005	3.589	14.93	21.38
Single ASS control	50.26	27.86	1.095	3.455	17.25	23.55
Integrated system control	42.13	21.91	0.982	3.398	14.67	21.11

Table 7.8   Response for ride comfort.

Performance index	Control method	Average	Root mean square
	Single EPS Control	0.0130	0.9115
Acceleration z_c (m.s^{-2})	Single ASS Control	0.0087	0.8273
	Integrated System Control	0.0070	0.7757
	Single EPS Control	0.0923	0.0747
Roll angle φ (rad)	Single ASS Control	0.0586	0.0458
	Integrated System Control	0.0521	0.0349
	Single EPS Control	0.0069	0.0097
Pitch angle θ (rad)	Single ASS Control	0.0065	0.0085
	Integrated System Control	0.0061	0.0081
	Single EPS Control	0.0222	0.0115
Suspension deflection f_d (m)	Single ASS Control	0.0189	0.0083
	Integrated System Control	0.0176	0.0076

7.5   Integrated Control of Active Suspension System (ASS) and Electric Power Steering System (EPS) using the Predictive Control Method

Predictive control (or model predictive control (MPC)) theory is an advanced control method developed from an industrial process control used in the 1980s. The principle behind predictive control is to use the past and current system states to predict the future change of the system output. The system's optimal control is achieved by minimizing the error between the controlled variables and the targets by applying an iterative, finite time-horizon optimization approach. The predictive control method is applied to the integrated control of the ASS and EPS systems in this chapter[29] since the predictive control has advantages in dealing with both soft and hard constraints, and uncertainties in a complex multivariable control framework.

7.5.1   Designing a Predictive Control System

As developed in the previous chapter, the same 7-DOF vehicle dynamic model is used. To apply the iterative, finite time-horizon optimization approach, the control system model must be represented by a discrete state equation[29]. The system predictive width is set

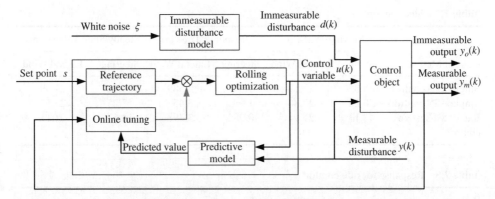

Figure 7.48   Block diagram of the predictive control system.

as P, and the control width C must follow the condition $C \leq P$. The optimization objective of the system between the reference trajectory $r(k)$ and the model predictive output $y(k)$ is given in a quadratic form:

$$\min J(k) = \sum_{i=0}^{p-1} \sum_{j=1}^{n_u} \left\{ \left| R_{i,j}^u u_j (k+i \mid k) \right|^2 + \left| R_{i,j}^{\Delta u} \Delta u_j (k+i \mid k) \right|^2 \right\}$$
$$+ \sum_{i=0}^{P-1} \sum_{j=1}^{n_y} \left| Q_{i+1,j}^y \left(y_j (k+i+1 \mid k) - r_j (k+i+1) \right) \right|^2 \tag{7.78}$$

where R and Q correspond to the weighting matrices of the control variables and output variables, respectively, and $R^u = diag(r_{11}, r_{21}, r_{31}, r_{41}, r_{51})$, $R^{\Delta u} = diag(r_{12}, r_{22}, ..., r_{52})$, $Q = diag(q_1, q_2, ..., q_{14})$; n_u and n_y are the dimensions of the control variables and output variables; $r_{11}, ..., r_{51}$ are the weighting coefficients of the control variables, $r_{12}, ..., r_{52}$ are the weighting coefficients of the variation rates of the control variables, $q_1, ..., q_{14}$ correspond to the weighting coefficients of system outputs; and $y(k+i+1 \mid k)$ are the predicted outputs at time k and step $i+1$.

The predictive control is based on the iterative, finite time-horizon optimization of the system model. At every sample time, the constrained optimization problem defined in equation (7.78) is solved online. Only the first term of the control sequence $u(k)(u(k-1)+\Delta u(k))$ is implemented to the control variables, then the system's states are sampled again and the optimization process is repeated starting from the new current states. The prediction horizon keeps being shifted forward, and for this reason MPC is also called receding horizon control. The block diagram of the proposed predictive control system is shown in Figure 7.48[30].

7.5.2   Boundary Conditions

One of the advantages of predictive control is the ability to explicitly handle the boundary conditions of the control variables in a multivariable control framework, and then predict the future output and take the control actions accordingly by applying online the iterative, finite horizon optimization approach. The major boundary conditions are defined as follows by considering the control requirements of the integrated EPS and ASS control system[31]:

1. The collision between the suspension and the frame/body should be avoided. The dynamic travel of the suspension should be constrained by its mechanical structure:

$$\left| z_{2i} - z_{3i} \right| \le f_{d\max} \quad \left(i = FL, FR, RL, RR \right) \tag{7.79}$$

$z_{2i} - z_{3i}$ is the suspension deflection at each suspension; and $f_{d\max}$ is the maximum dynamic deflection of the suspension. It is usually selected as 7–9 cm for sedans, 5–8 cm for buses, and 6–9 cm for commercial vehicles.

2. Tyre–road contact must be ensured in order to provide enough lateral and longitudinal forces to the vehicle. Hence, the dynamic load of the tyre does not exceed the static load.

$$\left| k_{ti} \left(z_{1i} - z_{2i} \right) \right| \le mg \quad \left(i = FL, FR, RL, RR \right) \tag{7.80}$$

$z_{1i} - z_{2i}$ is the dynamic displacement of each tyre; and k_{ti} is the tyre stiffness.

3. When the vehicle lateral acceleration reaches 0.4 g, the roll angle is selected as 2.5 ~ 4° for sedans, and not greater than 6 ~ 7° for commercial vehicles.

4. The vehicle lateral acceleration should not exceed 0.6 g, the yaw rate should not exceed 0.6 rad/s, and the pitch rate should not exceed 0.3 rad/s.

5. The active suspension force and the steering torque T_m of the EPS is also constrained:

$$\left| f_i \right| \le f_{\max} \tag{7.81}$$

$$\left| T_m \right| \le T_{m\max} \tag{7.82}$$

7.5.3 Simulation Investigation

To demonstrate the effectiveness of the developed integrated control system, a simulation investigation is performed. It is assumed that the vehicle speed u_c is 20 m/s; the steering wheel input is a step function with an amplitude of $\pi/2$; the predictive width P selected as 10, and the control width C as 4; and the sampling time is 0.005 s. After tuning, the weighting coefficients are selected as: $q1 = q2 = 10^3$, $q3=100$, $q4=500$, $q5= q6=1$, $q7 = q8 = q9 = q10 = 400$, $q11 = q12 = q13 = q14 = 10^3$, $r11 = r21 = r31 = r41 =10^{-3}$, $r51=10^{-2}$, $r12 = r22 = r32 = r42 = r52=1$. The integrated control is compared with the non-integrated control (i.e., the EPS and ASS subsystem controllers work independently), and the passive system to demonstrate the performance of the integrated control system.

It is observed in Figure 7.49(a–e) that the vehicle multiple performance indices for the integrated controlare reduced to various extents compared with the non-integrated control. The results indicate that the integrated control system based on the proposed predictive control is able to improve the overall vehicle performance, including handling stability and ride comfort, by coordinating the interactions between the ASS and EPS.

In addition, a sensitive study is performed to investigate the influence of the predictive width and control width. Figure 7.50 illustrates that control stability and robustness are

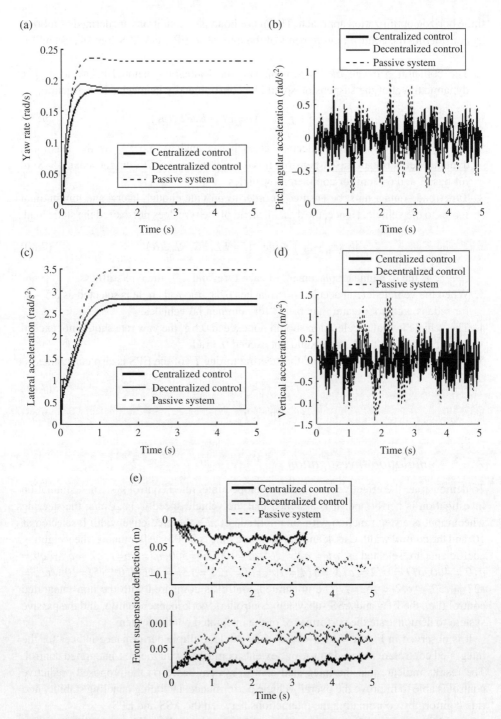

Figure 7.49 Comparisons of the responses for the three systems. (a) Yaw rate. (b) Pitch angular acceleration. (c) Lateral acceleration. (d) Vertical acceleration. (e) Dynamic deflection of the front suspension (left and right).

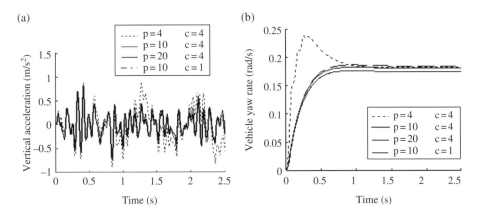

Figure 7.50   Influence of the predictive width and control width. (a) Predictive width variation. (b) Control width variation.

increased as the predictive width P increases, but the dynamic response is slow. Moreover, as the control width C increases, the dynamic response becomes faster, and the control sensitivity is improved correspondingly, but the system stability is reduced.

7.6   Integrated Control of the Active Suspension System (ASS) and Electric Power Steering System (EPS) using a Self-adaptive Control Method

In practice, numerous uncertainties exist in vehicle dynamics models, including stochastic excitations from the road surface, time-varying physical parameters of the vehicle, disturbances of the lateral wind, and measurement noises of on-vehicle sensors. In recent years, self-adaptive control has been identified as an attractive and effective control method to control systems with uncertainties. Self-adaptive control, which has a long history in the field of control engineering, is an advanced control method where the controller is able to adapt to a controlled system with uncertainties. The SISO self-tuning regulator was first proposed in 1973, and then the self-tuning regulator and controller were extended to MIMO in the 1980s. As a result, a number of multivariable self-tuning control methods were developed by combining the recursive parameter estimation method and minimum variance regulation law, generalized minimum variance regulation law, pole placement control law, feed forward control law, and so on. In this chapter, self-adaptive control is applied to the integrated EPS and ASS control by combining the recursive least square estimation method and the generalized minimum variance control law[32,33]. The block diagram of the self-adaptive integrated control system is illustrated in Figure 7.51.

7.6.1   Parameter Estimation of a Multivariable System

The foundation of the self-adaptive control method is parameter estimation. The parameter estimation of the multivariable integrated control system is performed according to the following steps. First, the integrated control system is simplified as a linear multivariable

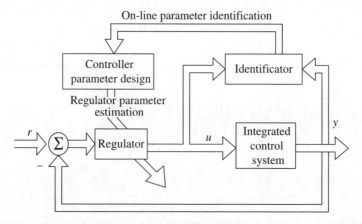

Figure 7.51 Block diagram of the self-adaptive control system for the integrated EPS and ASS control.

system, and the outputs of the two ASS and EPS subsystems are relatively independent. The outputs of the two subsystems have effects on vehicle performance including vertical acceleration, roll angle, yaw rate, etc. Therefore, if the system noise $\xi(t)$ given in equation (7.83) is an irrelevant measured noise with a mean of zero and variance of σ^2, the multivariable integrated control system is considered as a system consisting of m independent SIMO (single input multiple outputs) subsystems[21]. Thereafter, the parameter estimation equation of the multivariable system is decomposed into a set of parameter estimation equations, each of them corresponding to a SIMO subsystem. Finally, the recursive modified least square method is applied to the parameter estimation equations for the multiple SIMO subsystems, and hence the model parameters are obtained.

7.6.2   Design of the Multivariable Generalized Least Square Controller

The ARMAX model of the deterministic linear multivariable system is given as:

$$A\left(z^{-1}\right)y(t) = B\left(z^{-1}\right)u(t) + d_a + C\left(z^{-1}\right)\xi(t) \tag{7.83}$$

where $A\left(z^{-1}\right), B\left(z^{-1}\right)$ and $C\left(z^{-1}\right)$ are the polynomial matrices of the unit backward shift operator z^{-1}; u(t) and y(t) are the input and output vectors with dimension of n; d_a is the steady state error vector with dimension of n; and $\xi(t)$ is the system noise. The generalized output vector is defined as:

$$\varphi(t+k) = P\left(z^{-1}\right)T(z)y(t)$$

And the generalized ideal output vector is defined as:

$$y^*(t+k) = R\left(z^{-1}\right)p(t) - Q\left(z^{-1}\right)u(t) \tag{7.84}$$

where $P(z^{-1})$, $Q(z^{-1})$, and $R(z^{-1})$ are the weighted polynomial matrices of z^{-1}; $T(z)$ is the lower triangle matrix of z; and p(t) is the n-dimensional reference input vector with known boundary.

If the optimal prediction of the generalized output vector $\varphi^*(t+k|t)$ equals the generalized ideal output $y^*(t+k)$, the generalized minimum variance control law is obtained as:

$$\varphi^*(t+k|t) = R(z^{-1})p(t) - Q(z^{-1})u(t) \tag{7.85}$$

And the system performance factor J reaches the minimum:

$$J_{min} = E\left[\left\|e(t+k)\right\|^2 / F_t\right] = E\left[\left\|F(z^{-1})\xi(t+k)\right\|^2 / F_t\right] = \gamma^2 \tag{7.86}$$

where F_t denotes the non-descending σ-algebraic group; and $F(z^{-1})$ is the polynomial matrix of z^{-1}.

7.6.3   Design of the Multivariable Self-adaptive Integrated Controller

The multivariable self-adaptive integrated controller is designed according to the following steps[34]:

1. Measure the real output y(t) and external input p(t) at time t;
2. Compute $\hat{y}(t+k-1)$;
3. Construct the vectors $\hat{\phi}(t)$ and $\hat{\phi}(t-k)$;
4. Calculate the estimate parameter matrix $\hat{\Theta}(t)$;
5. Compute the new control input u(t);
6. Go back to step (1), add one to the time counter and repeat the steps above.

7.6.4   Simulation Investigation

To demonstrate the effectiveness of the developed integrated control system, a simulation investigation is performed. The vehicle physical parameters are the same as those defined in Section 7.3.12. The maneuver of step steering input is applied and the vehicle speed u_c is assumed to be 10 m/s and 20 m/s, respectively. The generalized ideal output of the system can be obtained as follows: The roll angle ϕ is expected to be as small as possible, and the yaw rate r is expected to reach an expected steady state value r_d. The self-adaptive integrated control is compared with the single EPS control and the single ASS control to demonstrate the performance of the integrated control system.

It is observed in Tables 7.9 and 7.10 that the self-adaptive integrated control system performs the best among the three control systems on the performance indices of the peak value and settling time of both the vehicle yaw rate and roll angle. The results indicate that both the handling stability and ride comfort are improved through applying the self-adaptive control method to the integrated EPS and ASS control. In addition, the results show that the application of the self-adaptive control method is able to reduce effectively

Table 7.9 Comparison of the peak value of responses.

Performance index	EPS control	ASS control $u_c = 10\,\text{m/s}$	Integrated control	EPS control	ASS control $u_c = 20\,\text{m/s}$	Integrated control
Yaw rate/r (rad · s^{-1})	0.17	0.15	0.12	0.22	0.20	0.18
Roll angle/ϕ (rad)	0.41	0.032	0.029	0.11	0.068	0.016

Table 7.10 Comparison of settling time responses.

Settling time of performance index	EPS control	ASS control $u_c = 10\,\text{m/s}$	Integrated control	EPS control	ASS control $u_c = 20\,\text{m/s}$	Integrated control
Yaw rate (s)	0.43	0.37	0.20	0.9	0.58	0.30
Roll angle (s)	1.7	1.2	1.0	1.6	1.2	1.0

the effects of both the model uncertainties and stochastic disturbances on the system. Moreover, comparisons between the single EPS and ASS demonstrate that the single ASS is able to maintain the vehicle attitude more effectively than the single EPS control under the maneuver of step steering input.

7.7 Integrated Control of an Active Suspension System (ASS) and Electric Power Steering System (EPS) using a Centralized Control Method

This section studies the integrated control of electric power steering system (EPS) and active suspension system (ASS) to achieve the goal of function integration of the control systems[35]. The nonlinear centralized control theory is applied to design a centralized controller in order to solve the system couplings between the ASS and EPS, and eliminate the disturbances from the road excitations. Moreover, a centralized PD controller is designed based on the centralized vehicle dynamic system in order to improve the dynamic responses of steering conditions.

7.7.1 Centralized Controller Design

7.7.1.1 Centralization of System Inertial Term

This study uses the same models as those developed in Section 7.4, including the nonlinear vehicle dynamic model, EPS system model, and ASS model.

The nonlinear dynamic model is transformed into the affine nonlinear form described in reference[36] through order reduction of the differential equation and centralized of the inertial terms. The state equation of the system is given as:

$$\dot{x} = Ax + Bu + \Delta f(x,t) + Pw \qquad (7.87)$$

where $\mathbf{A}_{n \times n} \ (n = 19)$ is the coefficient matrix; $\mathbf{B}_{n \times m} = [\mathbf{B}_1, \mathbf{B}_2, \mathbf{B}_3, \mathbf{B}_4, \mathbf{B}_5] \ (m = 5)$ is the input matrix; $\Delta \mathbf{f}(\mathbf{x}, t)$ is the affine nonlinear term of the state variables; and $\mathbf{P}$ is the disturbance coefficient matrix. The output equation is given as

$$y = Cx \tag{7.88}$$

where $C = [C_1, C_2, C_3, C_4, C_5]^T$ is the output coefficient matrix. The control variables include the control forces of the active suspension $f_i \ (i = 1,2,3,4)$, and the assist torque provided by the motor of the EPS. Therefore, the control input vector is defined as:

$$u = [f_1, f_2, f_3, f_4, T_m]^T \tag{7.89}$$

The multiple vehicle performance indices are considered to evaluate both the vehicle handling stability and ride comfort. These performance indices can be measured by the following physical terms: the vertical acceleration of the sprung mass $\ddot{Z}_c$; the roll angle Φ; the suspension dynamic deflection f_d; the yaw rate r; and the sideslip angle β. Therefore, the output vector is defined as:

$$y = [\ddot{Z}_c, \Phi, f_d, r, \beta]^T \tag{7.90}$$

For the above system with the same number of input variables and output variables, the static state feedback decouple method is used. For brevity, the detailed derivations of the matrices $A_{19 \times 19}$, $B_{19 \times 5}$, $P_{19 \times 5}$, $C_{5 \times 19}$, $D_{5 \times 5}$, x and the affine nonlinear term $\Delta f(x, t)$ are not presented here.

7.7.1.2   Centralized System

Let $\phi_j^i(x) = L_f^{j-1} h_i(x)$. It can be proven that the other $19 - \sum r_i = 6$ transformations can be found to construct the following mapping:

$$\Phi = col\left(\varphi_1^1, \ldots, \varphi_{r_5}^3, \varphi_{14}, \ldots, \varphi_{19}\right)$$

where $r_i \ (i = 1,2,3,4,5)$ is the relative degree of the system; $L_f \lambda(x)$ is a scalar function; $f(x)$ and $\lambda(x)$ are the functions of x. Please refer to Section 7.3 or reference[36] for the definitions of these functions. Thus, the transformation of the local coordinates at $x = 0$ is constructed as:

$$z(x) = [z_1(x) \ z_2(x) \ \cdots \ z_{11}(x)]^T$$

The defined system output is the last row of the first five sub-matrix. It is given as:

$$\dot{z}_i(x) = \begin{bmatrix} 0 & 1 & \cdots & 0 & 0 \\ 0 & 0 & 1 & \cdots & 0 \\ & & & 1 & \\ & & & & 1 \\ 0 & 0 & \cdots & & 0 \end{bmatrix}_{r_i} z_i(x) + \begin{bmatrix} 0 \\ 0 \\ 0 \\ 0 \\ b_i(x) \end{bmatrix} + \begin{bmatrix} 0 \\ 0 \\ 0 \\ 0 \\ E_i(x)u \end{bmatrix} + \begin{bmatrix} 0 \\ 0 \\ 0 \\ 0 \\ P_i w \end{bmatrix}$$

$$z_i(x) = \left[z_{i1}(x) \ \cdots \ z_{ir_i}(x) \right]^T \tag{7.91}$$

Combining this with the feedback control law u, the subsystem is given as:

$$\dot{z}_{ij} = z_{i\,(j+1)}, i \le 5, j \le r_i - 1$$
$$\dot{z}_{ir_i} = b_i(x) + E_i(x)u + p(x)w \tag{7.92}$$

Therefore, the matrix constructed by $\dot{Z}_{ir_i}$ is expressed as:

$$\left[\dot{z}_{1r_1} \ \dot{z}_{2r_2} \ \dot{z}_{3r_3} \ \dot{z}_{4r_4} \ \dot{z}_{5r_5} \right]^T = b(x) + E(x)u + p(x)w$$
$$= b(x) + E(x)\left[E^{-1}(x)b(x) + E^{-1}(x)v \right] + p(x)w = v + p(x)w$$

i.e.

$$\dot{z}_{ir_i} = v_i + \sum_{j=1}^{4} p_{ij} w_j \tag{7.93}$$

and the system output $y_i = z_{i1}$. Therefore, the centralized of the control channels is fulfilled since every control signal v_i only controls the system output $y_i\,(i \le 5)$ through a series of integrators with an order of r_i. However, it should be noted that the system output y_i is not only affected by v_i, but also by the road excitations. In fact, the control signal v_i does not have any physical meaning, in contrast to the feedback control law u as defined in equation (7.89).

Based on the centralized control theory, the system should be separated into the subsystems with independent control channels to derive the feedback control law. Therefore, the relative degree of the affine nonlinear system is needed. Then, the system can be transformed into a serial structure that consists of a number of subsystems with the degree of $r_i - 1$ and the integrators. Refer to Section 7.3 or reference[35] for the detailed derivation of the relative degree, the design of the input–output centralized controller, and the design of the disturbance centralized controller.

7.7.1.3   Centralized PD Controller Design

The control channels become independent and the disturbance from the road excitations are subdued after the system is centralized. However, the response quality of system is not improved significantly since the control signal v_i is not tuned. The response quality of the system includes the settling time, overshoot, and response error. To further improve the response quality of the system, a PD control law is introduced, and the control signal v_i becomes:

$$v = K_p diag(e) + K_d diag(de) \tag{7.94}$$

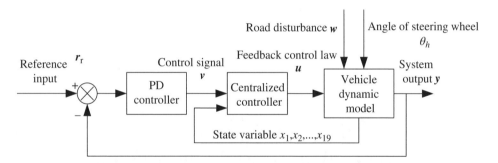

Figure 7.52   Block diagram of the centralized PD control system.

where e and de are the error of the output signal and the difference of the error, respectively; K_p and K_d are the vector of the proportional coefficient and differential coefficient, respectively.

Finally, the block diagram of the centralized PD control system is shown in Figure 7.52. In the figure, the reference input for the system is defined as:

$$r_r = \{0, 0.071, 0.0302, 0.1135, -0.0222\}$$

The above steady-state parameters are calculated by using equation (7.90) when the step steering is selected as the steering wheel input.

7.7.2   Simulation Investigation

A simulation investigation is performed to demonstrate the performance of the proposed centralized PD control system. We assume that the vehicle travels at a constant speed $u_c = 20\,\mathrm{m/s}$, and the step steering input $\theta_h = 90°$. A filtered white noise signal[37] is selected as the road excitation to the vehicle. The vector of proportional coefficient and differential coefficient is selected as $K_p = [1.8, 1.3, 1.8, 1.7, 1.5]$ and $K_d = [1.3, 2.5, 0.4, 4, 3]$. For comparison, two simulation studies are performed. The first is to compare to the system using the PD control and centralized control, and the other is to compare to the non-integrated systems, i.e., the ASS-only system and EPS-only system. The following discussions are made from the simulation results shown in Figures 7.53–7.59 and Table 7.11:

1. It is clearly observed from Figure 7.53 and Figure 7.54 that both the vertical acceleration of the sprung mass and the dynamic deflection of the front-right suspension for the proposed centralized PD control system are greatly reduced compared to that for both the PD control system and the centralized control system. It can be obtained through a quantitative analysis on the simulation results that the percentage decrease of the vertical acceleration of the spring mass is 65.1 and 27.6, and of the dynamic deflection is 44.2 and 30.8. It should be noted that the dynamic deflection of the front-right suspension is taken as an example since similar patterns can be observed for the other three

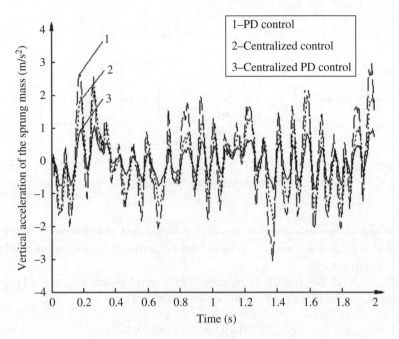

Figure 7.53   Vertical acceleration of the sprung mass (Front-right suspension).

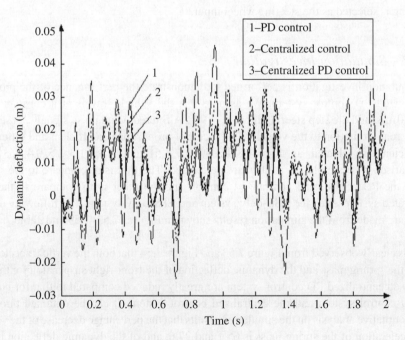

Figure 7.54   Dynamic deflection.

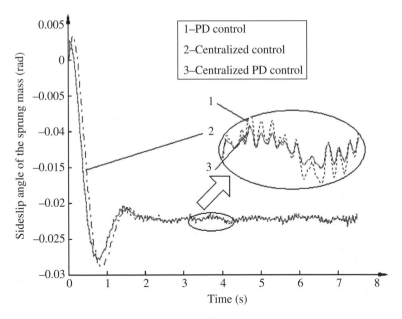

Figure 7.55   Sideslip angle of the sprung mass.

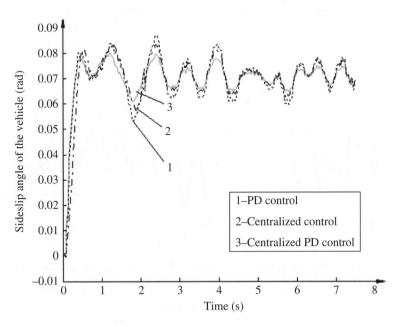

Figure 7.56   Sideslip angle of the vehicle.

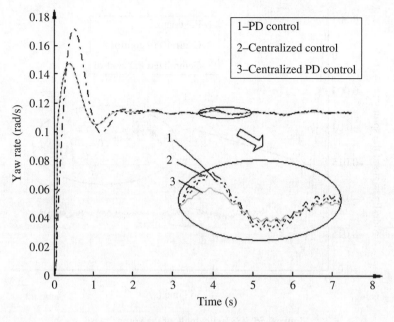

Figure 7.57   Yaw rate.

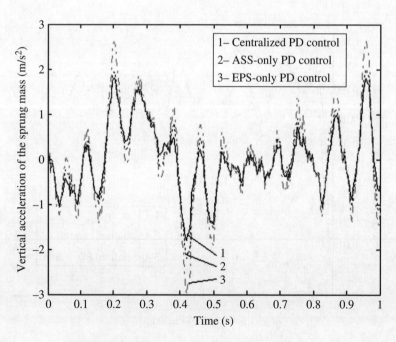

Figure 7.58   Vertical acceleration of the sprung mass.

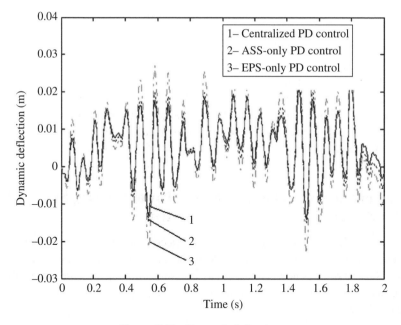

Figure 7.59 Dynamic deflection.

Table 7.11 Simulation results.

Control method		Peak value of response	Overshoot %	Response time (s)
Sideslip angle of vehicle	PD control	0.1446	27.36	0.783
	Centralized control	0.1484	30.76	0.834
	Centralized PD control	0.1421	25.18	0.745
Sideslip angle of sprung mass	PD control	0.0800	12.68	0.436
	Centralized control	0.0082	15.49	0.482
	Centralized PD control	0.0078	9.86	0.412
Yaw rate	PD control	−0.0286	28.63	0.423
	Centralized control	−0.0338	52.42	0.482
	Centralized PD control	−0.0282	26.87	0.423

suspensions. The results indicate that vehicle ride comfort is improved significantly by the proposed centralized PD control system in comparison with the other two control systems since the proposed control system integrates the advantages of the PD control and centralized control to eliminate effectively the disturbance from the road excitation.

2. The performance indices on lateral stability are also shown in Figures 7.55–7.57 and Table 7.11. Compared to both the PD control system and the centralized control system, the sideslip angle of the sprung mass, the sideslip angle of the vehicle, and the yaw rate

are reduced by the proposed centralized PD control system. The results indicate that lateral stability is improved by the proposed centralized PD control system.

3. It is observed clearly from Figure 7.58 and Figure 7.59 that both the vertical acceleration of the sprung mass and the dynamic deflection of the suspension for the proposed centralized PD control system are reduced significantly compared to that for both the ASS-only PD control system and the EPS-only PD control system. The results indicate that vehicle ride comfort is improved by the proposed centralized PD control system since the proposed control system is able to coordinate the EPS and ASS to achieve integration between the two systems.

7.8   Integrated Control of the Electric Power Steering System (EPS) and Vehicle Stability Control (VSC) System

This section studies the integrated control of the electric power steering system (EPS) and the vehicle stability control (VSC) system to achieve the goal of integrating the two control systems. The aim of the study is to design a new control strategy to compensate the return torque of the EPS system by considering the interactions of the VSC system under critical driving conditions.

7.8.1   Interactions Between EPS and VSC

The interactions between the EPS and VSC arise from the lateral forces of the tyres provided by the road surface. When we design the EPS separately, i.e., without taking into account the interactions of the VSC, the assist torque is determined mainly through calculating the return torque of the two front tyres. However, when the vehicle is under critical driving conditions, the lateral forces between the tyres and the road reach saturation, and the VSC intervenes to change the longitudinal forces of the tyres and thus change the lateral forces of the tyres. This results in the change of the return torque from the tyres. Therefore, it is necessary to design a new control strategy to compensate for the return torque by considering the intervention of the VSC.

7.8.2   Control System Design

The block diagram of the integrated control system is shown in Figure 7.60. The work principle of the proposed integrated control system is described as follows: in the VSC control unit, the reference inputs for the vehicle dynamic system, including the expected longitudinal speed of the vehicle u_d, expected lateral speed v_d, and expected yaw rate r_d, are calculated from the vehicle 2-DOF reference model. Then, a nonlinear sliding mode controller is used to calculate the expected control force F_{ud} in order to track the desired vehicle motions. The expected control force $F_{ud} = \begin{pmatrix} F_{xd} & F_{yd} & M_{zd} \end{pmatrix}^T$ is defined as a vector of the expected forces for vehicle stability control, where F_{xd}, F_{yd}, M_{zd} is the expected longitudinal force, expected lateral force, and expected yaw moment,

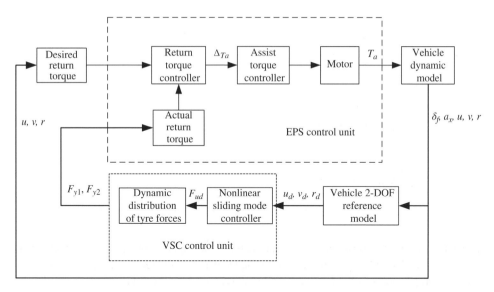

Figure 7.60   Block diagram of the integrated control system.

respectively. Thereafter, the expected control force F_{ud} is distributed optimally to the four wheels. The actual return torque is then generated based on the lateral forces of the two front wheels. In the EPS control unit, the compensated assist torque is obtained by comparing the desired return torque with the actual return torque in the return torque controller. The assist torque T_a is finally calculated by the assist torque controller. Therefore, the lateral stability of the vehicle is improved by compensating the return torque and considering the intervention of the VSC under critical driving conditions.

7.8.3   Dynamic Distribution of Tyre Forces

As mentioned above, the expected control force F_{ud} must be distributed optimally to the four wheels. The distribution of the tyre forces is a multivariable constrained optimization problem, subjected to the distribution accuracy among the four wheels and the control energy. A dynamic distribution method is applied to the distribution of the tyre forces according to the error between the actual value and the expected value of the yaw moment[38].

The compositions of the longitudinal and lateral forces and the yaw moment should have certain limits since the tyre–road friction forces cannot exceed the adhesion limits. Moreover, the feasible regions for the composition of forces and yaw moment are not rectangles since the longitudinal and lateral forces are coupled. When the composition of the longitudinal forces is determined, the yaw moment varies with the different distributions of the longitudinal forces of the four wheels. If all the possible distributions of the longitudinal forces are defined as a set, the yaw moment can be defined as a function of the set. The values of such a function can be constructed into a set with certain limits. When the composition of the longitudinal forces is determined, the upper limit of the functional set can be obtained. Therefore, a curve can be determined to describe the relation that the upper limits of the functional set vary with the different compositions of the longitudinal forces. The curve is defined as the feasible region of the composition of the longitudinal forces and the yaw moment. When the vehicle drives at high speeds, the steering angles of the front wheels are small and hence can be ignored. Thus, the compositions of forces and yaw moment are obtained as follows:

$$\begin{cases} F_x = F_{x1} + F_{x2} + F_{x3} + F_{x4} \\ F_y = F_{y1} + F_{y2} + F_{y3} + F_{y4} \\ M_z = d_F\left(F_{x2} - F_{x1}\right) + d_R\left(F_{x4} - F_{x3}\right) + a\left(F_{y1} + F_{y2}\right) - b\left(F_{y3} + F_{y4}\right) \end{cases} \tag{7.95}$$

where the subscripts 1 through 4 represent the left-front, right-front, left-rear, and left-front wheels, respectively; d_F, d_R represent half of the front and rear wheel track, respectively; a, b represent the distance of the front axle and rear axle between the C.G., respectively; F_x, F_y, and M_z represent the compositions of the longitudinal and lateral forces, and the yaw moment, respectively.

The distributions of the compositions of forces and the yaw moment are given as follows. First, the composition of the longitudinal forces is distributed. The error of the yaw moment is defined as:

$$e_M = M_{zd} - M_z = M_{zd} - I_z \dot{r} \tag{7.96}$$

The additional yaw moment resulting from the longitudinal forces is obtained from equation (7.95),

$$M_x = d_F\left(F_{x2} - F_{x1}\right) + d_R\left(F_{x4} - F_{x3}\right) \tag{7.97}$$

Let the additional yaw moment be:

$$M_x = k_1 e_M \tag{7.98}$$

Therefore, the error of the yaw moment e_M can be compensated by the additional yaw moment by selecting a suitable coefficient k_1. A constrained quadratic programming problem is defined as follows:

$$\begin{cases} \min_{F_{xi}} \quad J = \dfrac{F_{x1}^2}{a_{x1}^2} + \dfrac{F_{x2}^2}{a_{x2}^2} + \dfrac{F_{x3}^2}{a_{x3}^2} + \dfrac{F_{x4}^2}{a_{x4}^2} \\ s.t. \quad M_x = k_1 e_M \\ \qquad F_x = k_2 F_{x1} + k_3 F_{x2} + k_4 F_{x3} + k_5 F_{x4} \\ \qquad k_2 + k_3 + k_4 + k_5 = 1 \end{cases} \tag{7.99}$$

where $a_{xi}(i=1,2,3,4)$ are the maximum tyre road friction forces; $a_{xi}=\mu F_{zi}$, where F_{zi} is the vertical load of the tyre; and k_2,k_3,k_4,k_5 are the function distribution coefficients for the longitudinal forces. However, the above equation cannot compensate accurately for the error of the yaw moment since the longitudinal and lateral forces of the front wheels are coupled. Hence, it is necessary to adjust the lateral forces of the front wheels to further improve the track error of the yaw moment.

It can be observed that the additional yaw moment resulting from the lateral forces of the two front wheels is actually related to the composition of the two forces. Therefore, let the composition of the two forces be $F_{yF}=F_{y1}+F_{y2}$, and F_{yF} at the k-th sampling time is given as:

$$F_{yF}(k)=F_{yF}(k-1)+k_6 e_M \tag{7.100}$$

where k_6 is a constant. Finally, the lateral force of each front tyre is determined according to its proportion of the vertical load:

$$F_{yi}=\frac{F_{zi}}{F_{z1}+F_{z2}}F_{yF} \quad i=1,2 \tag{7.101}$$

7.8.4 Design of a Self-aligning Torque Controller

In order to design a self-aligning torque controller, an analysis of the relationship between the vehicle self-aligning torque and sideslip angle is performed. Figure 7.61 shows the relationship for the 7-DOF vehicle dynamic model by assuming that the vehicle speed is a constant of 72 km/h and the road adhesion coefficient is 0.3. It is observed in Figure 7.61 that the self-aligning torque is approximately proportional to the sideslip angle when the sideslip angle is small, and it reaches the maximum when the sideslip angle is around 4°–6°. As the sideslip angle keeps increasing, the self-aligning torque is decreased instead.

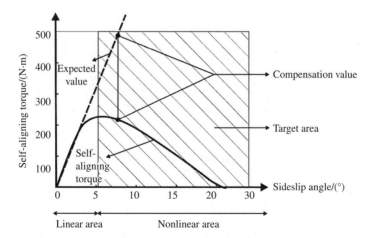

Figure 7.61 Relationship between self-aligning torque and sideslip angle.

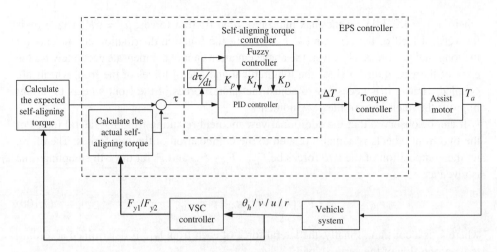

Figure 7.62 Control system chart of the self-aligning torque.

Under this circumstance, the hand torque required by the vehicle driver is increased, and also the road feel is lost. Therefore, the compensation of the self-aligning torque is required through adjusting the EPS based on the vehicle states obtained from VSC system.

The compensation value of the self-aligning torque for the EPS varies with the lateral force. The reason is that the self-aligning torque varies with respect to the sideslip angle. Moreover, the lateral forces also change when the VSC intervenes. Therefore, the PID fuzzy control is used to regulate the compensation value of the self-aligning torque in real time to track the expected value in order to design an appropriate self-aligning torque controller. As shown in Figure 7.62, the fuzzy controller has two input variables: the tracking error of the self-aligning torque τ, and the difference of the error $\dot{\tau}$. The three tuning parameters K_P, K_I, and K_D are selected as the outputs of the fuzzy controller, and they are the inputs of the PID controller. The application of the PID controller is to guarantee that the actual self-aligning torque tracks the expected one in real time.

The actual self-aligning torque is calculated as:

$$M = M_1 + M_2 = F_{y1}e_1 + F_{y2}e_2 \tag{7.102}$$

where M_1, M_2 are the self-aligning torques of the vehicle front-left and front-right wheels, respectively; and e_1, e_2 are the pneumatic trails of the two front tyres, respectively. The expected self-aligning torque is derived as:

$$M_c = \left(\frac{\partial M_z}{\partial \alpha}\right) \times \alpha_i \tag{7.103}$$

where M_c is the expected self-aligning torque; $\dfrac{\partial M_z}{\partial \alpha}$ is the slope rate of the self-aligning torque with respect to the sideslip angle when the linear relationship between them holds; and α_i is the actual sideslip angle of the two front wheels. The rule bases of the fuzzy controller are developed for the three output variables K_P, K_I, and K_D with respect to the input variable τ and its difference $\dot{\tau}$ as follows:

1. For a relatively large τ, a relatively larger K_P is selected to eliminate deviations and increase the response speed as soon as possible. In the meantime, K_I is set as zero to avoid large overshoots.
2. For a relatively small τ, K_P must be reduced and K_I should be a relatively small value to continue to reduce the deviations and also to prevent large overshoots and oscillations.
3. For an extremely small τ, K_P should keep decreasing and K_I should remain constant or increase a little to eliminate steady errors, and overcome the overshoots and stabilize the system as soon as possible.
4. When the signs of τ and $\dot{\tau}$ are the same, it means that the controlled variables vary away from the target values. Therefore, when the controlled variables approach the target values, the proportional parameter with the opposite sign counteracts the effects of the integral parameter to avoid the overshoots and the subsequent oscillations resulted from the integral parameter. While the control variables are far from the target values and vary towards them, the control process slows down due to the opposite signs of τ and $\dot{\tau}$. Therefore, when τ is relatively large and the signs of $\dot{\tau}$ and τ are opposite, K_I is selected to be zero or negative to accelerate the control process.
5. The value of $\dot{\tau}$ shows the change rate of the tracking error. As $\dot{\tau}$ becomes larger, K_P should become smaller and K_I larger, and vice versa. At the same time, the value of τ should also be taken into consideration.
6. The differential parameter K_D is used to improve the system dynamic characteristics and prevent the variation of τ. The differential parameter K_D is beneficial to reduce overshoots, eliminate oscillations, and shorten the settling time. Therefore, K_D should be increased to reduce the system steady state error and hence improve the control accuracy. When τ is relatively large, K_D is selected to be zero, and the controller becomes a PI controller; when τ is relatively small, K_D is selected to be medium, resulting in a PID control.

To design the fuzzy controller, τ and $\dot{\tau}$ are selected as the input language variables, the fuzzy subset of each variable is set as {Negative Big, Negative Medium, Negative Small, Zero, Positive Small, Positive Medium, Positive Large}, which is denoted as{NB,NM,NS,ZO,PS,PM,PB}. The discourse domains of both τ and $\dot{\tau}$ are defined as $\{-3,-2,-1,0,1,2,3\}$. The triangular full overlap function is adopted as the membership function, and the Sum-Product rule is used for fuzzy inference. The membership function of each rule is denoted as $w_j(\tau,\dot{\tau})(j=1,2,...,7)$. Therefore, the three output variables are obtained by the weighted mean method:

$$K_P = \frac{\sum_{j=1}^{7} w_j\left[(\tau,\dot{\tau}) \times K_{Pj}\right]}{\sum_{j=1}^{7} w_j(\tau,\dot{\tau})} \qquad K_I = \frac{\sum_{j=1}^{7} w_j\left[(\tau,\dot{\tau}) \times K_{Ij}\right]}{\sum_{j=1}^{7} w_j(\tau,\dot{\tau})} \qquad K_D = \frac{\sum_{j=1}^{7} w_j\left[(\tau,\dot{\tau}) \times K_{Dj}\right]}{\sum_{j=1}^{7} w_j(\tau,\dot{\tau})}$$

$$(7.104)$$

where K_{Pj}, K_{Ij}, and K_{Dj} are the weighed parameters of K_P, K_I, and K_D under different conditions. Therefore, the output of the fuzzy PID controller, i.e., the compensation of the self-aligning torque, is obtained as:

$$\Delta T_a = K_P \left(M - M_c \right) + K_I \int (M - M_c) dt + K_D \left(\dot{M} - \dot{M}_c \right) \tag{7.105}$$

7.8.5 Simulation Investigation

To validate the effectiveness of the compensation strategy of the self-aligning torque, a simulation is performed by comparing the integrated control of the VSC and EPS with the non-VSC control, and VSC control. The vehicle initial speed is set as 72 km/h, and a low road adhesion coefficient of 0.3 is selected. Two driving conditions are performed, including the step steering input and single lane change.

It is observed in Figures 7.63 and 7.64 that the integrated control system of VSC and EPS performed the best amongst the three systems considered on both the yaw rate and the

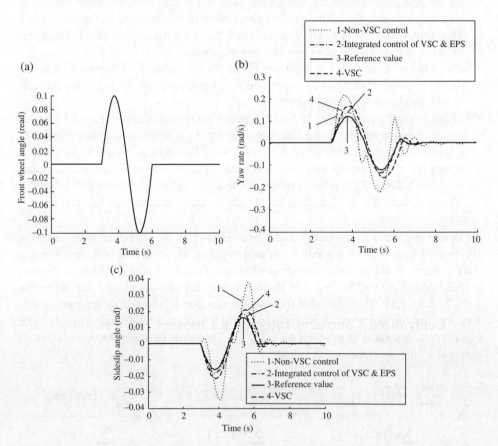

Figure 7.63 Comparison of responses for the maneuver of a single lane change. (a) Single lane change steering input to the front wheel. (b) Yaw rate. (c) Sideslip angle.

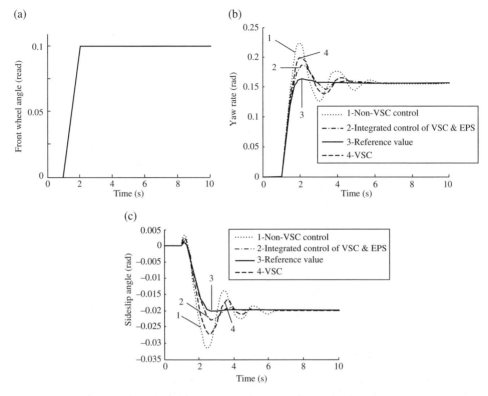

Figure 7.64 Comparison of responses for the maneuver of a step steering input. (a) Step steering input to the front wheel. (b) Yaw rate. (c) Sideslip angle.

sideslip angle. In addition, the application of the compensation strategy of the self-aligning torque is able to track the reference value faster and, hence, the settling time is shorter compared with the VSC control. Therefore, vehicle handling stability is improved by the integrated control of VSC and EPS.

7.9   Centralized Control of Integrated Chassis Control Systems using the Artificial Neural Networks (ANN) Inverse System Method

In recent years, intelligent control methods have been applied widely to the centralized control of multivariable systems to deal with nondeterministic and complex control problems. This study takes advantage of both the neural network and the centralized linearization of the inverse system. As a result, the neural network inverse system method is applied to the integrated control of vehicle Active Front Steering (AFS), Direct Yaw moment Control (DYC), and Active Suspension System (ASS)[26,39].

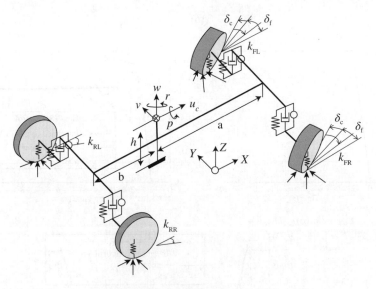

Figure 7.65   Vehicle dynamic model.

7.9.1   Vehicle Dynamic Model

As described above in Section 7.3, the nonlinear centralized method was applied to the integrated control of VSC and ASS. The overall vehicle performance, including handling stability and comfort, was improved by regulating the vertical load distribution and adjusting the brake forces. In this section, the AFS system is designed to serve as a steering correction system by applying an additional steer angle to the driver's steering input in the linear handling region. In this way, the AFS system is able to improve steerability by assisting the driver in handling the vehicle and preventing extreme handling situations. In this study, the integrated control VSC and ASS system is integrated further with the AFS system, and the neural network inverse system method is used to design the integrated centralized control of the AFS, DYC and ASS. As shown in Figure 7.65, the 7-DOF vehicle dynamic model is used, and the lateral, yaw, and roll motions are considered. In Figure 7.65, u_c, β, r, and ϕ are the vehicle longitudinal speed, sideslip angle, yaw rate, and roll angle, respectively; m, m_s, m_f, and m_r are the vehicle mass, sprung mass, unsprung mass of the front axle, unsprung mass of the rear axle, respectively; a and b are the distances between the center of gravity of the vehicle to the front and rear axles, respectively; I_z, I_x, I_{xz} are the yaw moment of inertia, roll moment of inertia, and product of inertia of the sprung mass about the roll and yaw axes, respectively ; $k_F \left(= K_{FL} + K_{FR} \right)$ and $k_R \left(= K_{RL} + K_{RR} \right)$ are the lateral stiffnesses of the front and rear tyres, respectively; δ_F, δ_c, T_z, and T_ϕ are the steering angle of the front wheel provided by the driver, additional steering angle to the front wheel provided by the AFS, corrective yaw moment, and suspension roll moment; K_ϕ and D_ϕ are the roll stiffness coefficient and damping coefficient of the suspension.

The state variable is defined as $x = (\beta \quad r \quad \phi \quad \dot{\phi})^{\mathrm{T}}$, and the control input variable $u = (\delta_c \quad T_z \quad T_\phi)^{\mathrm{T}}$, the system output variable $y = (\beta \quad r \quad \phi)^{\mathrm{T}}$, then the state equation of the integrated system is derived as:

$$\begin{cases} M\dot{x} = Kx + Nu + Q\delta_F \\ y = Cx \end{cases} \tag{7.106}$$

where,

$$M = \begin{pmatrix} mu_c & am_f - bm_r & 0 & m_s h \\ (am_f - bm_r)u_c & I_z & 0 & I_{xz} \\ 0 & 0 & 1 & 0 \\ m_s hu_c & I_{xz} & 0 & I_x \end{pmatrix}$$

$$K = \begin{pmatrix} -2(k_F + k_R) & -mu_c + \dfrac{2(bk_R - ak_F)}{u_c} & 0 & 0 \\ 2(bk_R - ak_F) & (bm_r - am_f)u_c - 2\dfrac{a^2 k_F + b^2 k_R}{u_c} & 0 & 0 \\ 0 & 0 & 0 & 1 \\ 0 & m_s hu_c & m_s gh - K_\phi & -D_\phi \end{pmatrix}$$

$$N = \begin{pmatrix} 2k_F & 0 & 0 \\ 2k_F & 1 & 0 \\ 0 & 0 & 0 \\ 0 & 0 & 1 \end{pmatrix}, \quad Q = \begin{pmatrix} 2k_F \\ 2ak_F \\ 0 \\ 0 \end{pmatrix}, \quad C = \begin{pmatrix} 1 & 0 & 0 & 0 \\ 0 & 1 & 0 & 0 \\ 0 & 0 & 1 & 0 \end{pmatrix}$$

It is obvious that equation (7.106) is a typical multivariable system with three inputs and three outputs. Due to the interactions among the tyre's longitudinal, lateral, and vertical forces, as well as the conflicts among the vehicle translational and rotational motions, the integrated control system of AFS, DYC, and ASS systems is highly coupled. Therefore, it is required to decouple the vehicle multivariable system into three independent SISO control systems in order to design the controller of the integrated system.

7.9.2 Design of the Centralized Control System

7.9.2.1 Analysis of System Invertibility

As described in Section 7.3.6, the Interactor algorithm is used to calculate the relative degree of the integrated control system given in equation (7.106). The detailed calculation of the relative degree is demonstrated as follows.

$$\dot{y}_1 = \dot{\beta} = -2(k_F + k_R)\beta / mu_c + 2r(bk_R - ak_F) / mu_c^2$$
$$- m_s h\ddot{\phi} / mu_c - (am_f - bm_r)\dot{r} / mu_c - r + 2k_F(\delta_F + \delta_c) / mu_c \tag{7.107}$$

Let $Y_1 = \dot{y}_1$, the rank of the corresponding Jacobian matrix with respect to input $\boldsymbol{u}$ is:

$$t_1 = rank\left[\frac{\partial Y_1}{\partial u^{\mathrm{T}}}\right] = rank\left[\begin{matrix} \dfrac{\partial \dot{y}_1}{\partial \delta_c} & \dfrac{\partial \dot{y}_1}{\partial T_z} & \dfrac{\partial \dot{y}_1}{\partial T_\phi} \end{matrix}\right] = 1 \tag{7.108}$$

$$\dot{y}_2 = \dot{r} = 2\left(bk_R - ak_F\right)\beta/I_z - 2\left(a^2k_F + b^2k_R\right)r/I_z u_c$$
$$-I_{xz}\ddot{\phi}/I_z - \left(am_f - bm_r\right)\left(\dot{\beta} + r\right)u_c/I_z + T_z/I_z + 2ak_F\left(\delta_F + \delta_c\right)/I_z \tag{7.109}$$

Let $Y_2 = \begin{bmatrix} \dot{y}_1 & \dot{y}_2 \end{bmatrix}^{\mathrm{T}}$, the rank of the corresponding Jacobian matrix with respect to $\boldsymbol{u}$ is:

$$t_2 = rank\left[\frac{\partial Y_2}{\partial u^{\mathrm{T}}}\right] = rank\left[\begin{matrix} \dfrac{\partial \dot{y}_1}{\partial \delta_c} & \dfrac{\partial \dot{y}_1}{\partial T_z} & \dfrac{\partial \dot{y}_1}{\partial T_\phi} \\[2mm] \dfrac{\partial \dot{y}_2}{\partial \delta_c} & \dfrac{\partial \dot{y}_2}{\partial T_z} & \dfrac{\partial \dot{y}_2}{\partial T_\phi} \end{matrix}\right] = 2 \tag{7.110}$$

$$\dot{y}_3 = \dot{\phi}$$
$$\ddot{y}_3 = \ddot{\phi} = -D_\phi\dot{\phi}/I_x - \left(K_\phi - m_s gh\right)\phi/I_x - I_{xz}\dot{r}/I_x - m_s h\left(\dot{\beta} + r\right)u_c/I_x + T_\phi/I_x \tag{7.111}$$

Let $Y_3 = \begin{bmatrix} \dot{y}_1 & \dot{y}_2 & \ddot{y}_3 \end{bmatrix}^{\mathrm{T}}$, the rank of the corresponding Jacobian matrix with respect to $\boldsymbol{u}$ is:

$$t_3 = rank\left[\frac{\partial Y_3}{\partial u^{\mathrm{T}}}\right] = rank\left[\begin{matrix} \dfrac{\partial \dot{y}_1}{\partial \delta_c} & \dfrac{\partial \dot{y}_1}{\partial T_z} & \dfrac{\partial \dot{y}_1}{\partial T_\phi} \\[2mm] \dfrac{\partial \dot{y}_2}{\partial \delta_c} & \dfrac{\partial \dot{y}_2}{\partial T_z} & \dfrac{\partial \dot{y}_2}{\partial T_\phi} \\[2mm] \dfrac{\partial \ddot{y}_3}{\partial \delta_c} & \dfrac{\partial \ddot{y}_3}{\partial T_z} & \dfrac{\partial \ddot{y}_3}{\partial T_\phi} \end{matrix}\right] = 3 \tag{7.112}$$

Therefore the relative degree of the system is $\boldsymbol{r} = \begin{bmatrix} r_1 & r_2 & r_3 \end{bmatrix}^{\mathrm{T}} = \begin{bmatrix} 1 & 1 & 2 \end{bmatrix}^{\mathrm{T}}$, and $\sum_{i=1}^{3} r_i = n = 4$ (n is the system order). According to the implicit function theorem, it is known that the inverse system of the original integrated control system exists, and the output of the inverse system $\boldsymbol{u}$ (i.e., the input of the original system) is given as:

$$\boldsymbol{u} = \varphi\left(\boldsymbol{x}, \dot{y}_1, \dot{y}_2, \ddot{y}_3\right) = \varphi\left(\boldsymbol{x}, \boldsymbol{v}\right) \tag{7.113}$$

where $\boldsymbol{v} = \begin{bmatrix} v_1 & v_2 & v_3 \end{bmatrix}^{\mathrm{T}} = \begin{bmatrix} \dot{y}_1 & \dot{y}_2 & \ddot{y}_3 \end{bmatrix}^{\mathrm{T}}$, $\varphi(\cdot)$ is the nonlinear relationship between the input and output of the inverse system. Let $z_1 = y_1$, $z_2 = y_2$, $z_{31} = y_3$, $z_{32} = \dot{z}_{31} = \dot{y}_3$, $v_1 = \dot{y}_1$,

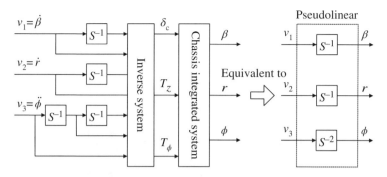

Figure 7.66   Structure of the pseudo-linear system.

$v_2 = \dot{y}_2$, $v_3 = \ddot{y}_3$ be the inputs to the inverse system, the standard form of the inverse system is given as:

$$
\begin{cases}
\dot{z}_1 = v_1 \\
\dot{z}_2 = v_2 \\
\dot{z}_{31} = z_{32} \\
\dot{z}_{32} = v_3 \\
u = \overline{\varphi}\left(z_1 \quad z_2 \quad z_{31} \quad z_{32} \quad v_1 \quad v_2 \quad v_3 \right)
\end{cases} \tag{7.114}
$$

As shown in equation (7.114), a pseudolinear system is constructed by connecting the inverse system ahead of the original one in series, which is equivalent to two first-order integral linear subsystems and one second-order integral linear subsystem. The structure of the pseudo-linear system is illustrated in Figure 7.66.

7.9.2.2   Design of a Neural Network Inverse System

The analytical expression of the inverse system given in equation (7.114) is based on the accurate mathematical model. Therefore, the centralized characteristics of the pseudolinear system are realized only if the original system parameters are known, accurate, and time-invariant. To improve the self-adaptability to the parameter variations and the robustness with respect to external disturbances, the static neural network and the integrator is applied to construct the inverse system[25].

It is known from equation (7.114) that the input layer includes β, r, ϕ, $\dot{\phi}$, $\dot{\beta}$, $\dot{r}$, and $\ddot{\phi}$; and the output layer includes δ_c, T_z, and T_ϕ. Therefore, the number of neurons on the input layer is set to be seven, and three for the output layer. The number of neurons on the implicit layer is determined to be 15 by the trial-and-error method. The nonlinear mapping of the inverse system is approached by using the static neural network. Four integrators are used to represent the dynamic characteristics of the inverse system. The structure of the developed neural network inverse system is 7-15-3, which is shown in Figure 7.67.

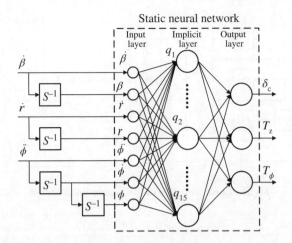

Figure 7.67   Structure of the BP neural network inverse system.

In the structure, the function *tansig* is used as the transfer function for the input layer and the implicit layer of the BP network, and the function *purelin* for the output layer.

It is assumed that the vehicle travels at a speed of 100 km/h, and the driver's steering input to the front wheels is a sinusoidal curve with a varying amplitude. As shown in Figure 7.68, the stimulating input signals used to train the neural network include the additional angle to the front wheel δ_c, corrective yaw moment T_z, and roll moment of the suspension T_ϕ. The dynamic responses of the yaw rate r, sideslip angle β, and roll angle ϕ are sampled with a period of 5 ms. Then $\dot{r}$, $\dot{\beta}$, $\dot{\phi}$ and $\ddot{\phi}$ are calculated according to the five-point numerical differential method. Two hundred sets of training data sets $[\beta, \dot{\beta}, r, \dot{r}, \phi, \dot{\phi}, \ddot{\phi}]$ and $[\delta_c, T_z, T_\phi]$ are obtained by combining the above-obtained data. The *premnmx* function in the neural network toolbox in MATLAB is used to normalize the network input and output data. In the network simulation test, the new data are preprocessed by the function *tramnmx* in the same way, and finally normalized by the function *postmnmx*. The BP neural network is established by the function *newff*, and the function *trainlm* of the Levenberg-Marquardt algorithm is selected as the training function. The established BP network has been trained by the selected training function for 500 times with a learning efficiency of 0.05, and a network target error of 10^{-3}. The required training accuracy is achieved after 72 sessions training in the simulation.

7.9.2.3   Design of the PD Controller

By performing the centralized process, the original integrated system is transformed into the three independent SISO systems, including the first-order AFS subsystem, the first-order DYC subsystem, and the second-order ASS subsystem. In order to improve the response quality of the integrated system, a compound controller shown in Figure 7.69 is designed by combining the PD controller and the neural network reverse system.

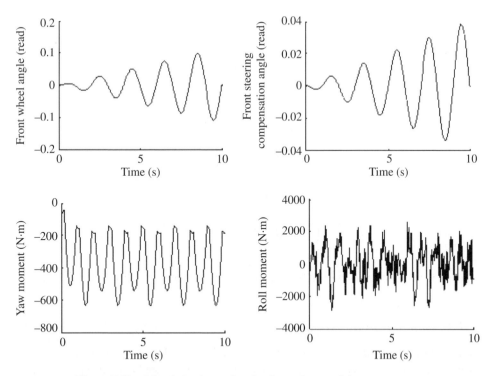

Figure 7.68   Stimulating input signals of neural network inverse system.

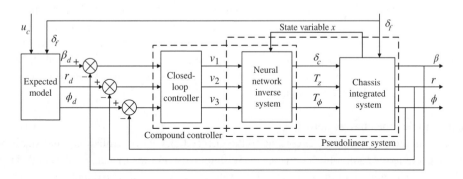

Figure 7.69   Block diagram of the centralized integrated control system.

As the input and output of the centralized pseudolinear system has a one-to-one linear relationship, a number of control methods for single variable linear systems can be applied to the design of the controller, including PID control, pole placement, and quadratic optimal. Here, the PD closed loop controller is designed as follows:

$$v = K_{\mathrm{p}}\mathrm{diag}\{e\} + K_{\mathrm{d}}\mathrm{diag}\{de\} \tag{7.115}$$

where e and de are the tracking error of the system output signals, and its difference, respectively; K_p and K_d are the proportional and differential coefficient, respectively. The vehicle reference model included in Figure 7.69, generates the expected dynamic response according to the driver's input, and the expected sideslip angle and yaw rate are obtained as:

$$\begin{cases} \beta_d = \dfrac{\dfrac{2k_F\left(2a^2k_F+2b^2k_R\right)}{2ak_F-2bk_R+mu_c^{\,2}}-2bk_F}{\dfrac{\left(2k_F+2k_R\right)\left(2a^2k_F+2b^2k_R\right)}{2ak_F-2bk_R+mu_c^{\,2}}+2bk_R-2ak_F}\delta_f \\[4mm] r_d = \min\left\{\left|\dfrac{u_c\delta_F}{(a+b)(1+Ku_c^{\,2})}\right|,\left|\dfrac{\mu g}{u_c}\right|\mathrm{sgn}\left(\delta_f\right)\right\} \end{cases} \qquad (7.116)$$

where μ is the road adhesion coefficient, and K is the understeer coefficient.

7.9.3   Simulation Investigation

To demonstrate the effectiveness of the developed integrated control system, a simulation investigation is performed by comparing the developed integrated control system with the integrated control system with PD control, and the system with the decentralized control (i.e., the three stand-alone controllers of AFS, ASS and DYC). The vehicle initial speed is set as $u_c = 80\,\mathrm{km/h}$, the road adhesion coefficient of $\mu = 0.85$ is selected. Two driving conditions are performed, including single lane change and step steering input. The vehicle physical parameters in Table 7.5 are used. After tuning, the PD coefficients of the closed loop controller are set as $K_p = \mathrm{diag}\{18\,52\,5000\}$ and $K_d = \mathrm{diag}\{4.5\,3\,10\}$. The roll stiffness coefficient of the suspension is $K_\phi = 65590\,\mathrm{N\cdot m/rad}$, and the suspension roll damping coefficient is $D_\phi = 2100\,\mathrm{N\cdot m\cdot s/rad}$.

(1)   Single lane change maneuver

The simulation is performed according to the GB/T6323.1-94 controllability and stability test procedure for automobiles – pylon course slalom test. For the maneuver of a single lane change, the amplitude of the front wheel steering angle is set as 0.08 rad and the frequency as 0.5 Hz. It is illustrated clearly in Figure 7.70 and Table 7.12 that the peak value of the yaw rate for the developed integrated control is reduced greatly by 12% and 29% respectively, compared with the integrated control system with PD control, and the system with the decentralized control. A similar pattern can be observed for the sideslip angle and roll angle. The results indicate that the application of the centralized PD control is able to track effectively the expected vehicle states and hence improve its handling stability.

(2)   Step steering input manoeuvre

The simulation is conducted according to the GB/T6323.2-94 controllability and stability test procedure for automobiles – steering transient response test (steering wheel angle

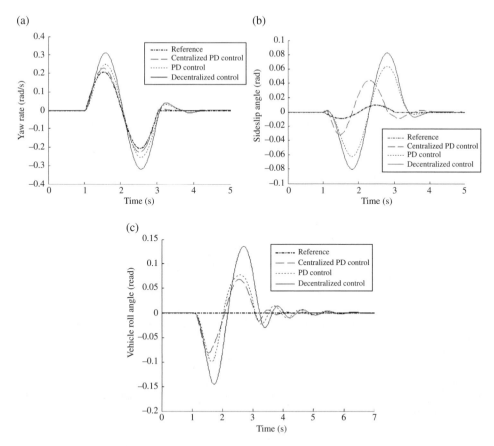

Figure 7.70 Comparison of responses for the single lane change maneuver. (a) Yaw rate. (b) Sideslip angle. (c) Roll angle.

Table 7.12 Comparison of the peak value of the single line change simulation results.

Control Method	Yaw rate r/rad. s^{-1}	Sideslip angle β/rad	Roll angle ϕ/rad
Expectation	0.2	0.01	0
Centralized PD control	0.22	0.041	0.075
PD control	0.25	0.062	0.096
Decentralized control	0.31	0.081	0.147

step input). The step steering input to the wheel is set at 1.57 rad. It is illustrated clearly in Figure 7.71 and Table 7.13 that the peak value of the yaw rate for the developed integrated control is reduced significantly by 19.1% and 26.4% respectively, compared with the integrated control system with PD control, and the system with the decentralized control. A similar pattern can be observed for the sideslip angle and roll angle. The results

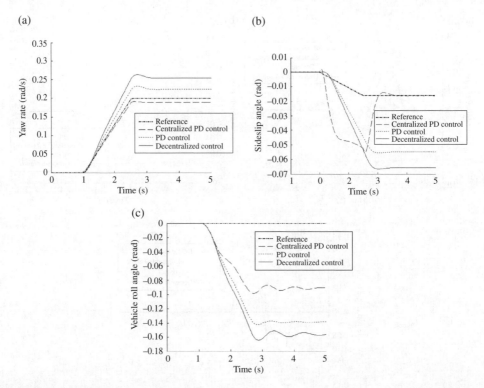

Figure 7.71 Comparison of responses for the maneuver of step steering input. (a) Yaw rate. (b) Sideslip angle. (c) Roll angle.

Table 7.13 Comparison of step steering simulation result peak values.

Control method	Yaw rate r (rad/s)	Sideslip angle β (rad)	Roll angle ϕ (rad)
Reference	0.2	0.017	0
Centralized PD Control	0.195	0.055	0.096
PD Control	0.241	0.056	0.142
Decentralized Control	0.265	0.066	0.164

indicate that the application of the centralized PD control is able to track effectively the expected vehicle states and hence improve the vehicle handling stability. In addition, the amplitudes of these performance indices for the combined control system are relatively larger since the conflicts among the three subsystems are unable to coordinate. Finally, these three performance indices for the integrated control system with PD control are deviated from the expected states since the PD controller fails to deal with the coupling effects of the control loop and, hence, it is inevitable that they negatively influence each other.

References

[1] Fruechte R D, Karmel A M, Rillings J H, Schilke N A., Boustany N M, Repa B S. Integrated vehicle control. Proceedings of the 39th IEEE Vehicular Technology Conference, May 1–3, 1989, San Francisco, CA, Vol. 2, 868–877.

[2] Gordon T J, Howell M, Brandao F. Integrated control methodologies for road vehicles. Vehicle System Dynamics, 2003, 40(1–3), 157–190.

[3] Yu F, Li D F, Crolla D A. Integrated vehicle dynamic control – State-of-the art review. Proceedings of IEEE Vehicle Power and Propulsion Conference (VPPC), September 3–5, 2008, Harbin, China, 1–6.

[4] Falcone P, Borrelli F, Asgari J, Tseng H E, Hrovat D. Predictive active steering control for autonomous vehicle systems. IEEE Transactions on Control Systems Technology, 2007, 15(3), 566–580.

[5] Chen W W, Xiao H S, Liu L Q, Zu J W. Integrated control of automotive electrical power steering and active suspension systems based on random sub-optimal control. International Journal of Vehicle Design, 2006, 42(3/4), 370–391.

[6] Hirano Y, Harada H, Ono E, Takanami K. Development of an Integrated System of 4WS and 4WD by H Infinity Control. SAE Technical Paper 930267, 1993, 79–86.

[7] Li D F, Du S Q, Yu F. Integrated vehicle chassis control based on direct yaw moment, active steering, and active stabiliser. Vehicle System Dynamics, 2008, 46(1), 341–351.

[8] Nwagboso C O, Ouyang X, Morgan C. Development of neural network control of steer-by-wire system for intelligent vehicles. International Journal of Heavy Vehicle Systems, 2002, 9(1), 1–26.

[9] Trächtler A. Integrated vehicle dynamics control using active brake, steering, and suspension systems. International Journal of Vehicle Design, 2004, 36(1), 1–12.

[10] Gordon T J. An integrated strategy for the control of a full vehicle active suspension system. Vehicle System Dynamics, 1996, 25, 229–242.

[11] Rodic A D, Vukobratovie M K. Design of an integrated active control system for road vehicles operating with automated highway systems. International Journal Computer Application Technology, 2000, 13, 78–92.

[12] Karbalaei R, Ghaffari A, Kazemi R, Tabatabaei S H. A new intelligent strategy to integrated control of AFS/DYC based on fuzzy logic. International Journal of Mathematical, Physical and Engineering Sciences, 2007, 1(1), 47–52.

[13] Chang S, Gordon T J. Model-based predictive control of vehicle dynamics. International Journal of Vehicle Autonomous Systems, 2007, 5(1–2), 3–27.

[14] Koehn P, Eckrich M, Smakman H, Schaffert A. Integrated Chassis Management: Introduction into BMW's Approach to ICM. SAE Technical Paper 2006-01-1219, 2006.

[15] Liu X Y. Research on direct yaw moment control for vehicle stablity control. Ph.D. Thesis, Hefei University of Technology, Hefei, China, 2010.

[16] Chen W W, Liu X Y, et al. Dynamic boundary control of sideslip angle for vehicle stablity control through considering the effects of road surface friction. Journal of Mechanical Engineering, 2012, 48(14), 112–118 (in Chinese).

[17] van der Steen R. Tyre/road Friction Modeling. Literature Survey, Department of Mechanical Engineering, Eindhoven University of Technology, 2007.

[18] Rabhi A, Amiens C R E A, et al. Estimation of contact forces and wheel road friction. Proceedings of the 15th Mediterranean Conference on Control and Automation, Athens, June 27–29, 2007, 1–6.

[19] Grip H, Imsland L, Johansen T A, et al. Nonlinear vehicle velocity observer with road-wheel friction adaptation. Proceedings of the 45th IEEE Conference on Decision and Control, San Diego, USA, December, 2007, 1080–1085.

[20] Yu F, Lin Y. Vehicle System Dynamics. China Machine Press, Beijing, 2008 (in Chinese).

[21] Zhao J, Lin H. Simulation of target attribute recognition based on BP neural network model. Journal of System Simulation, 2007, 19(11), 2571–2573 (in Chinese).

[22] Li M Y, Du Y L. Composite neural networks adaptive control of temperature based on genetic algorithm learning. Control Theory & Applications, 2004, 21(2), 242–246 (in Chinese).

[23] Dai W Z, Lou H C, Yang A P. An overview of neural network predictive control for nonlinear systems. Control Theory & Applications, 2009, 26(5), 521–528 (in Chinese).

[24] Zhou H Y, Zhang J, You L K, et al. Nonlinear predictive function control based on hybrid neural network. Control Theory & Applications, 2005, 22(1), 110–113 (in Chinese).

[25] Wang B W, Shen Y J, He T Z. Multi-layer feedforward neural network based on multi-output neural model and its applications. Control Theory & Applications, 2004, 21(4), 611–613 (in Chinese).

[26] Zhu M F. Research on centralized control for integrated chassis control system and time delay control for chassis key subsystems. Ph.D. Thesis, Hefei University of Technology, Hefei, China, 2011.

[27] Dai X Z. The Inverse Neural Network Control Method for Multivariable Nonlinear Systems. Beijing: Science Press, 2005 (in Chinese).

[28] Chen W W, Sun Q Q, et al. Integrated control of steering and suspension systems based on disturbance suppression. Journal of Mechanical Engineering, 2007, 43(11), 98–104 (in Chinese).

[29] Wang Q D, Wu B F, Chen W W. Integrated control of active suspension and electrical power steering based on predictive control. Transactions of the Chinese Society for Agricultural Machinery, 2007, 38(1), 1–5 (in Chinese).

[30] Zhu J. Intelligent Predictive Control and its Applications. Hangzhou: Zhejiang University Press, 2002 (in Chinese).

[31] Sun P Y, Chen H, Kang J, Guo K H, et al. Constrained predictive control of vehicle active suspension. Journal of Jilin University, 2002, 20(2), 47–52 (in Chinese).

[32] Chen W W, Wang Y W, Wang Q D. Integrated control of electrical power steering and active suspension using multivariable adaptive control. Journal of Vibration Engineering, 2005, 18(3), 360–365 (in Chinese).

[33] Chen W W, Xiao H S, Liu L Q, Zu J W. Integrated control of automotive electrical power steering and active suspension systems based on random sub-optimal control. International Journal of Vehicle Design, 2006, 42(3/4), 370–391.

[34] Cai T Y. Multivariable Adaptive Centralized Control and its Applications. Beijing: Science Press, 2001 (in Chinese).

[35] Chen W W, Xu J, Hu F. Input-output centralized and centralized PD control of vehicle nonlinear systems. Chinese Journal of Mechanical Engineering, 2007, 43(2), 64–70 (in Chinese).

[36] Xia X H, Gao W B. Nonlinear System Control and Decouple. Beijing: Science Press, 1993 (in Chinese).

[37] Yu F, Corolla D A. An optimal self-tuning controller for an active suspension. Vehicle System Dynamics, 1998, 29, 51–65.

[38] Liu Y, Fang M, Wang H B. Research on calculation and distribution of composition of forces for vehicle stability control. Proceedings of the 31st Chinese Control Conference, July 25–27, 2012, Hefei, China, 4652–4657 (in Chinese).

[39] Zhu M F, Chen W W, Xia G. Vehicle chassis centralized control based on neural network inverse method. Transactions of the Chinese Society for Agricultural Machinery, 2011, 42(12), 13–17 (in Chinese).

8

Integrated Vehicle Dynamics Control: Multilayer Coordinating Control Architecture

In recent years, multilayer control has been identified as the more effective control technique when compared with the centralized control discussed in Chapter 7. Application of the multilayer control brings a number of benefits, amongst which are: (1) facilitating the modular design of chassis control systems; (2) mastering complexity by masking the details of the individual subsystems at the lower layer; and (3) favoring scalability[1,2].

As described in Chapter 7, the design of the upper layer controller is crucial in the construction of the multilayer control structure. The upper layer controller determines the appropriate control commands and then distributes to the individual lower layer controllers through the coordinating control strategy. Therefore, this chapter mainly focuses on the development of the coordinating control strategy to create synergies among the different subsystems.

8.1 Multilayer Coordinating Control of Active Suspension System (ASS) and Active Front Steering (AFS)

In this section, the multilayer coordinating control of ASS and AFS is studied[3]. It is noted that the system models developed in Chapter 7 are used in the study, including the dynamic model equipped with ASS, the road excitation model, and the tyre model. The AFS model is developed as follows.

Integrated Vehicle Dynamics and Control, First Edition. Wuwei Chen, Hansong Xiao, Qidong Wang, Linfeng Zhao and Maofei Zhu.
© 2016 John Wiley & Sons Singapore Pte. Ltd. Published 2016 by John Wiley & Sons, Ltd.

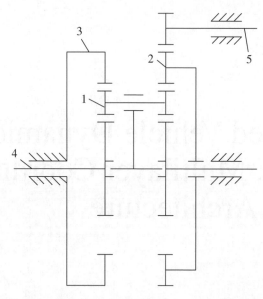

Figure 8.1 Structure of an active front steering system. (1) Steering valve (servo mechanism). (2) Electromagnetic locking unit. (3) Worm. (4) Servomotor. (5) Rack. (6) Pinion. (7) Planetary gear set.

8.1.1 AFS Model

Traditional steering systems are designed such that the ratio from the hand wheel angle to the steering angle of the front wheel is fixed. This design is unable to achieve ease at low speeds and agility at high speeds. To improve vehicle steerability, the AFS system was developed to realize a variable steering ratio by providing an additional angle to the steering wheel angle. The application of the AFS achieves a number of advantages, including improved steering comfort, enhanced steering response, and improved lateral stability through the steering correction of the additional steering angle.

As shown in Figure 8.1, the main components of the active front steering system mainly include a dual planetary gear set, a servomotor, and a transmission mechanism. The dual planetary gear set has two mechanical inputs, i.e., the steering angle input transmitted through the planet carrier and the servomotor angle input. The sun gear and the planet gear are the input and output, respectively. When the motor is static, the steering ratio is fixed. Once the motor turns according to a control command, the steering ratio is variable. The AFS controller collects the sensor signals and calculates the additional angle. The control command for the additional angle is in turn sent to the planetary gear set. Therefore, the purpose of the front active steering is realized. Let the transmission ratio from the motor angle to the steering angle of the front wheel be N_2, we then have:

$$\delta_1 = N_2 \delta_s \tag{8.1}$$

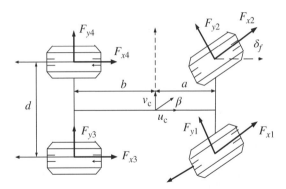

Figure 8.2   Vehicle steering model.

where δ_1 is the motor angle; and δ_s is the additional steering angle of the front wheel. The equation of motion is obtained by performing a dynamic analysis of the active steering axle, which is given as:

$$T_m = T_r + J_p \ddot{\delta}_1 + B_p \dot{\delta}_1 \qquad (8.2)$$

where T_m is the torque provided by the motor on the active steering axle; T_r is the aligning torque transferred from the tyres to the active steering axle; J_p is the equivalent moment of inertia of multiple parts reflected to the active steering axle; and B_p is the equivalent damping coefficient reflected to the active steering axle.

As shown in Figure 8.2, the steering model is established by considering the lateral, longitudinal, and yaw motions. The gradient resistance and wind resistance are ignored. The tyre–road forces are denoted as the longitudinal force F_{xi} and lateral force F_{yi} ($i = 1$–4) for each wheel. Then the equation of motion for the vehicle steering dynamics is derived as:

$$m\left(\dot{u}_c + v_c r\right) = F_{xf} \cos\delta_f + F_{xr} - F_{yf} \sin\delta_f \qquad (8.3)$$

$$m\left(\dot{v}_c + u_c r\right) = F_{xf} \sin\delta_f + F_{yr} + F_{yf} \cos\delta_f \qquad (8.4)$$

$$I_z \dot{r} = a\left(F_{xf} \sin\delta_f + F_{yf} \cos\delta_f\right) - bF_{yr} \qquad (8.5)$$

where $F_{xf} = F_{x1} + F_{x2}$; $F_{yf} = F_{y1} + F_{y2}$; $F_{xr} = F_{x3} + F_{x4}$; $F_{yr} = F_{y3} + F_{y4}$; u_c and v_c are the vehicle longitudinal speed and lateral speed, respectively; I_z is the yaw moment of inertia; and δ_f is the steering angle of the front wheel.

8.1.2   Controller Design

A rule-based method is proposed to design the upper layer controller in order to coordinate the two lower layer controllers of the AFS and ASS[3]. For the two lower layer controllers, the random optimal control method is applied to the ASS to regulate the suspension forces, and the sliding-mode variable-structure control method is applied to the AFS system to adjust the steering angle of the front wheel.

1. *Design of the ASS controller*

   As shown in Figure 8.3, the ASS controller is designed using the random optimal control method.

   The state variable is selected as $X = [z_s \ \dot{z}_s \ z_{u1} \ z_{u2} \ z_{u3} \ z_{u4} \ \dot{z}_{u1} \ \dot{z}_{u2} \ \dot{z}_{u3} \ \dot{z}_{u4} \ \theta \ \varphi \ \dot{\theta} \ \dot{\varphi}]^T$. The mean of the random initial state is given as m_0, and the variance matrix P_0. The output variable is defined as:

   $$Y = \begin{bmatrix} \ddot{z}_s & z_{u1} & z_{u2} & z_{u3} & z_{u4} & \theta & \varphi \end{bmatrix}^T$$

   Therefore, the state equation is derived as:

   $$\begin{cases} \dot{X} = AX + BU + FW \\ Y = CX + DU \end{cases} \tag{8.6}$$

   where U is the control input vector, and $U = [f_1 \ f_2 \ f_3 \ f_4]^T$, where f_1, f_2, f_3, and f_4 are the control forces provided by the active suspension actuators; W is the road excitation vector, and $W = [w_1(t) \ w_2(t) \ w_3(t) \ w_4(t)]^T$. A white noise with zero mean and a covariance matrix R_0 is selected as the road excitation for each wheel; A, B, C, D, and F are the coefficient matrices.

   The performance indices on both handling stability and ride comfort are selected, specifically: the displacement of each tyre z_{u1}, z_{u2}, z_{u3}, and z_{u4}; the dynamic deflection of each suspension $(z_{s1} - z_{u1}), (z_{s2} - z_{u2}), (z_{s3} - z_{u3})$, and $(z_{s4} - z_{u4})$; the vertical acceleration of the sprung mass $\ddot{z}_s$; the pitch rate $\dot{\theta}$; the roll rate $\dot{\varphi}$; and the control forces provided by each active suspension actuator f_1, f_2, f_3, f_4. Therefore, the performance index J is defined as:

   $$J = \lim_{T \to \infty} \frac{1}{T} \Big[q_1 z_{u1}^2 + q_2 z_{u2}^2 + q_3 z_{u3}^2 + q_4 z_{u4}^2 + q_5 \left(z_{s1} - z_{u1} \right)^2 + q_6 \left(z_{s2} - z_{u2} \right)^2 + q_7 \left(z_{s3} - z_{u3} \right)^2$$
   $$+ q_8 \left(z_{s4} - z_{u4} \right)^2 + q_9 \dot{\theta}^2 + q_{10} \dot{\varphi}^2 + r_1 f_1^2 + r_2 f_2^2 + r_3 f_3^2 + r_4 f_4^2 + q_{11} \ddot{z}_s^2 \Big] dt$$

   $$\tag{8.7}$$

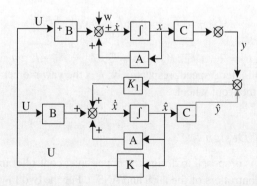

Figure 8.3 Block diagram of an active suspension control system.

where $q_1, ..., q_{11}$ and $r_1, ..., r_4$ are the weighting coefficients. Equation (8.7) can be rewritten as the following in matrix form:

$$J = \lim_{T \to \infty} \frac{1}{T} \int_0^T \left(X^T Q X + U^T R U + 2 X^T N U \right) dt \tag{8.8}$$

where Q, R, and N are the weighting matrices. According to the separation principle, the optimal control law of the random output feedback regulator is actually the optimal control law for the random state regulator. The only difference is that the system state X is replaced by the minimum variance estimated $\hat{X}$ provided by a Kalman filter. Therefore, the control law and filter can be designed separately. The control law is given as:

$$U_1 = -K\hat{X} \tag{8.9}$$

where K is the output feedback gain matrix, which can be derived from the following *Riccati* equation:

$$KA + A^T K + Q - KBR^{-1}B^T K + FWF^T = 0 \tag{8.10}$$

while $\hat{X}$ is obtained from the Kalman filter equation:

$$\dot{\hat{X}} = A\hat{X} + BU + K_1 \left(Y - C\hat{X} \right) \tag{8.11}$$

where the Kalman gain matrix K_1 is given as:

$$K_1 = P_1 C^T R^{-1} \tag{8.12}$$

and P_1 satisfies the Riccati equation:

$$AP_1 + P_1 A^T - P_1 C^T R_0^{-1} C P_1 + Q = 0 \tag{8.13}$$

2. *Design of the AFS controller*

As discussed in Chapter 7, a vehicle is a highly nonlinear system with numerous uncertainties. It requires the proposed controller to be able to provide robustness to deal with parameter variations and external disturbances. Therefore, the sliding-mode variable-structure control strategy based on the proportional switch function is applied to the design of the AFS because of its advantage in robustness control[3]. The control objective is to make the tracking error between the vehicle yaw rate and the expected yaw rate to approach zero. The vehicle state parameters are obtained by the sensors directly or indirectly, including the lateral velocity, lateral acceleration, yaw rate, yaw angular acceleration, etc. Thereafter, the sideslip angles of the front and rear wheels are solved in real time. In addition, the tyre lateral forces are calculated by using the tyre dynamic model. Hence, the tyre lateral forces are adjusted by regulating the steering angle of the front wheel and therefore fulfilling the yaw rate control through generating the additional yaw moment. For the AFS, the steering angle of the front wheel δ_f is calculated as:

$$\delta_f = \delta_c + \delta_s \tag{8.14}$$

where δ_c is the steering angle of the front wheel provided by the driver; and δ_s is the additional steering angle of the front wheel provided by the AFS. The real-time values of the yaw rate and yaw angular acceleration are r, $\dot{r}$, and the expected steady state values are r_d, and $\dot{r}_d$. Then, the error of the yaw rate and its difference are given as:

$$\begin{cases} e = r - r_d \\ \dot{e} = \dot{r} - \dot{r}_d \end{cases} \tag{8.15}$$

The switch function is defined as:

$$s = c_0 e + \dot{e} \tag{8.16}$$

According to the proportional switch control method, the control law is shown as:

$$U_s = \delta_s = \left(\alpha|e| + \gamma|\dot{e}|\right)\operatorname{sgn}(s) \tag{8.17}$$

where c_0 is the slope; α and γ are the positive constants.

3. *Design of the upper layer controller*

To design an upper layer controller, the performance indices on both handling stability and ride comfort are selected, specifically the vehicle vertical acceleration, roll angle, roll angular acceleration, and yaw rate. The upper layer controller monitors the driver's intentions and the current vehicle states. Based on these input signals, the upper layer controller is designed to coordinate the interactions between the ASS and AFS in order to achieve the desired vehicle states. Thereafter, the control commands are generated by the upper layer controller and distributed to the two lower layer controllers respectively. Finally, the individual lower layer controllers each execute their local control objectives to control the vehicle dynamics. The structure of the multilayer control system is illustrated in Figure 8.4. In the figure, U denotes the matrix of suspension regulating forces; U_s represents the controlled steering angle; and u_c is the vehicle speed. The variable δ_c is the steering angle of the front wheel provided by the driver; and r_d is the expected yaw rate.

The rule-based method is proposed to design the upper layer controller, which is described as follows.

1. The weighting parameters defined in the performance index of ASS $q_1, \ldots, q_{11}$ and $r_1, \ldots, r_4$ are determined according to the steering angle of the front wheel. Specifically, if $|\delta_f| \geq \delta_0$, the set of weighting parameters is determined by considering the effect of the steering on the roll of the suspension; otherwise the other set of the weighting parameters is used without considering the effect. δ_0 is the threshold of the steering angle of the front wheel.

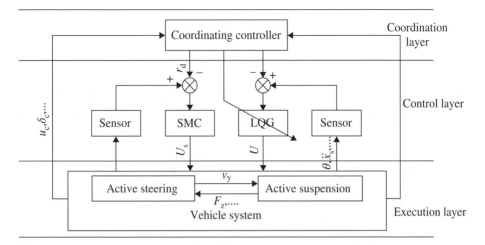

Figure 8.4 Block diagram of a multilayer control system for the ASS and AFS.

2. The aim of the upper layer controller is to track the expected yaw rate r_d, which is calculated as:

$$r_d = \frac{u_c \delta_f}{l\left(1 + K_s u_c^2\right)} \qquad (8.18)$$

where K_s is the vehicle stability factor; and l is the wheel base.

8.1.3 Simulation Investigation

A simulation study is performed to demonstrate the performance of the proposed multi-layer coordinating control system. We assume that the vehicle travels at a constant speed of $u_c = 20$m/s on a road with an unevenness coefficient of $5 \times 10^{-6} m^3$, and subjected to a step steering input with amplitude of $\pi/2$. After tuning, two sets of weighting parameters are selected for the ASS as discussed above: when $|\delta_f| \geq \delta_0$, $q_1 = q_2 = q_3 = q_4 = 0.9 \times 10^3$, $q_5 = q_6 = q_7 = q_8 = 1.1 \times 10^4$, $q_9 = 10^4 q_{10} = 1.2 \times 10^5$, $q_{11} = 5.0 \times 10^6$, $r_1 = r_2 = r_3 = r_4 = 1$; otherwise $q_1 = q_2 = q_3 = q_4 = 10^3$, $q_5 = q_6 = q_7 = q_8 = 10^4$, $q_9 = 2 \times 10^3$, $q_{10} = 10^5$, $q_{11} = 10^6$, $r_1 = r_2 = r_3 = r_4 = 1$. The parameters of the sliding mode variable structure controller are selected as $c_0 = 0.3$, $\alpha = 0.05$, and $\gamma = 0.01$.

The simulation results are shown in Figures 8.5(a–f). For comparison, the simulation for the decentralized control is also performed and in this case, we simply eliminate the upper layer controller. The following discussions are made by performing a quantitative analysis of the simulation results. First, it is obtained that the R.M.S. (Root-Mean-Square) values of the vertical acceleration, roll angular acceleration, and pitch angular acceleration of the sprung mass for the multilayer coordinating control are reduced significantly by 36.7%, 34.6%, and 18.8%, respectively, compared with those for the decentralized control.

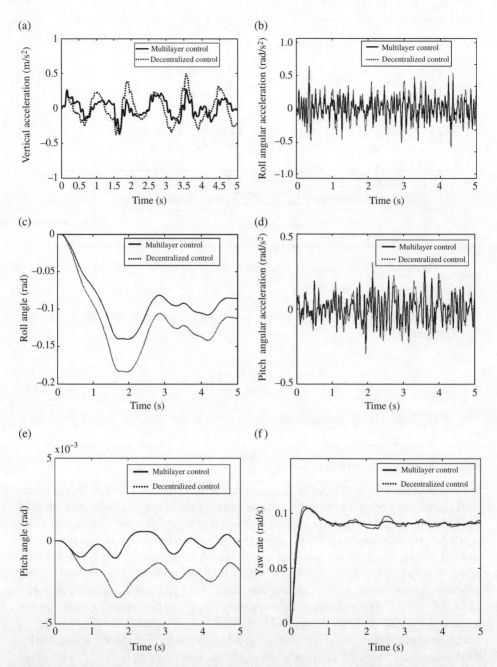

Figure 8.5 Simulation results. (a) Vertical acceleration. (b) Roll angular acceleration. (c) Roll angle. (d) Pitch angular acceleration. (e) Pitch angle. (f) Yaw rate.

In addition, a similar pattern can be observed for the performance index of the pitch angle and roll angle. Finally, the R.M.S. value of the yaw rate for the multilayer coordinating control is reduced by 11.3%, and the settling time is reduced by 0.15s, compared with the decentralized control. The results indicate that the multilayer coordinating controller is able to coordinate the interactions between the ASS and AFS. The application of the multilayer coordinating control improves the overall vehicle performance including ride comfort and handling stability.

8.2 Multilayer Coordinating Control of Active Suspension System (ASS) and Electric Power Steering System (EPS)

In general, the theories of Newtonian mechanics and analytical mechanics are applied to establish the vehicle dynamic models. As discussed in Chapter 1, classical mechanics may become difficult to deal with in complex dynamic systems. Vehicles are complex systems since they consist of numerous of rigid and/or flexible bodies interconnected by joints, and force elements (springs and dampers). To handle efficiently the complex vehicle dynamic systems, rigid multibody dynamics is applied in this chapter since it has the advantage of solving large-scale mechanical systems through using the standard calculation process[4,5].

8.2.1   System Modeling

1. *Vehicle rigid multibody dynamic model*
   Vehicles are complex dynamic systems in space with multiple degrees of freedom. As shown in Figure 8.6, the vehicle dynamic model equipped with ASS and EPS is established by applying rigid multibody dynamics in the Cartesian coordinates. Vehicles consist of seven components in total: a vehicle body, four wheel members, a tie rod, and a steering column. The body fixed coordinate system of each component is located at its center of gravity. The McPherson suspension is selected for the front suspension, and the trailing arm for the rear suspension. Each front wheel member connects with the vehicle body through the spherical rotation joint and pillar joint. The rear trailing arm suspension attaches to the vehicle body through rotational joints. In addition, the translational spring-damp-actuator (TSDA) in an ASS system is considered as the force element in the multibody dynamic model. The connection between the tie rod and the front wheel member is simplified as a spherical-spherical joint. The tie rod connects with the steering column through a rack-pinion mechanism. The assist torque of the EPS system is applied on the steering column. In summary, the system has 34 constraint equations and seven Euler parameter normalized constraint equations. The generalized Cartesian coordinates of the members in the space are defined as:

$$q_i = \begin{bmatrix} x_i & y_i & z_i & e_{0i} & e_{1i} & e_{2i} & e_{3i} \end{bmatrix}^T \quad (i = 1, 2 \cdots 7) \tag{8.19}$$

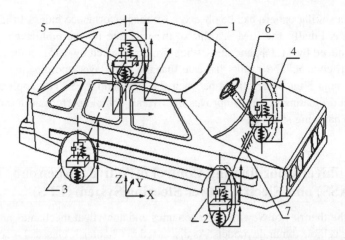

Figure 8.6 Vehicle rigid multibody model. (1) Vehicle body. (2–5) Wheel member. (6) Steering column. (7) Tie rod.

where x_i, y_i, and z_i are coordinates of the position of the members; and e_{0i}, e_{1i}, e_{2i}, and e_{3i} are the Euler quaternion of attitude coordinates of the members. The vehicle dynamic equation is derived as:

$$\begin{bmatrix} M & B^T \\ B & 0 \end{bmatrix}\begin{bmatrix} \dot{h} \\ -\lambda \end{bmatrix} + \begin{bmatrix} b \\ 0 \end{bmatrix} = \begin{bmatrix} g \\ \gamma^{\#} \end{bmatrix} \tag{8.20}$$

where $M = \begin{bmatrix} N & 0 \\ 0 & J' \end{bmatrix}$; $h = [\dot{s} \quad \omega']^T$; $b = [0 \quad \tilde{\omega}'J'\omega']^T$; $g = [f \quad n']$; N is the matrix of mass of member i; B is the Jacobian matrix, and the element of the matrix B is the partial derivative of the polynomial function on the left-hand side of the constraint equation with respect to the generalized coordinates. It is a function of both the system coordinates and time; λ is the Lagrange multiplier matrix; $\gamma^{\#}$ is the term on the right-hand side of the acceleration equation; s is the vector of the coordinates of the position; and J' is the moment of inertia. Due to the symmetry of the vehicle with respect to the longitudinal central line, only the product of inertia I_{xz} is taken into account, and the other products of inertia are ignored; and ω' is the vector of the angular velocity.

Let the vehicle angular velocities in x, y, and z directions be $[p, q, r]$, i.e., the roll, pitch, and yaw angular velocity. The three angular velocities given in the above equation are related with each other, and demonstrate the coupling effects of the three vehicle rotational motions. The dynamic equation of the angular velocity with respect to the Euler parameters is expressed as follows:

$$\omega' = 2G\dot{D} \tag{8.21}$$

where $G = \begin{bmatrix} -e_1 & e_0 & e_3 & -e_2 \\ -e_2 & -e_3 & e_0 & e_1 \\ -e_3 & e_2 & -e_1 & e_0 \end{bmatrix}$; $D = [e_0 \quad e_1 \quad e_2 \quad e_3]^T$; $\tilde{\omega}'$ expressed in equation

(8.20) and equation (8.21) is the antisymmetric matrix of the angular velocities; f and n' are the forces and moments applied on the members respectively, including the hand torque of the steering wheel, assist torque provided by the motor; tyre lateral forces, aligning moments, tyre vertical forces, TSDA forces of the suspensions, the gravitational forces of the members; and the frictional moment of the system. The detailed computation of the forces and moments is described in the following section.

2. *Force and moment*

In practice, the forces and moments applied on a vehicle are quite complex. It is necessary to simplify the actual forces and moments into a main vector and moment acting on the center of gravity of a rigid body by using the principle of force equivalence. In addition, the force resulting from tyre deformation is simplified into a force element acting on itself.

3. *Assist torque*

The assist torque provided by the motor is determined by the motor model, which is given as

$$T_a = \frac{i_t K_a}{R_0}\left(U_d - K_b i_t \dot{\delta}_1\right) \tag{8.22}$$

where T_a is the motor torque; i_t is the reduction ratio of transmission mechanism (i.e., the worm and worm shaft mechanism); K_a is the torque coefficient of the motor; K_b is the constant of the back electromotive force (EMF) of the motor; R_0 is the armature resistance; δ_1 is the steering angle of the pinion; and U_d is the terminal voltage of the motor.

4. *Suspension TSDA force*

It is assumed that the TSDA component connects with two members on the P_a and P_b points, respectively. The vector on the two points is given as:

$$d_{ab} = r_b + A_b s_b'^p - r_a - A_a s_a'^p \tag{8.23}$$

$$l^2 = d_{ab}^T d_{ab} \tag{8.24}$$

$$\dot{l} = \left(d_{ab}/l\right)^T \left(\dot{r}_b - A_b \tilde{s}_b'^p \omega_b' - \dot{r}_a + A_a \tilde{s}_a'^p \omega_a'\right) \tag{8.25}$$

where l is the length of the TSDA component; r_a and r_b are the coordinates of the positions of members a and b, respectively; $s_a'^p$ and $s_b'^p$ are the coordinate vectors of points P_a and P_b in the body-fixed coordination system, respectively; $\tilde{s}_a'^p$ and $\tilde{s}_b'^p$ are the antisymmetric matrices of $s_a'^p$ and $s_b'^p$, respectively; ω_a' and ω_b' are the vectors of the angular velocities of members a and b, respectively; A_a and A_b are the direction cosine matrices of members a and b, respectively. The antisymmetric matrix and the direction cosine matrix are derived as:

$$\tilde{S}'^p = \begin{bmatrix} 0 & -z & y \\ z & 0 & -x \\ -y & x & 0 \end{bmatrix} \quad A = 2\begin{bmatrix} e_0^2 + e_1^2 - \dfrac{1}{2} & e_1e_2 - e_0e_3 & e_1e_3 + e_0e_2 \\ e_1e_2 + e_0e_3 & e_0^2 + e_2^2 - \dfrac{1}{2} & e_2e_3 + e_0e_1 \\ e_1e_3 - e_0e_2 & e_2e_3 - e_0e_1 & e_0^2 + e_3^2 - \dfrac{1}{2} \end{bmatrix}$$

The suspension force is given as:

$$f = k(l - l_0) + c\dot{l} + F \tag{8.26}$$

where the first term on the right-hand side of the equation is the sprung force; the second term is the damping force; and the third term is the ASS control force. Therefore, the generalized force of the TSDA component applied on the member a is calculated as:

$$Q_a = \frac{f}{l}\begin{bmatrix} d_{ab} \\ 2G_a^T \tilde{s}_a'^p A_a^T d_{ab} \end{bmatrix} \tag{8.27}$$

5. *Tyre vertical force*
   The tyre vertical force Q_{zj} is given as:

$$Q_{zj} = k_{tj}\left(r_0 - z_{uj} + z_{0j}(t)\right) \quad (j = 1, 2, 3, 4) \tag{8.28}$$

where $z_{0j}(t)$ is the road excitation; k_{tj} is the tyre vertical stiffness; r_0 is the tyre free radius; and z_{uj} is the tyre vertical displacement. The filtered white noise described in Chapter 4 is selected as the road excitation in the time domain.

6. *Tyre lateral force and aligning moment*
   The Pacejka nonlinear tyre model is used to determine the dynamic forces of the tyre. The inputs of the tyre model include the tyre vertical force, tyre sideslip angle, and slip ratio; and the outputs include the lateral force F_y and aligning moment M_z. For the rigid multibody model, the front and rear sideslip angles are calculated as:

$$\begin{cases} \alpha_f = \arctan\dfrac{\dot{y}_f}{\dot{x}_f} - 2\arccos e_{0f} \\ \alpha_r = \arctan\dfrac{\dot{y}_r}{\dot{x}_r} \end{cases} \quad (0 \le 2\arccos e_{0f} < 2\pi) \tag{8.29}$$

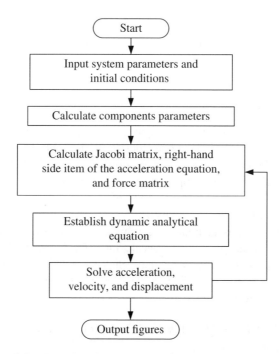

Figure 8.7   Flow chart for solving the dynamic equations of motion.

7. *Solution of the equations of motion*

The equations of motion for the rigid multi-body expressed in equation (8.20) are a set of nonlinear differential-algebraic equations. Numerical methods to solve these equations include the direct method and the coordination separation method. Here, the direct method is used and programmed to obtain s and $\dot{\omega}'$ in equation (8.20). Then, $\dot{D}$ is calculated by using equation (8.21) and the integral of $\dot{D}$ is performed with respect to the generalized coordinates of the Euler parameters and the corresponding coordinates of the positions. Hence, the state variables, including the angular velocities and accelerations of the members, are obtained by solving equation (8.20). The flowchart showing how to solve these equations is shown in Figure 8.7.

8.2.2   Controller Design

A rule-based method is proposed to design the upper layer controller in order to coordinate the two lower layer controllers of the AFS and ASS. For the two lower layer controllers, two different working modes are designed for each controller according to the different control objectives under different driving conditions.

The upper layer controller monitors the driver's intentions and the current vehicle state. Based on these input signals, the upper layer controller is designed to coordinate the interactions between the ASS and EPS in order to achieve the desired vehicle state. Thereafter, the control commands of switching to the appropriate working mode are generated by the upper layer controller and distributed to the two lower layer controllers respectively. Finally, each individual lower layer controllers executes their local control objectives to control the vehicle dynamics.

8.2.1.1 Design of the ASS Controller

Two working modes are considered in designing the ASS controller: (1) when the vehicle drives straight or the steering angle is relatively small, the control objective of the ASS is to improve the vehicle ride comfort; (2) when cornering, the control objective is to control or maintain the vehicle attitude to improve handling stability.

1. *Mode 1: Improving vehicle ride comfort*

   The ASS controller aims at improving vehicle ride comfort by controlling vehicle vertical acceleration and pitch angular acceleration. The increment PID control strategy is applied. The inputs to the PID controller include the tracking error of the vehicle pitch angular acceleration and the vertical acceleration, while the output is the ASS control force. The control law is obtained as:

$$\Delta U(k) = K_p \left[e(k) - e(k-1) \right] + K_i e(k) + K_d \left[e(k) - 2e(k-1) + e(k-2) \right] \quad (8.30)$$

$$U(k) = U(k-1) + \Delta U(k) \quad (8.31)$$

   where $U(k)$ is the control force; $e(k)$ is the tracking error; K_p, K_d, and K_i are the proportional, differential, integral coefficients of the PID controller, respectively.

2. *Mode 2: Regulating vehicle position during cornering*

   In this working mode, the ASS controller regulates the roll and yaw motions during cornering. The appropriate control force of each ASS is obtained by applying the sliding mode variable structure control strategy developed in Section 8.2, and the switch function and control law are given in equation (8.17) and equation (8.18), respectively.

   When regulating the roll motion, the control forces of the left-hand and right-hand side suspensions are regulated respectively. The input variables are selected as $e^p = p$ and $\dot{e}^p$, and the output is the adjustment of the control force of the suspension ΔF_n^p ($n = 1, 2, 3, 4$).

   When regulating the yaw motion, the additional yaw moment is generated indirectly by adjusting the ASS control force, resulting in the regulation of the tyre forces. The inputs are defined as $e^r = r - r_d$ and $\dot{e}^r$, where r_d is the expected yaw rate given in equation (8.6). The output is the adjustment of the control force of the suspension ΔF_n^r ($n = 1, 2, 3, 4$).

   In the rigid multibody dynamic model, the steering angle of the front wheel δ_f is derived according to the Euler parameter definition, by assuming that the tyre plane is vertical.

$$\delta_f = \left(2\arccos e_{0fl} + 2\arccos e_{0fr} \right) / 2 = \arccos e_{0fl} + \arccos e_{0fr} \quad (8.32)$$

   where e_{0fl} and e_{0fr} are the Euler parameters of the front-left and front-right wheels, respectively.

In addition, a weighting factor $\varepsilon(0 < \varepsilon < 1)$ is defined to take into account the coupling effects of the roll and yaw motions, and ε is determined in real time according to the values of r and p. Hence, the corresponding suspension control force F_n' is defined as:

$$F_n' = F_n + \varepsilon\Delta F_n^p + (1-\varepsilon)\Delta F_n^r \quad (n = 1,2,3,4) \tag{8.33}$$

8.2.1.2 Design of the EPS Controller

Similarly, two working modes are considered when designing the EPS controller: (1) when steering in steady-state, the objective of the EPS controller is to improve the ease of steering; (2) when the vehicle loses stability, e.g., oversteering, the assist torque must be reduced in order to weaken the driver's ability to quickly change the steering angle of the front wheel, and hence increase the road feel.

1. *Mode 1: Improving steering ease*
   The increment PID control strategy is applied to the design of the EPS controller. The input to the PID controller is selected as the tracking error between the target current and the actual feedback current of the motor; while the output is the voltage of the motor. The target current is determined by the characteristic curve of the motor [5]. The control law has a similar expression as equation (8.30) and equation (8.31).
2. *Mode 2: Improving handling stability*
   The reduction of the assist torque is fulfilled by multiplying the steady-state target torque by a gain coefficient that is less than 1.

8.2.1.3 Design of an Upper Layer Coordinating Controller

A rule-based control method is adopted to design the upper layer coordinating controller. The structure of the multilayer control system is illustrated in Figure 8.8.

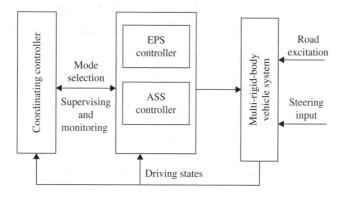

Figure 8.8   Block diagram of the multi-layer coordinating control system.

The control rule is developed as follows:

Rule 1: When the vehicle drives straight or the steering angle is relatively small, the ASS controller works in mode 1 and the EPS system is idle.

Rule 2: When cornering, the coordinating controller sends out the control command to the ASS controller to switch it to mode 2. Meanwhile, the coordinating controller recognizes the vehicle steering characteristics according to the tracking error e^r between the actual and expected yaw rate, combining the value of the steering angle of the front wheel. For example, when the vehicle steers to the left, $e^r > 0$ indicates the oversteering phenomenon. The coordinating controller sends out the control command to the EPS controller to work in mode 2. The gain coefficient values in mode 2 are then determined to reduce the extent of the oversteering. Otherwise, the EPS controller works in mode 1.

8.2.3   Simulation Investigation

A simulation study is performed in MATLAB to demonstrate the performance of the proposed multilayer coordinating control system. Two driving conditions are investigated, and the simulation results are illustrated in Figure 8.9. It is noted that the vehicle physical parameters presented in Table 7.5 are also used in the simulation. The following discussions are made:

Driving condition 1: When the vehicle travels straight (or the steering angle is relatively small) at a speed of 20m/s, the upper layer controller works according to rule 1, i.e., the ASS works and EPS is idle. The vehicle ride comfort is the major control objective. It is observed clearly in Figure 8.9(a, b) that both the RMS and peak values of the vertical acceleration and pitch angular acceleration for the multilayer coordinating control system improve significantly compared with the passive suspension. Specifically, the RMS value of the vertical acceleration for the multilayer coordinating control system is reduced from 0.582m/s^2 to 0.316m/s^2, and the RMS value of the pitch angular acceleration from 0.673 rad/s^2 to 0.496 rad/s^2. The results indicate that the vehicle ride comfort is effectively enhanced.

Driving condition 2: The vehicle travels at a speed of 10m/s and is subjected to a step steering input with amplitude of $\frac{\pi}{2}$. The upper layer controller works according to rule 2.

It is observed clearly in Figure 8.9(c) that the RMS value of the roll angular acceleration for the multilayer coordinating control system is reduced from 0.0389 rad/s^2 to 0.0225 rad/s^2, compared with the decentralized control system. As shown in Figure 8.9(d), both the peak value and settling time of the yaw rate for the multilayer coordinating control system are decreased greatly. In summary, the simulation results indicate that the multilayer coordinating controller is able to coordinate the interactions between the ASS and EPS. The application of multilayer coordinating controls effectively improves the overall vehicle performance including ride comfort and handling stability. In addition, the application of rigid multibody dynamics is successful in constructing and solving complex vehicle dynamic systems.

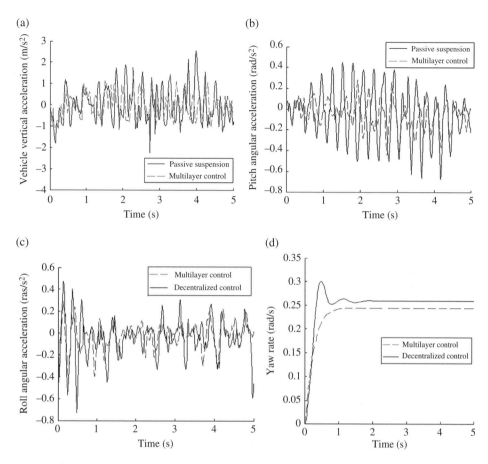

Figure 8.9 Simulation results. (a) Vehicle vertical acceleration. (b) Pitch angular acceleration. (c) Roll angular acceleration. (d) Yaw rate.

8.3 Multilayer Coordinating Control of an Active Suspension System (ASS) and Anti-lock Brake System (ABS)

This section studies the multilayer coordinating control of an active suspension system (ASS) and anti-lock brake system (ABS) to achieve the goal of function integration of the control systems[6,7]. The architecture of the proposed multilayer coordinating control system is shown in Figure 8.10. In the figure, a_x is longitudinal deceleration and $a_x = \dfrac{du_c}{dt}$; ΔT_s and ΔT_b represent the changes of the pitch torque and the braking torque, respectively; and u_c is the wheel speed. The definitions of the other symbols in Figure 8.10 can be found in the previous section. The control system consists of three layers, i.e., coordinating layer, control layer, and target layer. The coordinating controller monitors the driver's intentions and the current vehicle conditions including the pitch angle θ and longitudinal deceleration a_x. Based on these input signals, the coordinating controller computes the desired vehicle

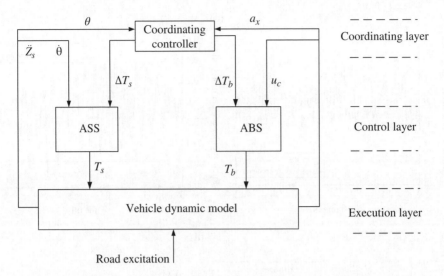

Figure 8.10   Block diagram of the multilayer coordinating control system.

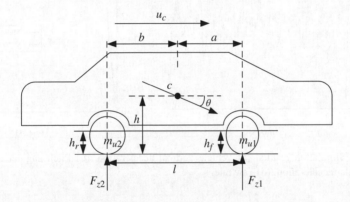

Figure 8.11   Half vehicle model.

motions in order to achieve an optimal overall performance criterion. Thereafter, the coordinating controller generates the control commands ΔT_s and ΔT_b to the two controllers, i.e., the ASS and the ABS, respectively. The ASS and the ABS in turn execute their local control objectives to apply (increase, decrease, or hold) the pitch torque T_s and the braking torque T_b, respectively.

8.3.1   Coordinating Controller Design

It is known that interactions exist between the suspension system and brake system, especially when the brakes are applied. When the brakes are applied, the dynamic loads of the front wheel and rear wheel can be calculated by equation (8.34), based on the half vehicle model shown in Figure 8.11.

$$F_{ZMi} = \pm \left(m \frac{du_c}{dt} h + c_\theta \theta + m_{u1} h_f \frac{du_c}{dt} + m_{u2} h_r \frac{du_c}{dt} \right) / l \quad (i = 1, 2) \tag{8.34}$$

where c_θ is pitch angular stiffness and $c_\theta = k_{s1} a^2 + k_{s2} b^2$; h_f and h_r are the height of the C.G. for front and rear unsprung mass, respectively; h is the height of the C.G. of the vehicle; k_{s1} and k_{s2} are the stiffness of front suspension and rear suspension, respectively; and l is the wheelbase. A simple rule-based control strategy for the coordinating controller is designed as follows by using the two vehicle states, i.e., the pitch angle θ and longitudinal deceleration a_x.

1. If $\left(\frac{du_c}{dt} > a_0 \ \& \ \theta > \theta_0 \right)$, the coordinating controller monitors the vehicle states and does not generate control commands. In this case, the ABS is not applied, and only the EPS system executes its local control objective.

2. If $\left(\frac{du_c}{dt} < a_0 \ \& \ \theta < \theta_0 \right)$, the coordinating controller coordinates the ASS system and the ABS in order to achieve an optimal overall performance criteria. Based on the pitch angle θ and longitudinal deceleration a_x, the coordinating controller generates the control commands ΔT_s and ΔT_b to the two controllers, i.e., the ASS and the ABS, respectively. The ASS and the ABS in turn execute their local control objectives to apply the pitch torque T_s and the braking torque T_b, respectively.

Where a_0 is the logic threshold of the deceleration; θ_0 is the logic threshold of the pitch angle. It should be noted that these parameters have negative values. The pitch torque T_s and the braking torque T_{b1} for front wheel and T_{b2} for rear wheel are calculated as:

$$T_s = a F_{ZM1} - b F_{ZM2} \tag{8.35}$$

$$T_{b1} = \left(F_{ZM1} + \frac{mgb}{l} \right) \phi_x R_1 \tag{8.36}$$

$$T_{b2} = \left(F_{ZM2} + \frac{mga}{l} \right) \phi_x R_2 \tag{8.37}$$

The proposed rule-based control strategy shows that only the deceleration and pitch angle exceed their thresholds, the coordinating controller generates the control commands to the ASS and ABS. Otherwise, the coordinating controller acts as a state monitor.

8.3.2 Simulation Investigation

A simulation study is performed to demonstrate the performance of the proposed multilayer coordinating control system. We assume that the vehicle travels at a constant speed $u_c = 20$m/s, and the logic thresholds are selected as: $a_0 = -8 m / s^2$, and $\theta_0 = -0.0689 rad$. The simulation results are shown in Figure 8.12(a–f). For comparison, the simulation for the decentralized control is also performed. In this case, we simply eliminate the upper

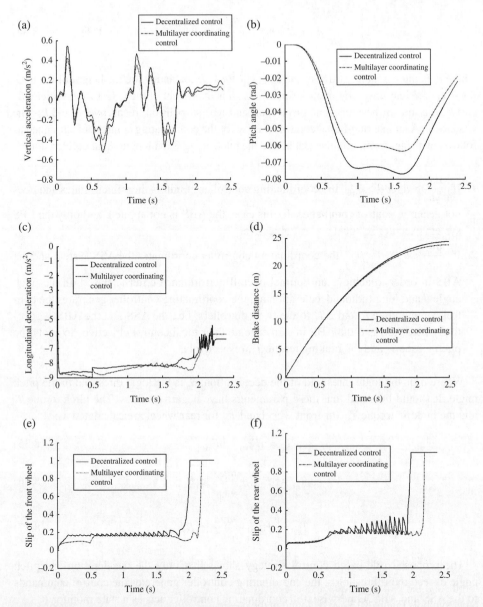

Figure 8.12 Simulation results. (a) Vertical acceleration of the sprung mass. (b) Pitch angle. (c) Longitudinal deceleration. (d) Brake distance. (e) Slip of the front wheel. (f) Slip of the rear wheel.

layer controller. In addition, a quantitative analysis of the simulation result is also performed. The following discussions are made from the simulation results:

First, it is observed in Figure 8.12(a) that the peak values of the vertical acceleration of the sprung mass for the multilayer coordinating control is reduced compared to that of the decentralized control. Moreover, the R.M.S. (Root-Mean-Square) value of the performance

Table 8.1   Comparison of the R.M.S. values of the responses of the multilayer coordinating control and decentralized control.

Performance index	Decentralized control	Multilayer coordinating control	Improvement
Vertical acceleration of sprung mass (m/s^2)	0.2320	0.1762	24.05%
Pitch angle (rad)	0.0635	0.0526	17.17%
Longitudinal deceleration(m/s^2)	7.9714	8.1498	2.24%
Max. brake distance(m)	24.306	23.734	0.572

index for the multilayer coordinating control shown in Table 8.1 is reduced greatly by 24.05%. A similar pattern can be observed for the performance index of the pitch angle. The results indicate that the ride comfort of the vehicle is improved significantly when the multilayer coordinating control is applied.

In addition, it is shown in Figure 8.12(c) and (d) that both the longitudinal deceleration and the brake distance are improved slightly for the multilayer coordinating control compared to those of the decentralized control.

Finally, it is noted in Figure 8.12(e) that there is not much difference on the slip of the front wheel for the two control systems. However, the slip of the rear wheel for the multilayer coordinating control, as shown in Figure 8.12(f), is reduced clearly compared to that of the decentralized control. The results indicate that the brake swerve can be avoided, and hence the lateral stability is improved.

In summary, the multilayer coordinating controller is able to coordinate the interactions between the ASS and ABS. The application of the multilayer coordinating control improves the overall vehicle performance under critical driving conditions: the vehicle ride comfort is improved and the lateral stability is improved, and the braking performance is ensured.

8.4   Multilayer Coordinating Control of the Electric Power Steering System (EPS) and Anti-lock Brake System (ABS)

This section studies the multilayer coordinating control of an electric power steering system (EPS) and anti-lock brake system (ABS) to achieve the goal of function integration of the control systems[8,9]. Similar to Section 8.4, the architecture of the proposed multilayer coordinating control system is shown in Figure 8.13. Here, ΔT_m and ΔT_b represent the changes of the assist torque and the braking torque, respectively; and u_c is the wheel speed. The definitions of the other symbols Figure 8.13 can be found in the previous section. The control system consists of three layers: coordinating layer, control layer, and target layer. The coordinating controller monitors the driver's intentions and the current vehicle conditions including the torque applied on the steering wheel T_c, the vehicle speed u_c, the slip λ, and the yaw rate γ. Based on these input signals, the coordinating controller computes the desired vehicle motions in order to achieve an optimal overall performance criterion. Thereafter, the coordinating controller generates the control commands ΔT_m and ΔT_b to the two controllers, i.e., the EPS system and the ABS, respectively. The EPS system and the ABS in turn execute their local control objectives to apply (increase, decrease, or hold) the assist torque T_m and the braking torque T_b, respectively.

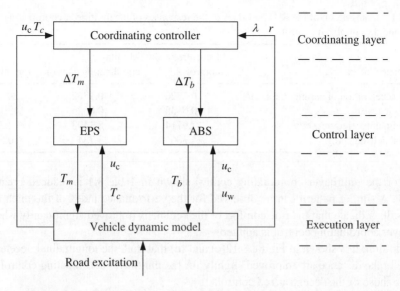

Figure 8.13   Block diagram of the multilayer coordinating control system.

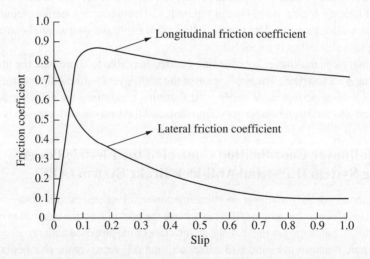

Figure 8.14   Relationships of the slip and the friction coefficients (on a road surface with high adhesion).

8.4.1   Interactions between the EPS System and ABS

To design the coordinating controller, we must first investigate the interactions between the EPS system and the ABS. An adhesion-slip curve of a tyre is used to reveal the interactions. Figure 8.14 illustrates the relationship between the slip and the lateral friction coefficient, and the relationship between the slip and the longitudinal friction coefficient.

Figure 8.14 shows that when the slip is less than 0.05, the lateral friction coefficient is relatively large, while the longitudinal friction coefficient is relatively small. Therefore, the vehicle is able to achieve a good lateral stability but a bad braking performance. When the slip is in the range between 0.1 and 0.3, the lateral friction coefficient drops dramatically while the longitudinal friction coefficient stays at a relatively large value. Therefore, the vehicle could have a good braking performance but a bad lateral stability. However, the performance conflicts between the steering system and the braking system have not been taken into account when designing the EPS system and the ABS separately. Therefore, it is necessary to coordinate the conflicts between the two control systems in order to obtain an optimal overall performance. The coordinating controller is designed as follows.

8.4.2 Coordinating Controller Design

A simple control strategy for the coordinating controller is designed as follows.

Three performance indices are considered in designing the coordinating controller, including the driver's steering torque T_c, the slip λ for both the front wheel and rear wheel, and the yaw rate γ. The overall performance index J is defined as:

$$J = \sqrt{\frac{W_1 J_1^{\,2} + W_2 J_2^{\,2} + W_3 J_3^{\,2}}{W_1 + W_2 + W_3}} \tag{8.38}$$

where J_1, J_2 and J_3 are the variances of the driver's steering torque T_c; the slip λ, and the yaw rate γ, respectively; W_1, W_2 and W_3 are the weighting parameters. The following values are assigned to these weighting parameters, considering the relatively higher importance of the braking performance: $W_1 = 0.2$, $W_2 = 0.5$, and $W_3 = 0.3$. For the first two indices, the driver's steering torque T_c and the slip λ, three cases are considered:

1. If the vehicle speed is less than 20km/h, the coordinating controller does not generate any control commands. In this case, the ABS is not applied and only the EPS system executes its local control objective.
2. If the vehicle speed is at the range of 20–40km/h, the coordinating controller coordinates the EPS system and the ABS to minimize the overall vehicle performance index defined in equation (8.38). In this case, the coordinating controller sends the control command of actuation to the EPS to adjust the assist torque according to the variation of the measured driver's steering torque. In the meantime, the coordinating controller generates the control command of actuation according to the measured wheel angular acceleration and the slip. The ABS then adjusts the brake pressure of the wheel cylinders to guarantee the brake safety and the handling stability.
3. If the vehicle speed is more than 40km/h, the coordinating controller does not generate any control commands. In this case, the EPS system is not applied, and only the ABS executes its local control objective.

The above control strategy is fulfilled through a decision-making controller. The main function of the decision-making controller is to monitor the driving conditions of the

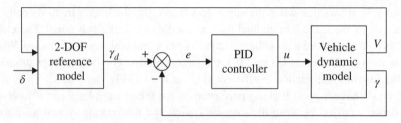

Figure 8.15   Block diagram of the PID controller.

vehicle, and then decide whether or not the EPS system and ABS are applied. If both are applied, the decision-making controller coordinates the interactions between the two controllers.

For the third index, the yaw rate γ, a PID controller is designed to make the vehicle track the reference yaw rate. As shown in Figure 8.15, the reference yaw rate γ_d is obtained from the 2-DOF vehicle reference model derived in equation (8.6) since the 2-DOF reference model reflects the desired relationship between the driver's steering input and the vehicle yaw rate. The block diagram of the PID controller is shown in Figure 8.15.

In addition, it is noted that the controllers for the EPS system and ABS are designed in the same way as those in the previous chapter.

8.4.3   Simulation Investigation

A simulation study is performed to demonstrate the performance of the proposed multi-layer coordinating control system. We assume that the vehicle travels at a constant speed of $u_c = 36$ km/h, and is subjected to steering input from the steering wheel. In the meantime, an emergency brake is applied. The steering input is set as a step signal with an amplitude of $180°$. After tuning the parameter setting for the coordinating PID controller, we select $P = 86$, $I = 2$, and $D = 0.01$ for the upper layer PID controller, and the optimal slip $\lambda_{opt} = 0.18$ for the ABS.

The simulation results for the three performance indices are shown in Figure 8.16(a)–(d). For comparison, the simulation for the decentralized control is also performed. In this case, we simply eliminate the upper layer controller. The following discussion is made from the simulation results.

First, it is observed that the yaw rate for the multilayer control, as shown in Figure 8.16(a), follows the desired yaw rate much better than that for the decentralized control. Moreover, it is also reduced more quickly than that for the decentralized control. The results indicate that the vehicle lateral stability is well controlled when the multilayer coordinating control is applied. Moreover, the driver's steering torque for the multilayer coordinating control, as shown in Figure 8.16(b), is kept almost the same as that for the decentralized control. The results indicate that the steering agility is ensured when the multilayer coordinating control is applied. Finally, there is not much difference in wheel slip between the multilayer coordinating control and decentralized control, as can be seen in Figure 8.16(c) and (d). The results indicate that, inboth cases, the slip can be kept around the optimal value.

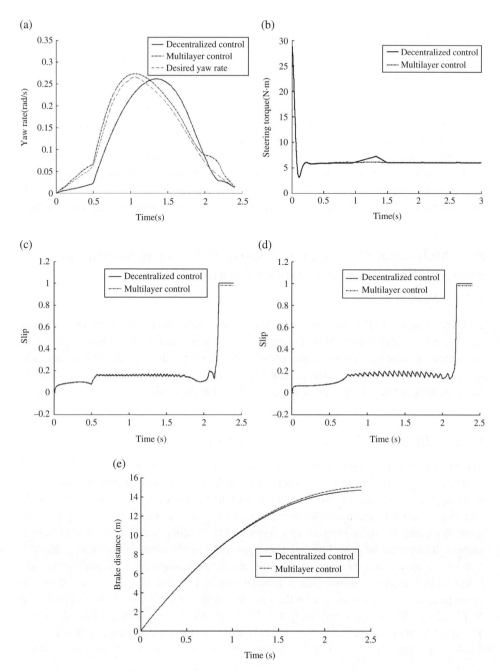

Figure 8.16 Simulation results. (a) Yaw rate. (b) Driver's steering torque. (c) Slip for the front wheel. (d) Slip for the rear wheel. (e) Brake distance.

In addition, it is interesting to investigate the braking performance for both cases. The brake distances for both cases are shown in Figure 8.16(e). We can see from the figure that the brake distance for the multilayer coordinating control is slightly longer than that for the decentralized control. The reason for the small loss on the braking performance is that the steering direction is one of the factors when adjusting the brake pressure of the diagonal brake wheel cylinders under the simultaneous steering and braking driving conditions. However, the loss on the brake distance for the multilayer coordinating control in this investigation is too small to be taken into account.

In summary, the multilayer coordinating controller is able to coordinate the interactions between the EPS system and ABS. The application of the multilayer coordinating control improves the overall vehicle performance under the critical driving condition: the vehicle lateral stability is improved, and the steering agility and the braking performance are ensured.

8.5   Multi-layer Coordinating Control of the Active Suspension System (ASS) and Vehicle Stability Control (VSC) System

8.5.1   System Model

In this section, the 7-DOF nonlinear vehicle dynamic model, developed in the previous section, is used to calculate the vehicle dynamics of the system. The Pacejka nonlinear tyre model is used to determine the dynamic forces of each tyre. Again, the linear 2-DOF reference model is used for designing the controllers and calculating the desired response to the driver's steering input. A filtered white noise signal is selected as the road excitation to the vehicle.

8.5.2   Multilayer Coordinating Controller Design

The architecture of the proposed multilayer coordinating control system is shown in Figure 8.17 [2,10]. The upper layer controller monitors the driver's intentions and the current vehicle states including the vehicle speed u_c, the steering angle of the front wheel δ_f, the sideslip angle β, the yaw rate r, and the lateral acceleration a_y. Based on these input signals, the upper layer controller computes the corrective yaw moment M_{zc} in order to track the desired vehicle motions. Thereafter, the upper layer controller generates the distributed torques M_{VSC} and M_{ASS} to the two lower layer controllers, i.e., the VSC and the ASS, respectively, according to a rule-based control strategy. Moreover, the distributed torques M_{VSC} and M_{ASS} are converted into the corresponding control commands for the two individual lower layer controllers. Finally, the VSC and the ASS execute respectively their local objectives to control the vehicle dynamics. The upper layer controller and the two lower layer controllers are designed as follows.

8.5.2.1   Upper Layer Controller Design

It is known that both the VSC and the ASS are able to develop corrective yaw moments (either directly or indirectly). To coordinate the interactions between the ASS and the VSC, a simple rule-based control strategy is proposed to design the upper layer

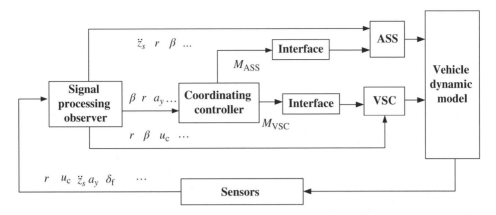

Figure 8.17 Block diagram of the multilayer coordinating control system. (Adapted from: *Integrated Control of Active Suspension System and Electronic Stability Program Using Hierarchical Control Strategy: Theory and Experiment*, by H. S. Xiao, W. W. Chen, H. H. Zhou, J. W. Zu. Reproduced with permission of the Taylor & Francis Group)

controller. The aim of the proposed control rule is to distribute the corrective yaw moment appropriately between the two lower layer controllers. The control rule is described as follows.

First, the corrective yaw moment M_{zc} is calculated by using the 2-DOF vehicle reference model based on the measured and estimated vehicle input signals.

Second, the braking/traction torque M_d and the pitch torque M_p are computed by using the following equations:

$$M_d = c_p \cdot p_w - 0.5 M_{zc} + I_w \cdot \dot{\omega}_w \tag{8.39}$$

$$M_p = \frac{k_\alpha \tan\alpha}{c_\lambda \lambda_w} M_d \tag{8.40}$$

where equation (8.39) is derived by considering the dynamics of one of the front wheels. It should be noted that, although a front-wheel-drive vehicle is assumed, the main conclusions of this study can be easily extended to vehicles with other driveline configurations. In general, the brake torque at each wheel is a function of the brake pressure p_w at that wheel, and c_p is an equivalent braking coefficient of the braking system, which is determined by using the equation $c_p = A_w \mu_b R_b$. The number "0.5" shows that the corrective yaw moment is evenly shared by the two front wheels.

Finally, the distributed torques M_{ESP} and M_{ASS} are generated by using a linear combination of the braking/traction torque M_d and the pitch torque M_p, which is given as:

$$\begin{cases} M_{ESP} = n_1 M_d + (1 - n_1) M_p \\ M_{ASS} = n_2 M_p + (1 - n_2) M_d \end{cases} \tag{8.41}$$

where n_1 and n_2 are the weighting coefficients, and $1 > n_1 > 0.5$, $1 > n_2 > 0.5$. Therefore, by tuning the weighting coefficients n_1 and n_2, the upper layer controller is able to coordinate with the two lower layer controllers and determine to what extent these are to be controlled.

8.5.2.2 Lower Layer Controller Design

8.5.2.2.1 ASS controller design

The LQG control method is used to control the active suspension system. The state variables are defined as $X = [z_s \ \dot{z}_s \ z_{u1} \ z_{u2} \ z_{u3} \ z_{u4} \ \dot{z}_{u1} \ \dot{z}_{u2} \ \dot{z}_{u3} \ \dot{z}_{u4} \ \theta \ \phi \ \dot{\theta} \ \dot{\phi}]^{\mathrm{T}}$; and the output variables are chosen as $[y = [\ddot{z}_s \ z_{u1} \ z_{u2} \ z_{u3} \ z_{u4} \ \theta \ \phi]^{\mathrm{T}}$. Therefore, the state equation and the output equation can be written as:

$$\begin{cases} \dot{X} = AX + BU \\ Y = CX + DU \end{cases} \tag{8.42}$$

where $U = [U_1 \ U_2]^{\mathrm{T}}$ is the control input vector, and $U_1 = [f_1 \ f_2 \ f_3 \ f_4]^{\mathrm{T}}$ is the control force vector, $U_2 = [z_{g1} \ z_{g2} \ z_{g3} \ z_{g4}]^{\mathrm{T}}$ is the road excitation vector. The multiple vehicle performance indices are considered to evaluate the vehicle handling stability, ride comfort, and safety. These performance indices can be measured by the following physical terms: vertical displacement of each wheel z_{u1}, z_{u2}, z_{u3}, z_{u4}; the suspension dynamic deflections $(z_{s1} - z_{u1})$, $(z_{s2} - z_{u2})$, $(z_{s3} - z_{u3})$, $(z_{s4} - z_{u4})$; the vertical acceleration of the sprung mass $\ddot{z}_s$; the pitch angular acceleration $\ddot{\theta}$; the roll angular acceleration $\ddot{\phi}$; and the control forces of the active suspension f_1, f_2, f_3, f_4. Therefore, the combined performance index is defined as:

$$\begin{aligned} J = \underset{T \to \infty}{Lim} \frac{1}{T} \int_0^T \Big[& q_1 z_{u1}^2 + q_2 z_{u2}^2 + q_3 z_{u3}^2 + q_4 z_{u4}^2 + q_5 \left(z_{s1} - z_{u1} \right)^2 \\ & + q_6 \left(z_{s2} - z_{u2} \right)^2 + q_7 \left(z_{s3} - z_{u3} \right)^2 + q_8 \left(z_{s4} - z_{u4} \right)^2 + q_9 \ddot{\theta}^2 \\ & + q_{10} \ddot{\phi}^2 + q_{11} \ddot{z}_s^2 + r_1 f_1^2 + r_2 f_2^2 + r_3 f_3^2 + r_4 f_4^2 \Big] dt \end{aligned} \tag{8.43}$$

where $q_1, \dots, q_{11}$, and $r_1, \dots, r_4$ are the weighting coefficients. The above equation can be rewritten as the following matrix form:

$$J = \underset{T \to \infty}{Lim} \frac{1}{T} \int_0^T \left(X^T Q X + U^T R U + 2 X^T N U \right) dt \tag{8.44}$$

where Q, R, N are the weighting matrices.

The state feedback gain matrix K is derived using the optimal control method, and it is the solution of the following *Riccati* equation:

$$K A + A^T K + Q - K B_1 R^{-1} B_1^T K + B_2 U_2 B_2^T = 0 \tag{8.45}$$

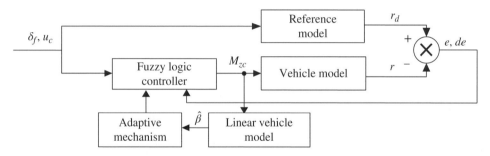

Figure 8.18 Block diagram of the adaptive fuzzy logic controller for the VSC.

8.5.2.2.2 *VSC controller design*

In this study, an adaptive fuzzy logic (AFL) method is applied to the design of the VSC controller [2]. The fuzzy logic controller (FLC) has been identified as an attractive control method in vehicle dynamics. This method has advantages when the following situations are encountered: (1) there is no explicit mathematical model that describes how the control outputs functionally depend on the control inputs; (2) there are experts who are able to incorporate their knowledge into the control decision-making process. However, traditional FLC with a fixed parameter setting cannot adapt to changes in the vehicle operating conditions or in the environment. Therefore, an adaptive mechanism must be introduced to adjust the controller parameters in order to achieve satisfactory vehicle performance in a wide range of changing conditions.

As shown in Figure 8.18, the AFL controller consists of a FLC and an adaptive mechanism. To design the AFL controller, the yaw rate and the sideslip angle of the vehicle are selected as the control objectives. The yaw rate can be measured by a gyroscope, but the sideslip angle cannot be measured directly and thus has to be estimated by an observer. The observer is designed by using the 2-DOF vehicle reference model. The linearized state space equation of the 2-DOF vehicle reference model is derived as follows, with the assumptions of having a constant forward speed and a small sideslip angle.

$$\begin{cases} \dot{X} = A_E \cdot X + B_E \cdot U \\ Y = C_E \cdot X + D_E \cdot U \end{cases} \tag{8.46}$$

where,

$$X = \begin{bmatrix} \beta \\ \omega_z \end{bmatrix}, U = \begin{bmatrix} \delta_f \\ M_{zc} \end{bmatrix}, A_E = \begin{bmatrix} -\dfrac{C_f + C_r}{mv} & -1 - \dfrac{aC_f - bC_r}{mv^2} \\ -\dfrac{aC_f - bC_r}{I_z} & -\dfrac{a^2C_f + b^2C_r}{I_z v} \end{bmatrix}, B_E = \begin{bmatrix} \dfrac{C_f}{mv} & 0 \\ \dfrac{aC_f}{I_z} & \dfrac{1}{I_z} \end{bmatrix},$$

$$C_E = \begin{bmatrix} 1 & 0 \\ 0 & 1 \end{bmatrix}, D_E = \begin{bmatrix} 0 & 0 \\ 0 & 0 \end{bmatrix}.$$

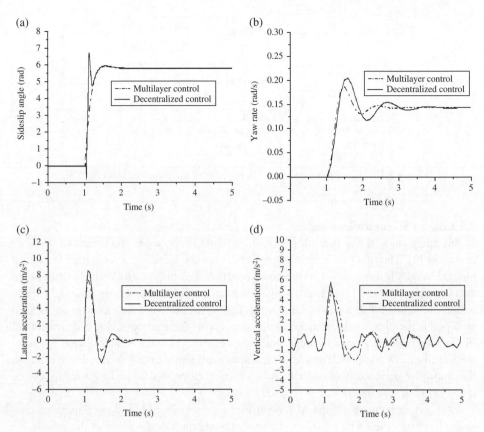

Figure 8.19 Comparison of responses for the step steering input maneuver. (a) Sideslip angle. (b) Yaw rate. (c) Lateral acceleration. (d) Vertical acceleration.

The aim of the AFL is to track both the desired yaw rate and the desired sideslip angle. The desired yaw rate can be calculated using equation (8.6). As shown in Figure 8.19, the FLC has two input variables: the tracking error of the yaw rate e and the difference of the error de. They are defined as follows, at the k-th sampling time:

$$e(k) = r(k) - r_d(k) \tag{8.47}$$

$$de(k) = e(k) - e(k-1) \tag{8.48}$$

The output variable of the FLC is defined as the corrective yaw moment M_{zc}. To determine the fuzzy controller output for the given error and its difference, the decision matrix of the linguistic control rules is designed and presented in Table 8.2. These rules are determined based on expert knowledge and the large number of simulation results performed in the study. In designing the FLC, the scaling factors k_e and k_{de} have great effects on the performance of the controller. Therefore, the adaptive mechanism is applied to adjust the

Table 8.2 Fuzzy rule bases for VSC control.

e \ de	PB	PM	PS	O	NS	NM	NB
PB	NB	NB	NB	NB	NM	O	O
PM	NB	NB	NB	NB	NM	O	O
PS	NM	NM	NM	NM	O	PS	PS
PO	NM	NM	NS	O	PS	PM	PM
NO	NM	NM	NS	O	PS	PM	PM
NS	NS	NS	O	PM	PM	PM	PM
NM	O	O	PM	PB	PB	PB	PB
NB	O	O	PM	PB	PB	PB	PB

parameters in order to achieve a satisfactory control performance when there are changes in the vehicle operating conditions or in the environment. The adaptive law is given as in reference [2].

$$\beta\left(k_e\right) = \beta_0 + k_e \int_0^t \left(\frac{a_y}{u_c} - r\right) dt \tag{8.49}$$

$$\dot{\beta}\left(k_{de}\right) = -r + k_{de} \frac{1}{u_c}\left(a_y \cos\beta - a_x \sin\beta\right) \tag{8.50}$$

where $\beta_0 = 0$.

8.5.3 Simulation Investigation

In order to evaluate the performance of the developed multilayer coordinating control system, a simulation investigation is performed. We assume that the vehicle travels at a constant speed of u_c = 90 km/h. Two driving conditions are performed: (1) step steering input, and (2) double lane change. For the first case, the vehicle is subjected to a steering input from the steering wheel and the steering input is set as a step signal with an amplitude of 120°. The road excitation is assumed to be independent for the four wheels.

After tuning the parameter setting for the multilayer coordinating control system, we select the set of weighting parameters for the ASS: $r_1 = r_2 = r_3 = r_4 = 1$, $q_1 = q_2 = q_3 = q_4 = 10^3$, $q_5 = q_6 = q_7 = q_8 = 10^4$, $q_9 = 2 \times 10^3$, $q_{10} = 10^5$, and $q_{11} = 10^6$. Moreover, the weighting parameters for the upper layer controller are selected as: $n_1 = 0.80$ and $n_2 = 0.85$.

The simulation results for the multiple performance indices are shown in Figure 8.19 and Figure 8.20 (for brevity, only some representative performance indices are presented here). For comparisons, the simulation investigation for the decentralized control is also performed. In the case, we simply eliminate the upper layer controller. The following discussion is made:

1. For the maneuver of step steering input, it can be seen that the peak value of the sideslip angle for the multilayer coordinating control, as shown in Figure 8.19(a), is reduced by

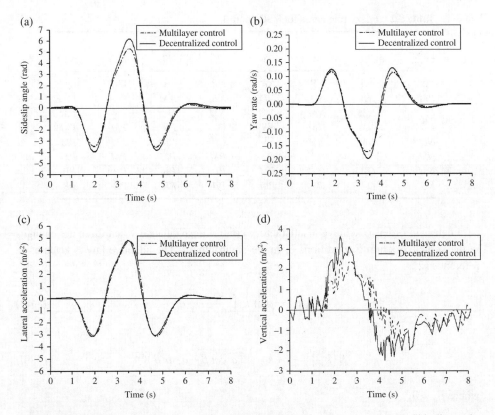

Figure 8.20 Comparison of responses for the double lane change maneuver. (a) Sideslip angle. (b) Yaw rate. (c) Lateral acceleration. (d) Vertical acceleration.

11.6% compared to that of the decentralized control. Moreover, the sideslip angle for the multilayer coordinating control is quickly damped and thus has less oscillation than that of the decentralized control. Similar patterns can be observed for the yaw rate and the lateral acceleration. The results indicate that the vehicle lateral stability is improved by the proposed multilayer coordinating control system compared to with the decentralized control system. In addition, the vertical acceleration of the spring mass, which is one of the ride comfort indices, is presented in Figure 8.19(d). It can be observed that the peak value of this performance index is decreased by 13.8% for the multilayer coordinating control compared to that of the decentralized control.

2. For the double lane change maneuver, it is observed that the peak value of the sideslip angle for the multilayer coordinating control is reduced by 15.3% compared to that of the decentralized control, as shown in Figure 8.20(a). Moreover, for the peak value of the yaw rate shown in Figure 8.20(b), the percentage of decrease is 7.9%. However, as shown in Figure 8.20(c), there is no significant difference on the lateral acceleration between the two control cases. However, for the vertical acceleration of the spring mass shown in Figure 8.20(d), it can be clearly seen that the peak value of this performance index for the multilayer coordinating control is reduced significantly by 30.5% compared

to that of the decentralized control. In addition, a quantitative analysis of the vertical acceleration shows that the R.M.S. (Root-Mean-Square) value of the vertical acceleration for the multi-layer coordinating control is reduced by 21.9% compared to that of the decentralized control.

In summary, the application of the multilayer coordinating control system improves the overall vehicle performance including the ride comfort and the lateral stability under critical driving conditions. The results show that the multilayer coordinating control system is able to coordinate the interactions between the ASS and the VSC and thus expand the functionalities of the two individual control systems.

8.6 Multilayer Coordinating Control of an Active Four-wheel Steering System (4WS) and Direct Yaw Moment Control System (DYC)

8.6.1 Introduction

As discussed in Chapter 5, the work principle of four-wheel steering (4WS) control is to simultaneously steer the rear wheels according to a specified relationship with the steering angles of the front wheels. Therefore, vehicle handling stability is improved by adjusting the tyre lateral forces. In general, the control strategy of the 4WS is designed based on the assumption that the tyre lateral force is proportional to the tyre steering angle. The assumption is valid only when the lateral acceleration is relatively small. However, when the lateral acceleration increases, the relationship among the tyre forces in the three directions and the steering angle of the wheels become nonlinear. This driving condition leads to difficulties in designing an appropriate control law to maintain vehicle stability.

There are quite a few control methods applied to the design of the 4WS, including adaptive control, robust control, and neural network. Adaptive control is able to achieve a great control performance when the lateral acceleration is relatively small. However, with adaptive control it is difficult to identify the real-time vehicle response parameters since the required steering angle is very small when a large lateral acceleration is reached. Hence, it is difficult to design an adaptive control system and maintain system stability. The application of the robust control method to the 4WS controller design ultimately results in solving linear matrix inequalities. The nonlinear error resulting from the decrease of the tyre lateral stiffness cannot be tolerated, and hence leads to system instability. In practice, this problem occurs under critical driving conditions including lane change (or cornering) combined with acceleration or deceleration. To overcome these difficulties, a control method based on neural network theory was proposed. The neural network-based control method is able to solve successfully the nonlinear problem caused by large tyre sideslip, and also adapt to vehicle parameter variations by tuning online the neural network coefficients.

At present, the application of the direct yaw moment control (DYC) exploits the limitations of the 4WS control systems. The DYC is able to handle effectively the saturation of the tyre lateral force and hence improve the vehicle lateral stability. In general, the yaw rate is selected as the control objective to adjust the tyre longitudinal forces since the yaw rate is more convenient to measure compared with the sideslip angle. However, it is more effective to maintain the vehicle trajectory through regulating directly the lateral tyre forces and therefore adjusting the sideslip angle. Therefore, the integrated control is able to improve significantly the vehicle lateral stability by combining the regulations of both the tyre lateral forces and longitudinal forces.

As discussed in Chapter 7, the keys to designing the integrated control system of the 4WS and DYC include: (1) coordinating the interactions between the two systems, and (2) exploiting the potentials of each subsystems. In this chapter, the integrated control of the 4WS and DYC regulates both the sideslip angle and the yaw rate to enhance overall vehicle performance over a large range of the lateral acceleration[11,12].

8.6.2   Coordinating Control of DYC and 4WS

8.6.2.1   Design of a Coordinating Control System

In this section the upper layer coordinating controller and two subsystem controllers for the DYC and 4WS are designed separately. In designing the 4WS controller, the sideslip angle is selected as the control objective since the 4WS adjusts directly the steering angle of the rear wheels. To simplify the design of the 4WS controller, the improvement of the steering sensitivity and lateral displacement are not taken into account. In addition, this simplification also results in a model with fewer degrees of freedom. Moreover, the yaw rate is selected as the control objective in designing the DYC controller. As illustrated in Figure 8.21, the upper layer coordinating controller monitors the driver's intentions and the current vehicle states. Based on these input signals, the upper layer controller is designed to coordinate the interactions between the 4WS and DYC in order to achieve the desired vehicle states. The expected adjustment of the tyre longitudinal forces and steering angles of the real wheels are calculated. Thereafter, the control commands are generated by the upper layer controller and distributed to the two lower layer controllers respectively. Finally, the individual lower layer controllers execute respectively their local control objectives to control the vehicle dynamics. A rule-based control method is proposed to design the upper layer controller, and it is described as follows.

The expected yaw rate r_d is an important performance index for the DYC system, which is defined as[12]:

$$|r_d| = \min\left\{\frac{u_c}{l\left(1 + Ku_c^2\right)}\delta_f, \left|\frac{\mu g}{u_c}\right|\right\}$$

(8.51)

where μ is the road adhesion coefficient; u_c is the vehicle speed; g is the gravitational acceleration; l is the wheel base; K is the vehicle stability factor; and δ_f is the steering angle of the front wheel. It is known from equation (8.51) that the road adhesion coefficient and vehicle speed are vital parameters to determine the expected yaw rate.

The control objective of the 4WS controller is to make the sideslip angle close to 0. Therefore, the feedback control law of the 4WS controller derived from the linear 2-DOF vehicle reference dynamic model is given as:

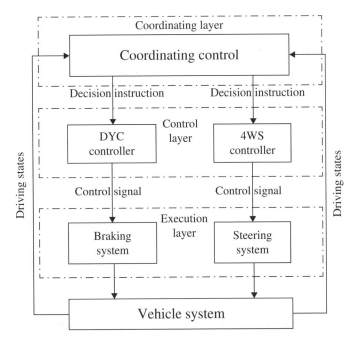

Figure 8.21   Coordinating system control configuration of the DYC and 4WS.

$$\delta_r = -c_1\delta_f + c_2u_cr \tag{8.52}$$

where $c_1 = 1$; $c_2 = \dfrac{mb}{k_Fl} + \dfrac{ma}{k_Rl}$; δ_f and δ_r are the steering angles of the front and rear wheels, respectively; r is vehicle actual yaw rate; m is the vehicle mass; a and b are the horizontal distance between the center of gravity of the vehicle and the front and rear axles, respectively; k_F and k_R are the cornering stiffness of the front and rear wheels, respectively. It is known from equation (8.52) that the vehicle speed and the yaw rate are important parameters to determine the control output of the 4WS controller.

It is clear from the above discussion that the yaw rate is the major parameter affecting both subsystems, and it is not only the control objective of the DYC system, but also the important parameters determining the control output of the 4WS controller. In addition, both the road adhesion coefficient and vehicle speed determine the control outputs of the two subsytems. Therefore, the control rules for designing the upper layer coordinating controller are based on three factors, i.e., the yaw rate r, the road adhesion coefficient μ, and the vehicle speed u_c.

It is known that the 4WS system is designed to have a specified speed u_a. When the vehicle speed is lower than u_a, the front and rear wheels steer in the opposite direction to reduce the vehicle steering radius and also improve steerability at low speeds. When the vehicle speed is higher than u_a, the front and rear wheels steer in the same direction at a relatively small angle range to improve the vehicle transient steering characteristics, and also reduce the roll angle to a certain extent. Therefore, the control rules of the upper layer controller are developed according to the above two working conditions of the 4WS system. The detailed control rules are described as follows.

1. When the vehicle drives straight or the steering angle is relatively small, the upper layer controller performs a supervision function, and no control command is sent out. In the meantime the DYC and 4WS systems are idle.
2. When cornering:
    a. if $u_c \le u_a$, the upper layer controller performs the supervision function, and no control command is sent out. In the meantime the DYC system is idle; and the 4WS performs its local control objective to improve steerability at low speeds.
    b. if $u_c > u_a$, two thresholds of the yaw rate $\nabla \omega_1$ and $\nabla \omega_2$ are determined to coordinate the two subsystems in different work domains. The two thresholds of the yaw rate $\nabla \omega_1$ and $\nabla \omega_2$ are both positive, and $\nabla \omega_2 > \nabla \omega_1$.

If $|r| \le \left| \dfrac{\mu g}{u_c} \right|$, and $\left\| |r| - |r_d| \right\| > \nabla \omega_1$, the upper layer controller generates the control commands and distributes those to the two subsystem controllers. The DYC controller works together with the 4WS controller and improves the steering sensitivity of the 4WS system. In the meantime, the 4WS controller performs its local control objective to improve the vehicle stability.

If $|r| > \left| \dfrac{\mu g}{u_c} \right|$ and $\nabla \omega_2 \ge \left\| |r| - |r_d| \right\| > \nabla \omega_1$, the upper layer controller generates the control commands and distributes those to the two subsystem controllers. The DYC controller works together with the 4WS controller to improve the steering sensitivity of the 4WS system. In the meantime, the 4WS controller performs its local control objective to improve the vehicle stability.

If $|r| > \left| \dfrac{\mu g}{u_c} \right|$ and $\left\| |r| - |r_d| \right\| > \nabla \omega_2$, the upper layer controller generates the control commands and distributes those to the two subsystem controllers. The DYC controller improves the vehicle stability and, in the meantime, the 4WS system is idle.

Else, the upper layer controller performs the supervision function, and no control command is sent out. In the meantime the DYC system is idle; and the 4WS performs its local control objective to improve the vehicle stability.

In conclusion, the function allocation of the upper layer controller is based on the above rules. The upper layer controller generates the control command of improving steering sensitivity of the 4WS and distributes it to the DYC system. The DYC system works together with the 4WS and tracks the expected steady-state yaw rate generated by the linear 2-DOF vehicle dynamic reference model. The work domains of the two subsystems are divided by the two thresholds of the yaw rate and the upper layer controller coordinates and performs the function allocation of the two subsystems.

8.6.2.2 Lower Level Controller Design

1. *DYC controller design*

   The PID control method is applied to the design of the DYC controller to regulate the braking forces of the wheels. The calculation process of the PID control law is shown in Figure 8.22.

   For simplification, the coordination between the ABS and the DYC is based on the slip ratio control approach. A target slip ratio of 0.2 is selected. When the slip ratio reaches or exceeds the target value, the ABS sends out the signal to reduce the braking pressure of the hydraulic cylinder in the brake system. If the tyre slip ratio is less than the target value, the proposed PID-based DYC controller works in this area.

2. *4WS controller design*

   The neural network theory is applied to the design of the 4WS controller. The neural network-based control method is able to overcome the nonlinear problem caused by large tyre sideslip, and in the meantime adapt to vehicle parameter variations through tuning online the neural network coefficients. The detailed development of the control method can be found in Section 5.4.

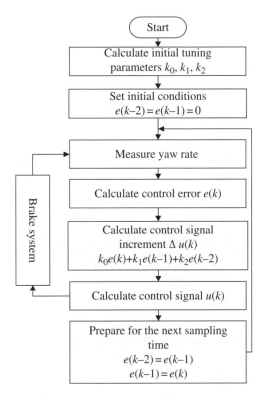

Figure 8.22   Flowchart of the PID control algorithm.

8.6.3   Simulation Investigation

To demonstrate the performance of the proposed multilayer coordinating control system, a simulation investigation is performed by combining ADAMS and MATLAB for the dynamic analysis and control of the integrated control system. In the simulation, the vehicle speed is set as 80 km/h, the road adhesion coefficient is selected as 0.8, and the steering input to the steering wheel is a sinusoidal signal. The simulation results are shown in Figure 8.23 and a quantitative analysis of the results is presented in Table 8.3. In the figures, 2WS denotes that only the front steering system is used, and 4WS represents that only the 4WS system is used; 4WS + DYC represents that the proposed multilayer coordinating control system is applied.

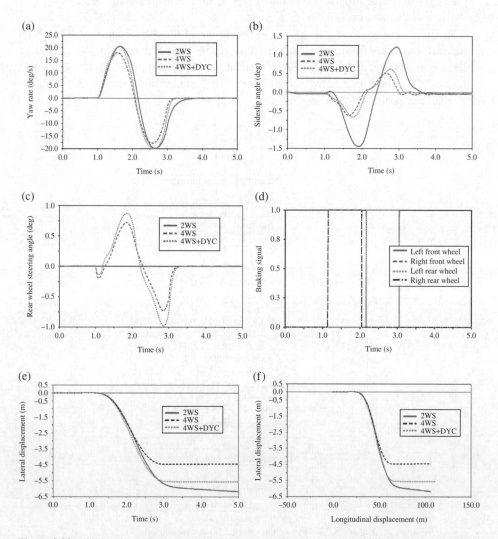

Figure 8.23   Comparison of responses. (a) Yaw rate. (b) Sideslip angle. (c) Steering angle of the rear wheel. (d) Braking signal on each wheel. (e) Lateral displacement. (f) Traveling trajectory.

Table 8.3   Comparison of the peak value of responses.

Peak value of performance index	2WS	4WS	4WS + DYC
Yaw rate (deg/s)	20.55	18.03	20.51
Sideslip angle (deg)	1.21	0.52	0.63
Lateral displacement (m)	6.17	4.47	5.56

It is shown in Figure 8.23(a) and Table 8.3 that the peak value of the yaw rate reaches 20.51 deg/s for the 4WS+DYC integrated control system. The value is closer to 20.55 deg/s for the 2WS system than 18.03 deg/s for the 4WS system. The results indicate that the proposed multilayer coordinating control system is able to track the expected yaw rate, and the steering sensitivity of the 4WS is improved. In addition, the settling time of the yaw rate for the multilayer coordinating control system is also reduced, compared with those for the other two systems. A similar pattern can also be observed for the sideslip angle.

It can be further concluded from Figure 8.23(e and f) that the lateral diaplacement response and traveling trajectory of the multilayer coordinating control system are close to those of the 2WS system. The results indicate that the multilayer coordinating control system is able to regulate the vehicle motions more accurately. The margin of the tyre lateral force is enlarged and, hence, the anti-sideslip ability is improved. While for the 4WS system, the deviations of the lateral displacement and traveling trajectory for the 2WS are significant even though the anti-sideslip ability of the 4WS is better. Moreover, both the lateral displacement and traveling trajectory of the 2WS demonstrate the phenomenon of sideslip to a certain extent.

In summary, the proposed multilayer coordinating control system is able to maintain the advantages of the 2WS, and also overcome its disadvantages through the coordination and function allocation of the 4WS and DYC systems. As a result, the vehicle stability is enhanced significantly.

8.7   Multilayer Coordinating Control of Integrated Chassis Control Systems

8.7.1   Introduction

As studied earlier in this chapter, the multilayer coordinating control method was applied to the coordination and function allocation of different combinations of two subsystem controllers, including ASS, DYC, ABS, EPS, AFS, and 4WS. However, it is known that the modern vehicle may be simultaneously equipped with more subsystem controllers on the steering, braking/traction, and suspension systems. Therefore, it is interesting to investigate the multilayer coordinating control of the overall vehicle dynamics in the three directions – lateral, longitudinal, and vertical – in order to enhance the overall vehicle performance including handling stability, ride comfort, acceleration and braking performance[13]. This is the aim of this section.

8.7.2   Controller Design

The architecture of the integrated control of chassis systems is illustrated in Figure 8.24. As analyzed in Chapter 6, the coupling effects among the three subsystems of steering, braking, and suspension are also presented in the figure. Based on the architecture, the structure of the multilayer coordinating control system is developed in Figure 8.25. Similarly, the upper layer controller monitors the driver's intentions and the current vehicle states. Based on these input signals, the upper layer controller is designed to coordinate the interactions

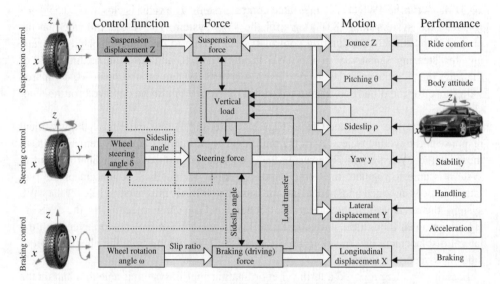

Figure 8.24   Architecture of the integrated chassis control system.

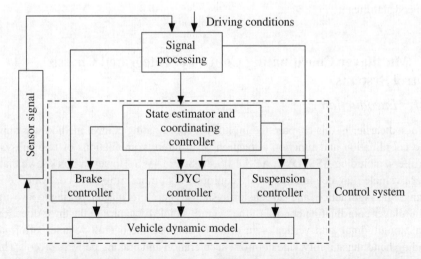

Figure 8.25   Block diagram of the multilayer coordinating control system.

amongst the three subsystems by identifying the driving conditions of the vehicle in order to achieve the desired vehicle states. Thereafter, the control commands are generated by the upper layer controller according to the developed control rules for a certain driving condition of the vehicle. Then, the control commands are distributed to the three lower layer controllers respectively. Finally, the individual lower layer controllers execute respectively their local control objectives to control the overall vehicle dynamics. The three low layer controllers and the upper layer coordinating controller are developed as follows. It is noted that the detailed development of the vehicle dynamic model can be found in reference[13].

8.7.2.1   Design of a Lower Level Controller

As shown in Figure 8.26, the PID control method is applied to the design of the controller of the suspension system. The PID controller is developed to adapt to the different driving conditions of the vehicle. When the vehicle drives in a normal condition, i.e., no emergency braking or cornering (the steering angle of the front wheel is relatively large) is applied, the vertical acceleration $\ddot{z}_{si}$ of the i-th suspension is selected as the control input. It is assumed at the k-th sampling time, the controller output is $U(k)$ and the corresponding vertical acceleration at that time is $\ddot{z}_{si}(k)$. The PID control law for the i-th suspension is given as:

$$U(k) = K_{pi}\left[\ddot{z}_{si}(k) - \ddot{z}_{si}(k-1)\right] + K_{ii}\ddot{z}_{si}(k) + K_{di}\left[\ddot{z}_{si}(k) - 2\ddot{z}_{si}(k-1) + \ddot{z}_{si}(k-2)\right] \quad (8.53)$$

where K_{pi}, K_{ii}, and K_{di} ($i = 1, 2, 3, 4$) are the proportional, differential, and integral coefficients, respectively. When the vehicle drives in the condition of either using emergency brakes or cornering, the aim of the PID controller is to regulate the pitch or the roll motions, and only the proportional coefficient K_{pi} is tuned while the other two are kept the same. Specifically, if regulating the pitch motion, the control inputs to the PID controller include the pitch angular velocity $\dot{\theta}$ and its difference $\ddot{\theta}$, and the outputs are the weighting parameters of the PID controller K_{pi}, K_{ii}, and K_{di} (the parameters of the two front suspensions 1, 2 and the two rear suspensions 3, 4 are tuned in pairs). If regulating the roll motion, the control inputs to the PID controller include the roll rate $\dot{\varphi}$ and its difference $\ddot{\varphi}$. The outputs are the weighting parameters of the PID controller K_{pi}, K_{ii}, and K_{di} (the parameters of the two left suspensions 1,3 and two right suspensions 2,4 are tuned in pairs). It is noted that the

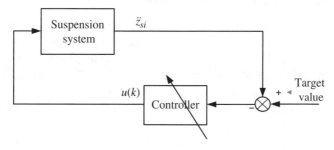

Figure 8.26   Block diagram of the suspension PID control system.

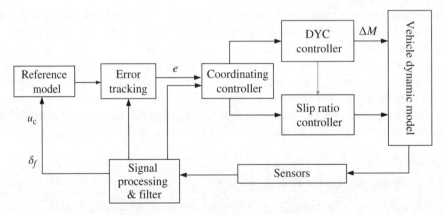

Figure 8.27   Block diagram of the DYC system.

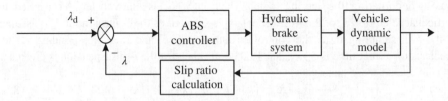

Figure 8.28   ABS control principle.

driving conditions are determined by the upper layer controller and also the three weight-ing parameters of the PID controller are assigned by the upper layer controller.

The sliding mode variable structure control method is applied to the design of the DYC controller. The structure of the DYC system is illustrated in Figure 8.27 and the detailed development of the controller can be found in Section 8.2.

The slip ratio control is applied to the design of the ABS, and the work principle of the slip ratio control is presented in Figure 8.28. The detailed development of the ABS controller refers to Section 3.3.

8.7.2.2   Design of the Upper Layer Controller

A rule-based control method is proposed to design the upper layer controller to coordinate the interactions among the three subsystems by identifying the driving conditions of the vehicle. The driving conditions of the vehicle can be divided into four types: (1) normal condition, i.e., no emergency brakes or cornering are applied; (2) normal condition and then emergency brakes; (3) normal condition and then sudden cornering; and (4) braking in turn. The different driving conditions are identified by calculating the measured signals from the sensors, including the vehicle longitudinal speed u_c, lateral speed v_c, longitudinal acceleration $\dot{u}_c$, lateral acceleration $\dot{v}_c$, pitch rate $\dot{\theta}$, roll rate $\dot{\varphi}$, and steering angle of the hand wheel. The control rules of the upper layer coordinating controller are designed as follows.

Driving condition 1: The upper layer controller generates the control commands and distributes those to the three subsystem controllers. Both the ABS and DYC system are idle. The weighting parameters for the PID controller of the ASS are assigned according to the normal driving condition developed in the above section. These weighting parameters are kept unchanged throughout driving condition 1.

Driving condition 2: In this case, ABS works according to the tyre slip ratio control method and while the DYC system is idle. The set of the weighting parameters for the ASS PID controller are assigned and adjusted according to the driving condition concerning the pitch motions. The specific adjustment strategy for the weighting parameters is shown as follows by assuming that the counter clockwise direction around the y-axis is positive.

If $\dot{\theta} > \alpha_1$ and $\ddot{\theta} > \alpha_2$, then $K_{p1} = K_{p2} = \sigma_1$, and $K_{p3} = K_{p4} = \sigma_2$;

If $-\alpha_1 < \dot{\theta} < \alpha_1$ and $\ddot{\theta} > \alpha_2$, then $K_{p1} = K_{p2} = \sigma_3$, and $K_{p3} = K_{p4} = \sigma_4$;

If $\dot{\theta} < -\alpha_1$ and $\ddot{\theta} > \alpha_2$, then $K_{p1} = K_{p2} = \sigma_5$, and $K_{p3} = K_{p4} = \sigma_6$;

If $\dot{\theta} > \alpha_1$ and $-\alpha_2 < \ddot{\theta} > \alpha_2$, then $K_{p1} = K_{p2} = \sigma_7$ and $K_{p3} = K_{p4} = \sigma_8$;

If $-\alpha_1 < \dot{\theta} < \alpha_1$ and $-\alpha_2 < \ddot{\theta} > \alpha_2$, then $K_{p1} = K_{p2} = \sigma_9$, and $K_{p3} = K_{p4} = \sigma_{10}$;

If $\dot{\theta} < -\alpha_1$ and $-\alpha_2 < \ddot{\theta} > \alpha_2$, then $K_{p1} = K_{p2} = \sigma_{11}$, and $K_{p3} = K_{p4} = \sigma_{12}$;

If $\dot{\theta} > \alpha_1$ and $\ddot{\theta} < -\alpha_2$, then $K_{p1} = K_{p2} = \sigma_{13}$, and $K_{p3} = K_{p4} = \sigma_{14}$;

If $-\alpha_1 < \dot{\theta} < \alpha_1$ and $\ddot{\theta} < -\alpha_2$, then $K_{p1} = K_{p2} = \sigma_{15}$, and $K_{p3} = K_{p4} = \sigma_{16}$;

If $\dot{\theta} < -\alpha_1$ and $\ddot{\theta} < -\alpha_2$, then $K_{p1} = K_{p2} = \sigma_{17}$, and $K_{p3} = K_{p4} = \sigma_{18}$.

In addition, the values of the two weighting parameters K_{ii} and K_{di} are adjusted according to the adjustment chart presented in Table 8.4.

In the rule, the parameters α_i and σ_i are obtained by tuning using enormous numbers of simulations, and in practice those can be calibrated in vehicle road tests. The parameters in the driving condition 3 are obtained by the same way.

Driving condition 3: In this case, the ABS is idle, and the DYC system generates the appropriate additional yaw moment to track the expected yaw rate. The set of the weighting parameters for the ASS PID controller are assigned and adjusted according to the driving condition

Table 8.4 Adjustment chart of the weighting parameters of the PID controller when pitching.

	$\ddot{\theta}$ (pitch motion)		
	$\ddot{\theta} < -\alpha_2$	$-\alpha_2 < \ddot{\theta} < \alpha_2$	$\ddot{\theta} > \alpha_2$
K_{i1}^{θ} and K_{i2}^{θ}	1.00	0.80	1.12
K_{i3}^{θ} and K_{i4}^{θ}	1.06	0.80	1.00
K_{d1}^{θ} and K_{d2}^{θ}	3.18	3.00	3.14
K_{d3}^{θ} and K_{d4}^{θ}	3.16	3.00	3.24

Table 8.5 Adjustment chart of the weighting parameters of the PID controller when rolling.

	$\ddot{\varphi}$ (Roll motion)		
	$\ddot{\varphi} < -\beta_2$	$-\beta_2 < \ddot{\varphi} < \beta_2$	$\ddot{\varphi} > \beta_2$
K_{i1}^{ϕ} and K_{i3}^{ϕ}	1.06	0.8	1.10
K_{i2}^{ϕ} and K_{i4}^{ϕ}	1.16	0.8	1.04
K_{d1}^{ϕ} and K_{d3}^{ϕ}	3.10	3.0	3.12
K_{d3}^{ϕ} and K_{d4}^{ϕ}	3.18	3.0	3.10

concerning the roll motion. The specific adjustment strategy for the weighting parameters is shown as follows assuming that the counter clockwise direction around x axis is positive.

If $\dot{\varphi} > \beta_1$ and $\ddot{\varphi} > \beta_2$, then $K_{p1} = K_{p3} = \lambda_1$, and $K_{p2} = K_{p4} = \lambda_2$;

If $-\beta_1 < \dot{\varphi} < \beta_1$ and $\ddot{\varphi} > \beta_2$, then $K_{p1} = K_{p3} = \lambda_3$, and $K_{p2} = K_{p4} = \lambda_4$;

If $\dot{\varphi} < -\beta_1$ and $\ddot{\varphi} > \beta_2$, then $K_{p1} = K_{p3} = \lambda_5$, and $K_{p2} = K_{p4} = \lambda_6$;

If $\dot{\varphi} > \beta_1$ and $\beta_2 < \ddot{\varphi} < \beta_2$, then $K_{p1} = K_{p3} = \lambda_7$, and $K_{p2} = K_{p4} = \lambda_8$;

If $-\beta_1 < \dot{\varphi} < \beta_1$ and $\beta_2 < \ddot{\varphi} < \beta_2$, then $K_{p1} = K_{p3} = \lambda_9$, and $K_{p2} = K_{p4} = \lambda_{10}$;

If $\dot{\varphi} < -\beta_1$ and $\beta_2 < \ddot{\varphi} < \beta_2$, then $K_{p1} = K_{p3} = \lambda_{11}$, and $K_{p2} = K_{p4} = \lambda_{12}$;

If $\dot{\varphi} > \beta_1$ and $\ddot{\varphi} < -\beta_2$, then $K_{p1} = K_{p3} = \lambda_{13}$, and $K_{p2} = K_{p4} = \lambda_{14}$;

If $-\beta_1 < \dot{\varphi} < \beta_1$ and $\ddot{\varphi} < -\beta_2$, then $K_{p1} = K_{p3} = \lambda_{15}$, and $K_{p2} = K_{p4} = \lambda_{16}$;

If $\dot{\varphi} < -\beta_1$ and $\ddot{\varphi} < -\beta_2$, then $K_{p1} = K_{p3} = \lambda_{17}$, and $K_{p2} = K_{p4} = \lambda_{18}$.

In addition, the values of the other two weighting parameters K_{ii} and K_{di} are adjusted according to the adjustment chart presented in Table 8.5.

Driving condition 4: In this case, both the ABS and the DYC system execute their functions based on the inputs of the measured signals. The set of weighting parameters for the ASS PID controller are assigned and adjusted according to the driving condition concerning the coupled roll and pitch motions. The specific adjustment strategy for the weighting parameters is shown as follows.

A weighting factor ε is defined in order to take into account the coupling effect of the roll and pitch motions during this driving condition. If $|\ddot{\theta} > \varepsilon_0|$, and $|\ddot{\varphi} > \gamma_0|$, the coupling effect is identified, and the proportional weighting parameter of the i-th suspension is regulated as:

$$K_{pi} = \varepsilon K_{pi}^{\theta} + (1 - \varepsilon) K_{pi}^{\varphi} \tag{8.54}$$

where ε_0 and γ_0 are the thresholds of the pitch and roll angular accelerations respectively, and their values are adjusted through tuning by simulations; K_{pi}^{θ} and K_{pi}^{φ} are the proportional weighting parameters of the i-th suspension obtained from the parameter assignment defined in the above pitch and roll driving conditions, respectively; and the value of ε is adjusted in real time by considering the values of $\ddot{\theta}$ and $\ddot{\varphi}$.

Otherwise, if the coupling effect of the roll and pitch motions are not identified, the maximum value of the pitch and roll angular accelerations, i.e., $\max\{\ddot{\theta}\quad\ddot{\phi}\}$, is considered to identify which driving condition the vehicle performs on and, hence, the ASS is adjusted accordingly based on the above rules. In addition, the integral and differential weighting parameters of the ASS controller are assigned as follows[2].

$$\text{If } |\ddot{\varphi}| > \beta_2 \text{ and } |\ddot{\theta}| > \alpha_2, \text{ then } K_i = 1.0, K_d = 3.16;$$
$$\text{else } K_i = 0.8, \text{ and } K_d = 3.0.$$

8.7.3   Simulation and Experiment Investigations

8.7.3.1   Simulation Investigation

To demonstrate the performance of the proposed multi-layer integrated control system, a simulation investigation is performed by considering the four different driving conditions. The vehicle speed is set to 70km/h, and the physical parameters used in the simulation can be found in Table 7-5. The following discussion is made on the simulation results.

Driving condition 1: In this normal driving condition, the vehicle ride comfort is the performance index of most concern. It is observed in Figure 8.29(a) that the vehicle vertical acceleration for the multilayer integrated control system is improved greatly compared with that of the passive suspension.

Driving condition 2: For the step steering input maneuver the vehicle lateral stability is the performance index of most concern. It is observed in Figure 8.29(b) that the yaw rate for the multilayer integrated control system is better at tracking the expected yaw rate compared with that of the standalone DYC. In addition, it is obvious in Figure 8.29(c) that the roll angular acceleration for the multilayer integrated control system is reduced compared with that of the passive suspension, and the standalone ASS.

Driving condition 3: For the emergency brake maneuver, it is shown in Figure 8.29(d) that the brake distance for the multilayer integrated control system is reduced compared with that of the standalone ABS. In addition, it is shown in Figure 8.29(e) that the pitch angular acceleration for the multilayer integrated control system is also reduced compared with that of the standalone ASS based on the ordinary PID control and the passive suspension.

Driving condition 4: For the braking in turn maneuver, it is observed in Figure 8.29 (f)–(h) that the RMS of the vehicle vertical acceleration for the multilayer integrated control system is reduced by 3.1% and 25.8%, compared with that of the standalone ASS based on the ordinary PID control and the passive suspension, respectively. Moreover, the peak value of the yaw rate for the multilayer integrated control system is also reduced compared with that of the standalone DYC system. However, the brake distance for the multilayer integrated control system is increased compared with the standalone ABS system because of the intervention of the DYC system.

In summary, the application of the proposed multilayer coordinating control system is able to improve the overall vehicle dynamics in the lateral, longitudinal, and vertical directions and, hence, enhance the overall vehicle performance including handling stability, ride comfort, and braking performance.

8.7.3.2 Experiment Investigation

To validate the effectiveness of the control strategy, a vehicle road test is conducted. The configuration of the experiment setting is shown in Figure 8.30. The upper layer controller monitors the driver's intentions and the current vehicle states measured by the sensors, including the speed, the steering angle of the front wheel, the yaw rate, the vertical, lateral, and longitudinal accelerations, and the pitch and roll angular accelerations. Based on these input signals, the upper layer controller generates control commands according to the above rule-based control method in order to control the overall vehicle dynamics. Then the control commands are distributed to the three lower

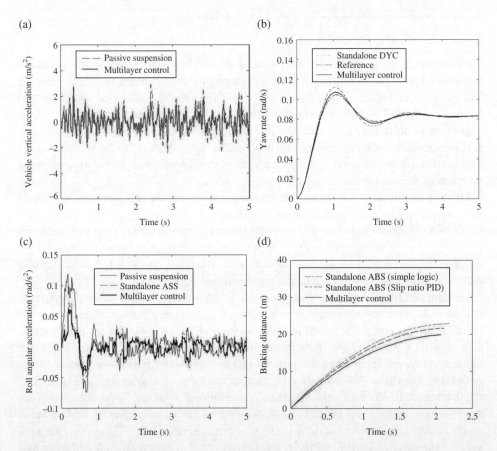

Figure 8.29 Comparison of responses for the four different driving conditions. (a) Vehicle vertical acceleration (normal driving). (b) Yaw rate (step steering). (c) Roll angular acceleration (step steering). (d) Braking distance comparison (emergency brakes). (e) Pitch angular acceleration (emergency brake). (f) Vertical acceleration (braking in turn). (g) Yaw rate (braking in turn). (h) Braking distance (braking in turn).

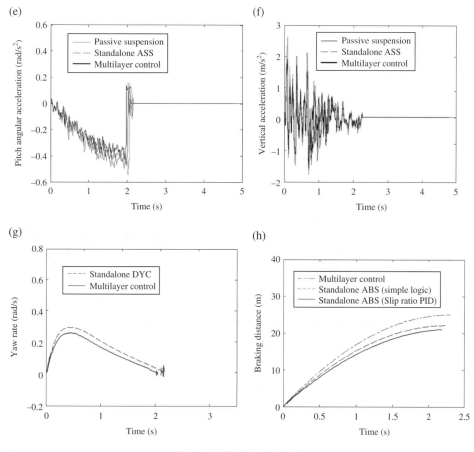

Figure 8.29   (Continued)

layer controllers respectively. Finally, the individual lower layer controllers execute respectively their local control objectives.

The experiment is performed by considering four different driving conditions. The following discussion is made on the simulation results. For driving condition 1, i.e., the normal driving condition, the RMS value of the vehicle vertical acceleration for the multilayer integrated control system is reduced from 1.02m/s^2 to 0.96m/s^2 compared with the passive suspension. For the emergency brake maneuver, the pitch angular acceleration of the multilayer integrated control system is also reduced from 0.41 rad/s^2 to 0.34 rad/s^2 compared with the standalone ASS based on the ordinary PID control. For the step steering input maneuver, the RMS value of the roll angular acceleration for the multilayer integrated control system is reduced from 0.034 rad/s^2 to 0.029 rad/s^2 compared with the standalone ASS. For the braking in turn maneuver, the peak value of the yaw rate for the multilayer integrated control system is also greatly reduced compared with that for the standalone DYC system. However, the brake distance for the multilayer integrated control system is increased compared with the standalone ABS system because of the intervention of the

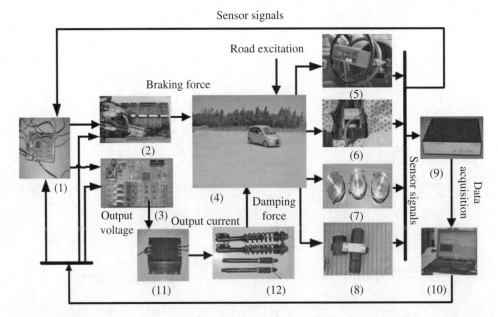

Figure 8.30 Configuration of the vehicle coordinating control test. (1) Upper layer controller. (2) Solenoid valve drive. (3) Suspension controller. (4) Test vehicle. (5) Steering wheel angle sensor. (6) Gyroscope. (7) Acceleration sensor. (8) Vehicle speed sensor. (9) Data acquisition. (10) PC. (11) Power source with constant current. (12) Magneto rheological damper.

DYC system. In summary, the experiment results demonstrate a good agreement with the simulation results. Therefore, the application of the proposed multilayer coordinating control system is effective in enhancing the overall vehicle performance including handling stability, ride comfort, and braking performance.

8.8 Multilayer Coordinating Control of Integrated Chassis Control Systems using Game Theory and Function Distribution Methods

Due to the large number of chassis subsystems, it is hard for simple control systems to guarantee the system's overall optimal control performance. When using hierarchy structures, however, the function of each level is clarified, i.e., the subcontrol system local performance is guaranteed by lower level controllers. The overall optimal performance of integrated control systems is guaranteed by upper level controllers. However, the functions of the chassis subsystems are different, for instance, the suspension subsystem can improve vehicle ride comfort and handling stability, and the braking subsystem can enhance the vehicle safety. To improve the chassis control system optimal performance, the function of each subsystem needs to be considered. An upper level controller needs to be designed to coordinate each subsystem and reach the optimal state of the chassis system overall control performance[14].

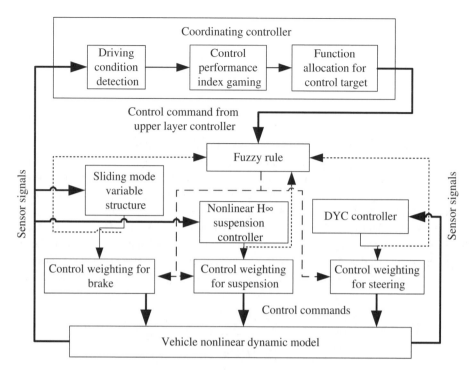

Figure 8.31   Block diagram of the multilayer coordinating control system.

8.8.1   Structure of the Chassis Control System

As discussed in the previous section, the control strategy of the upper layer controller is applied to coordinate the interactions of the three subsystem controllers, and then fulfill the function distributions of the three controllers in order to improve the overall vehicle dynamics. In this section, game theory is applied to coordinate the conflicts amongst the different performance indices in order to determine the appropriate overall performance indices. In addition, the fuzzy logic control method is applied to the real-time self-tuning of the weighting parameters of the outputs of the subsystems to reduce the tracking errors between the actual control objectives of the lower layer controllers and the expected control objectives of the upper layer controller. The structure of the proposed multilayer coordinating control system is illustrated in Figure 8.31.

8.8.2   Design of the Suspension Subsystem Controller

The H_∞ control method is applied to the design of the controller of the suspension system in order to suppress the input disturbances and hence ensure the stability and robustness of the close loop control system of the suspension. A nonlinear state equation of the suspension system is established as:

$$\dot{X} = A(X) + B_1 W + B_2 U \tag{8.55}$$

where $X = [\varphi \; \dot{\varphi} \; \theta \; \dot{\theta} \; r \; z_s \; \dot{z}_s \; z_{u1} \; \dot{z}_{u1} \; z_{u2} \; \dot{z}_{u2} \; z_{u3} \; \dot{z}_{u3} \; z_{u4} \; \dot{z}_{u4}]^T$ is the state vector; $W = [z_{01} \; z_{02} \; z_{03} \; z_{04}]^T$ is the disturbance vector; and $U = [f_{c1} f_{c2} f_{c3} f_{c4}]^T$ is the control input vector. The nonlinear output equation of the suspension system is obtained as:

$$Z = C_1(X) + D_{11}W + D_{12}U \tag{8.56}$$

$$Y = C_2(X) + D_{21}W + D_{22}U \tag{8.57}$$

where $Y = [\ddot{z}_s \; r \; \dot{\theta} \; \dot{\varphi}]^T$ is the measured output vector; $Z = [r \; \ddot{z}_s \; \ddot{\varphi} \; \ddot{\theta} \; z_{s1} - z_{u1} \; z_{s2} - z_{u2} \; z_{s3} - z_{u3}$ $z_{s4} - z_{u4} \; f_{c1} \; f_{c2} \; f_{c3} \; f_{c4}]^T$ is the controlled output vector; f_{c1} through f_{c4} are the suspension control forces; z_s is the displacement at the vehicle center of gravity; z_{s1} through z_{s4} are the suspension displacements; z_{u1} through z_{u4} are the wheel displacements; and z_{01} through z_{04} are the road displacements. The aim of the H_∞ control of the nonlinear suspension system is to design a stable close loop controller to generate the control forces and hence make the H_∞ norm of the transfer function T_{zw} of the disturbance input W and the controlled output Z reach the minimum, i.e., min $\|T_{zw}\|_\infty$. Based on the dissipation theory, the control law $U(Y)$ is obtained so that the constructed close loop control system of the suspension satisfies the inequality of the L_2 gain as follows:

$$\frac{\int_0^T \|y(t)\|^2 \, dt}{\int_0^T \|w(t)\|^2 \, dt} \leq p^2$$

where $0 \leq p \leq 1$. When $D = D_{12}^T D_{12}$ is a positive definite matrix, the necessary and sufficient condition that the system is a p dissipation is the existence of a smooth and differentiable semi-positive storage function $V(X(t))$ with $V(X_0) = 0$, and HJI inequalities are satisfied.

$$H[X, V_X(X), W, U] - p^2 W^T W \leq 0 \tag{8.58}$$

where,

$$H[X, V_X(X), W, U] = V_X(X)^T [A(X) + B_{11}W + B_{12}U]$$
$$+ [C_1 X + D_{11}W + D_{12}U]^T [C_1 X + D_{11}W + D_{12}U] \tag{8.59}$$

$$V_X(X) = \frac{\partial V(X)}{\partial X} \tag{8.60}$$

Therefore, solving the nonlinear H_∞ control problem is to find the control law U and the corresponding storage function $V(X)$ defined in the above HJI inequalities through the Taylor series expansion method.

8.8.3 Design of the Steering Subsystem Controller

The PID control method is applied to the design of the controller of the DYC system. The expected yaw rate r_d is defined in equation (8.51). After tuning, the weighting parameters of the PID controller are selected as $K_p = 0.2$, $K_i = 0.5$, and $K_d = 0.1$.

8.8.4   Design of the Braking Subsystem Controller

The sliding mode variable structure control method is applied to the design of the ABS in order to maintain the slip ratio around the target value to improve the system's braking performance. The tracking error between the actual vehicle slip ratio λ and the expected value λ_d is defined as $e = \lambda - \lambda_d$. The switch function of the sliding mode controller is selected as $s = c_0 e + \dot{e}$. Therefore, the control law is given as $U = (\alpha |e| + \gamma |\dot{e}|)\operatorname{sgn}(s)$, where c_0, α, and γ are the positive constants. After tuning, a good control performance is reached when $c_0 = 0.68$, $\alpha = 0.36$, and $\gamma = 0.06$.

8.8.5   Design of the Upper Layer Controller

Game theory is applied to the design of the optimal control strategy in order to coordinate the conflicts amongst the performance indices and therefore achieve a global optimal performance. The basic principle of the game theory-based control strategy is described as follows. The control objectives of each subsystem are considered as a player in the game. Then the game matrix is constructed by the reciprocal of each control objective. Therefore, the global optimal solution of the control objective game problem is obtained through using the cooperative game method.

The control objective of the suspension is selected as $J_1 = \|T_{zw}\|_\infty$ in order to suppress the disturbance input to the suspension system. Moreover, the control objective of the DYC system is selected as $J_2 = \int t |r - r_d| dt$ in order to track effectively the expected yaw rate. In addition, the control objective of the ABS is selected as $J_3 = \int t \left(|\lambda - \lambda_d| + \left| \dfrac{s}{s_0} \right| \right) dt$ to regu-

late the slip ratio around the optimal value and reach the shortest braking distance. Here, s is the actual distance, and s_0 is the braking distance when the simple logic control is applied. The payoff values of the game for the subsystem control objectives are defined as $m_1 = a_1 / J_1$, $m_2 = a_2 / J_2$, and $m_3 = a_3 / J_3$, where a_1, a_2, and a_3 are the coefficients. By considering the different driving conditions, including the emergency brakes, step steering input, and braking in turn, the corresponding payoff combinations of the game are obtained respectively as (m_1, m_3), (m_1, m_2), (m_1, m_2, m_3). Therefore, the global optimal solution of the game problem is obtained as $\left(m_1^*, m_3^*\right)$, $\left(m_1^*, m_2^*\right)$, and $\left(m_1^*, m_2^*, m_3^*\right)$ through the cooperative game of the different subsystems; and a_1 / m_1^*, a_2 / m_2^*, a_3 / m_3^* are the expected control objectives of the corresponding subsystems.

In addition, the fuzzy logic control method is applied to the real-time self-tuning of the weighting parameters of the outputs of the subsystems to reduce the tracking errors between the actual control objectives of the lower layer controllers and the expected control objectives of the upper layer controller. The control outputs of the subsystems include the control forces of suspension f_{ci} ($i = 1,2,3,4$), the additional yaw moment generated by the DYC system ΔM, and the braking torques of the brake system T_{bi}. The detailed strategy of the function distribution is developed as follows.

For the emergency brake maneuver, ABS works according to the slip ratio control method, while the DYC system is idle. In addition, there is a conflict on the function distribution between the ASS and ABS because of the pitch motion. To guarantee the overall optimal performance, the coefficients are selected as $a_1 = 0.8$ and $a_3 = 1.5$. Then, the global optimal solutions m_1^* and m_3^* are obtained through the cooperative game of the two subsystems. Therefore, the expected control objectives of the ASS and ABS, i.e., $0.8 / m_1^*$ and $1.5 / m_3^*$ respectively, are used to regulate in real time the control outputs of the ASS and ABS.

In addition, the fuzzy logic control method is applied to the real-time self-tuning of the weighting parameters of the outputs of the subsystems, and the fuzzy logic controller is developed as follows, assuming a counter-clockwise direction around the y-axis positive.

The input variables of the fuzzy logic controller are defined as $e_1 = 0.8 / m_1^* - 0.8 / m_1$, $e_2 = 1.5 / m_2^* - 1.5 / m_2$, $ec_i = \dot{e}_i$ $(i = 1,2)$; the discourse domain is defined as $e_i, ec_i = \{-5,-4,-3,-2,-1,0,1,2,3,4,5\}$; and the fuzzy subset is defined as $e_i, ec_i = \{$NS, NM, NS, ZO, PS, PM, PB$\}$. The elements in the subset represent large negative, medium negative, small negative, null, small positive, medium positive, and large positive, respectively. It is assumed that the input variables e_i, ec_i $(i = 1,2)$ and the output variable q_i are subject to Gaussian distribution. It is noted that the value of i is determined by the number of controlled outputs. For instance, for the ASS, $i = 1, 2, 3, 4$, for the DYC system $i = 1$, and for the ABS $i = 1, 2, 3, 4$. Therefore, the membership degree of each fuzzy subset can be determined. Based on the value assignment table of each subset membership degree and the fuzzy control model of each parameter, the fuzzy matrix of the output variables q_i is obtained by applying the compositional rule of inference; next, the corrected parameter is found and then it is substituted into the equation $q_i = 1 + \{e, ec\}_i$, where 1 is the initial value of the weighting parameter. In summary, the self-tuning of the output variables q_i is fulfilled through the calculation process shown in Figure 8.32. The fuzzy self-tuning rule bases of the output variables q_i of the ASS are shown in Table 8.6. A good control performance of the ASS is achieved when the fuzzy factors are selected as $k_{e1} = 10$ and $k_{ec1} = 0.3$, and the defuzzification factors as $k_{q1} = k_{q2} = 0.1$, $k_{q3} = k_{q4} = 0.08$.

For the step steering input maneuver, the ABS is idle, and there is a conflict on the function distribution between the ASS and DYC because of the roll motion. Similarly, to guarantee the overall optimal performance, the coefficients are selected as $a_1 = 0.9$ and $a_2 = 1.2$. Then the global optimal solutions m_1^* and m_2^* are obtained through the cooperative game of the two subsystems. Therefore, the expected control objectives of the ASS and DYC, i.e., $0.9 / m_1^*$ and $1.2 / m_2^*$ respectively, are used to regulate in real time the control outputs of the ASS and DYC.

For the braking in turn maneuver, the three subsystems perform the control functions according to the following rule. The expected slip ratio λ_d is regulated in real time by the upper layer controller. There are conflicts on the function distributions among the three subsystems because of the coupled pitch and roll motions. Similarly, to guarantee the overall optimal performance, the coefficients are selected as $a_1 = 0.79$, $a_2 = 1.16$, and $a_3 = 1.38$. Then, the global optimal solutions m_1^*, m_2^* and m_3^* are obtained through the cooperative game of the two subsystems. Therefore, the expected control objectives of the three subsystems ASS, DYC, and ABS, i.e., $0.79 / m_1^*$, $1.16 / m_2^*$, and $1.38 / m_3^*$ respectively, are used to regulate in real time the control outputs of the three subsystems.

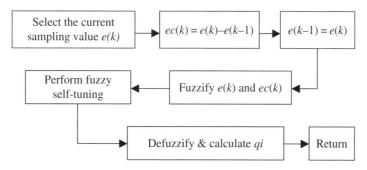

Figure 8.32 Flowchart of online fuzzy self-tuning.

Table 8.6 Rule bases of fuzzy self-tuning of the ASS outputs.

e \ ec	NB	NM	NS	ZO	PS	PM	PB
NB	PB	PB	PM	PM	PS	ZO	ZO
NM	PB	PB	PM	PS	PS	ZO	NS
NS	PM	PM	PM	PS	ZO	NS	NS
ZO	PM	PM	PS	ZO	NS	NM	NM
PS	PS	PS	ZO	NS	NS	NM	NM
PM	PS	ZO	NS	NM	NM	NM	NB
PB	ZO	ZO	NM	NM	NM	NB	NB

8.8.6   Simulation Investigation

To demonstrate the performance of the proposed multilayer integrated control system, a simulation investigation is performed by considering the three different driving conditions. The vehicle speed is set to 70km/h and the physical parameters used in the simulation can be found in Table 7.5. The following discussions are made based on the simulation results shown in Figures 8.33–8.35.

For the step steering input maneuver, it is observed in Figure 8.33(a) that the yaw rate for the multilayer integrated control system is better on tracking the expected yaw rate, compared with that of the decentralized control system (i.e., the three subsystems work independently). In addition, it is obvious in Figure 8.33(b) that the roll angular acceleration for the multilayer integrated control system is reduced significantly compared with that of the decentralized control system. The simulation results indicate that the proposed multilayer integrated control system is able to achieve a better handling stability and ride comfort than the decentralized control system.

For the emergency brakes maneuver, it is observed in Figure 8.34 that both the brake distance and the pitch angular acceleration for the multilayer integrated control system is reduced compared with that of the decentralized control system. The simulation results indicate that the proposed multilayer integrated control system is able to achieve a better braking performance and ride comfort than the decentralized control system.

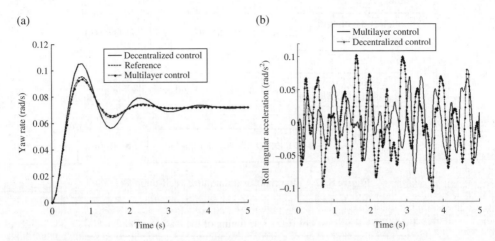

Figure 8.33 Comparison of responses for the step steering input maneuver. (a) Yaw rate. (b) Roll angular acceleration.

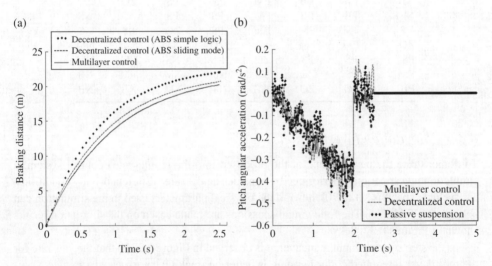

Figure 8.34 Comparison of responses for the emergence brake maneuver. (a) Brake distance. (b) Pitch angular acceleration.

For the braking in turn maneuver, it is observed in Figure 8.35(a) that the peak value and RMS of the vertical acceleration for the multilayer integrated control system is reduced compared with that of the decentralized control system. In addition, the settling time of the vertical acceleration for the multilayer integrated control system is also reduced. Moreover, as shown in Figure 8.35(b), the peak value of the yaw rate for the multilayer integrated control system is also reduced. However, the brake distance for the multilayer integrated control system is increased compared with that for the decentralized control system because of the intervention of the DYC system.

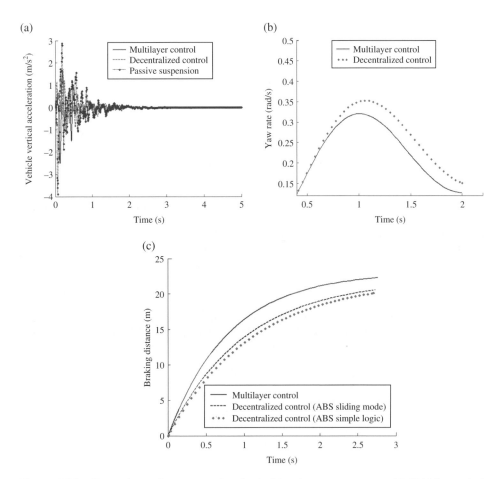

Figure 8.35 Comparison of responses for the braking in turn maneuver. (a) Vehicle vertical acceleration (b) Yaw rate. (c) Brake distance.

In summary, the application of the proposed multilayer coordinating control system is able to improve the overall vehicle dynamics in the three directions including lateral, longitudinal, and vertical and, hence, enhance the overall vehicle performance including handling stability, ride comfort, and braking performance.

References

[1] Gordon T J, Howell M, Brandao F. Integrated control methodologies for road vehicles. Vehicle System Dynamics, 2003, 40(1–3), 157–190.

[2] Xiao H S, Chen W W, Zhou H H, Zu J W. Integrated control of active suspension system and electronic stability program using hierarchical control strategy: Theory and experiment. Vehicle System Dynamics, 2011, 49(1/2), 381–397.

[3] Zhu H, Chen W W. Active control of vehicle suspension and steering systems based on hierarchical control method. Transactions of the Chinese Society for Agricultural Machinery, 2008, 39(10), 1–6 (in Chinese).

[4] Wang Q D, Qing W H, Chen W W. Integrated control of vehicle ASS and EPS based on Multi-degree of Freedom Model. Journal of System Simulation, 2009, 21(16), 5130–5137 (in Chinese).

[5] Qing W H. Integrated Chassis Control using Multi-body Model. PhD Thesis, Hefei University of Technology, Hefei, China, 2010 (in Chinese).

[6] Chu C B. Hierarchical Control of Chassis Systems. PhD Thesis, Hefei University of Technology, Hefei, China, 2008.

[7] ChuC B, ChenW W. Hierarchical control of chassis systems. Journal of Mechanical Engineering, 2008, 44(2), 157–162 (in Chinese).

[8] Chen W W, Chu C B. Hierarchical control of electrical power steering system and anti-lock brake system. Journal of Mechanical Engineering, 2009, 45(7), 188–193 (in Chinese).

[9] ChenW W, Xiao H S, Chu C B, Zu J W. Hierarchical control of automotive electric power steering system and anti-lock brake system: Theory and experiment. International Journal of Vehicle Design, 2012, 59(1), 23–43.

[10] Chen W W, Zhou H H, Liu X Y. Hierarchical control of ESP and ASS. Journal of Mechanical Engineering, 2009, 45(8), 190–196 (in Chinese).

[11] Song Y. Vehicle Stability Control System and Four-wheel Steering System and its Integrated Control. PhD Thesis, Hefei University of Technology, Hefei, China, 2012 (in Chinese).

[12] Song Y, Chen W W, Chen L Q. Simulation of vehicle stability control through combining ADAMS and MATLAB. Journal of Mechanical Engineering, 2011, 47(16), 86–92 (in Chinese).

[13] Chen W W, Zhu H. Coordination control of vehicle handling stability and ride comfort based on state identification. Journal of Mechanical Engineering, 2011, 47(6), 121–129 (in Chinese).

[14] Wang H B, Chen W W, Yang L Q, et al. Coordination control of vehicle chassis systems based on game theory and function distribution. Journal of Mechanical Engineering, 2012, 48(22), 105–112 (in Chinese).

9

Perspectives

Vehicle system dynamics and integrated control is a new subject that has emerged in recent years, and is a research hotspot in the field of vehicle engineering internationally. Over the past two decades, vehicle system dynamics and integrated control technologies have developed a relatively complete theoretical and technical foundation with the drive of various applications and techniques in relative fields. At present, this technology is leading towards the integration of dynamic system modeling, simulation, analysis, and control. The previous chapters of this book respectively describe the background on the modeling of vehicle system dynamics, tyre dynamics, full vehicle dynamics, longitudinal/lateral/vertical vehicle dynamics and control, centralized integrated control, and layered coordinated control. However, vehicles are complicated multivariable nonlinear dynamic systems, developed with an increasingly higher demand for their performance. Hence, when a complete analysis and integrated control to improve the overall vehicle performance is conducted, there are many key technical problems that require further in-depth studies.

9.1 Models of Full Vehicle Dynamics

Under the influence of complex operating conditions and various uncertain factors (due to the uncertainty of the environment and inaccuracy of design factors), the tyre load, road adhesion condition, and vertical/lateral/tangential forces change during vehicle travel. The uncertainty included in the vehicle ride comfort and handling stability are the functions of the dynamic models. Therefore, when performing the system's dynamical modeling and integrated control, the factors corresponding to the driver, vehicle, road, and environment

Integrated Vehicle Dynamics and Control, First Edition. Wuwei Chen, Hansong Xiao,
Qidong Wang, Linfeng Zhao and Maofei Zhu.
© 2016 John Wiley & Sons Singapore Pte. Ltd. Published 2016 by John Wiley & Sons, Ltd.

lead to extremely complicated models and numerous parameters. In order to solve the controlling problems based on integrity and relevance, it is important to propose effective calculation models, simulation algorithms, and control strategies.

9.2 Multi-sensor Information Fusion

With the development of the fields of microelectronics, field bus control, computer measuring and control, information processing, wireless communication, and drive-by wire technologies, there are more and more types of sensors installed on vehicles. Also, it has become a new research direction to explore the applications of multisensor information fusion technology in condition detection, fault diagnosis, and integrated control for vehicle systems. Multisensor information fusion is actually a kind of functional simulation of complicated problem-solving by the human brain. In the multisensor system, the information provided by a variety of sensors is likely to have different characteristics, including time-varying/time-invariant, real-time/unreal-time, fuzzy/exact, accurate/incomplete, and mutually supportive/complementary. By making the best use of multiple sensors and combining complementary and redundant information in both space and time based on some kinds of optimization criterions, a consistent interpretation or description of the observational environment can be created, which improves the performance and effectiveness of the whole sensor system and avoids the limit of a single or few sensors. Therefore, with the aid of multisensor information fusion and multisensor management, the optimum use of the limited sensor resources is achieved, as well as multiple goals and multiple scanning spaces of the vehicle system, and the values of each specific characteristic is also obtained. Information fusion technology refers to the theories and techniques in various aspects, such as signal processing, estimation theory, uncertainty theory, pattern recognition, optimization techniques, neural networks, and artificial intelligence; but each method developed according to the requirements of various applications is a subset of the fusion method. For the vehicle integrated control system, the research study is mainly focused on the comprehensive utilization of the fusion information provided by multiple sensors, the foundation of a more exact integrated control modeling, and the development of a desirable optimum control strategy, allowing the improvement of the comprehensive performance of a full vehicle[1].

9.3 Fault-tolerant Control

Fault-tolerant control is a practical interdisciplinary subject, and also a high-reliability technology. Starting in the 1980s, this technology has achieved great advances with numerous significant successes. Fault-tolerant control technology is especially applicable to a chassis integrated control system: a complicated system consisting of several subsystems including the braking/driving system, steering system, and suspension system. If the sensor or actuator in one of the subsystems fails while the vehicle is traveling, other subsystems or the whole integrated control system remain with a suboptimal performance, allowing a safe and stable motion. This is a novel way to improve the reliability of the complicated chassis integrated control system. As each system is inevitably broken down, fault-tolerant control can be regarded as the last defense to maintain the vehicle operating safety. Fault-tolerant control is classified as passive fault-tolerant control and active fault-tolerant control according to the design methodology. Passive tolerant control is able to make the system insensitive to failure, while the active one utilizes a fault accommodation or signal reconstruction to maintain the performance and stability of the system after failure. The design of the passive fault-tolerant control is always conservative, but the active fault-tolerant control, including measurement module, fault diagnosis module, execution module, and fault-tolerant processing module, is more appropriate to the complicated chassis integrated control.

The application and development of fault-tolerant control technology in vehicle electronic control systems is scarce compared to its wide application in the fields of aeronautics, astronautics, computer sciences, and nuclear energy. For example, the current electronic control units (ECU) in vehicle systems are in dual machine structure, and the actuators and sensors use hardware redundancy methods, which results in extremely redundant hardware, high-cost fault tolerance, and a complex system structure if a second-order system is applied to compensate the failure of the first subsystem. Hence, it is essential to explore novel fault diagnosis and fault-tolerant control methods applicable to chassis integrated control systems because they can improve the fault tolerance performance and avoid the high cost of hardware redundancy.

For the application of fault-tolerant control technology in chassis integrated control systems, the main research involves fault-tolerant control structure, control mode, control scheme, and controller design[2]. By utilizing fault-tolerant control technology of integration, networking, and intelligence, the resource-sharing of a complicated chassis control system is accomplished, along with the synthetic diagnosis and fault-tolerant control of multiple methods. Also, based on the properties of smart materials and smart structures, the mentioned method can develop smart fault-tolerant control structures of self-healing and self-compensating that meet the requirements of vehicle operation, and the fault-tolerant control of nonlinear systems and time-delay multivariable dynamic systems. The theoretical level and practical application values of the aforementioned studies are obviously of great importance.

9.4   Active and Passive Safety Integrated Control Based on the Function Allocation Method

Recently, the rapid development of electronic control technologies has played an essential role in promoting advances in vehicle technology, especially in security technology. The progression of vehicle security technology is moving in several directions, by using radar technology and vehicle-mounted photography technology to develop: lane departure warning systems, automatic collision avoidance systems, high-performance tyre comprehensive monitoring systems, adaptive cruise control (ACC) systems, ABS/ASR/VSC systems, driver identity recognition systems, seat belts, and airbags. With the wide applications of advanced intelligent sensors, fast-response actuators, high-performance electronic

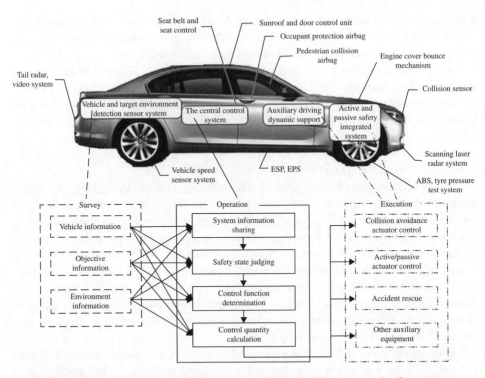

Figure 9.1 Arrangement and structure of an active and passive integrated safety system.

control units, advanced control strategies, vehicle networking technology and radar technology, modern vehicles will be developed towards the electromechanical integration of high intelligentization, automation, and informationalization for the purpose of maximizing driver and passenger safety with the effects of driving assistance systems, active and passive integrated control systems, combined occupant protection in all driving and traffic situations (Figure 9.1).

In different road conditions and driving cycles, it is the combined effect of the driver, driving assistance system, and integrated control system that improve the full vehicle ride comfort, handling stability, and driving safety. Generally speaking, drivers have great capacities of forecasting, solving non-programmed problems, and handling burst interference (i.e., lateral wind, bumpy shock); however, an integrated control system is capable of handling complex tasks and responding rapidly, something that is unmatched by drivers. Hence, in the design of an active and passive integrated control system, if the respective advantages of the driver and the integrated control system are analyzed, and

the functionalities are allocated to both based on a certain criterion and requirements after taking the overall performance of the vehicle into account, both the driver and the integrated control system can make full use of each advantage, improving the overall vehicle performances to the maximum. That is the man–machine function allocation problem at the decision-making level—the upper layer of the integrated control system. Moreover, due to the restrictions of the functionalities of each lower-level control execution subsystem and dynamic properties of tyres, each individual subsystem has its own active zone which enables it to mostly affect the vehicle performance in a specific direction or effective zone. If the system functions are allocated reasonably to each subsystem in its effective working zone, and the functional overlapping between the subsystems occurring during the integrated control is avoided, the overall performance of the overall integrated control system will certainly be improved. Thus, the function allocation matters between the low-level subsystems of the integrated control system[3]. Therefore, in the design of an active and passive safety integration control system, the "driver-vehicle-road" closed-loop system should be focused on, and each effective working zone and objective function difference of the driver, up-level decision-making system, and low-level control execution subsystems analyzed. In addition, by the advisable optimum allocation of the overall vehicle performance to the driver and the decision-making system, as well as the objective function of the decision-making level to each low-level subsystem, the respective advantages of the driver, up-level decision-making system, and each low-level control execution subsystem can be brought into full play, allowing the enhancement of the comprehensive performances of the integrated control system in different road and driving conditions.

In order to better accomplish an integrated control, more attention must be paid to some key issues, as discussed below.

1. *Real-time estimation algorithm of the vehicle driving situation and road parameters*
   The obtained vehicle driving situation and road parameters are the basis for designing an integrated control system. During the actual motion of a vehicle, it is hard to directly obtain many key parameters indicating the vehicle driving status from the vehicle-mounted sensors due to the influence of external disturbances and measuring conditions, such as the vehicle centroid sideslip angle, front and rear tyre sideslip angle, road adhesion coefficient, and tyre longitudinal/lateral/vertical forces; so it is necessary to consider how to know the vehicle motion status and road conditions via necessary information sharing and fusion technologies[4].

2. *Identification of the operating intention of the driver in different road and driving situations*
   In a "driver-vehicle-road" closed-loop system, the driver determines the actual operation intention by applying in a timely manner the steering angle, throttle, brake pedal displacement, and other physical parameters based on the vehicle dynamic performance and road information, such as the desired yaw rate, centroid sideslip angle, and traveling trajectory. Therefore, it is necessary to analyze the variation of such parameters as the driver's forward-viewing time, action lag time, system-following orders with respect to road conditions, and vehicle performance. It is also necessary to assess the operating behaviors of the driver, i.e., straight driving, curve steering, emergency steering, and fatigue. Thus the models of the driver operating intentions in different driving situations are built, hence the driving assistance system can better understand the driver operating behaviors, maintaining the vehicle motion as the driver expects.

3. *Analysis of the stability of the integrated control system consisting of multiple cooperative subsystems*

If the coupling between subsystems is strong, it is necessary to analyze the effect of the subsystems' coupling on stability by conducting the analysis of the interconnected stability between them; thus, it is necessary to consider whether the extended subsystems are stable, and whether the integrated control system after extension is still stable. It is the analysis of the interconnected stability between the subsystems and the stability of the integrated control system that plays an important role in the study of the vehicle instability mechanism in critical conditions.

4. *Study of the coordinated decisions based on objective function allocation*

In a wide range of driving situations, there is a great difference between the driver's operating intention and the objective function accomplished, where an effective work zone of each subsystem is limited. In order to ensure active and passive safety, ride comfort, handling stability of the vehicle in normal traveling and critical conditions, it is important to analyze the respective objective functions familiar to the driver, up-level coordinators, and low-level control execution subsystems in the design at the system decision-making level. Then, by the coordinated decision at the decision-making level, the man–machine function sharing and the function allocation for the low-level control execution subsystems, all the driver, driving assistance, coordinated and control execution subsystems can implement their respective functions in the effective working zones. Hence, the comprehensive performance of the vehicle is enhanced, especially for objectives under the critical conditions.

9.5   Design of System Integration for a Vehicle

In the design of the mechanical, control, information perception, and actuator units of a vehicle system, it is necessary to consider such factors as hardware integration, information integration, function integration, and driving situation of the system as a whole. Fault-tolerance and redundancy design are also important in the system design. Furthermore, attention should be paid to other matters, including topology expression of multiscale parameters and various domain parameters, objective optimization of singular modes and evolution indication, coupling of multiple active and passive control systems in the work process, coordinated design of multi-unit techniques, and so on[5].

The main purposes of advanced vehicle control technology, intelligent driving assistance, real-time collision early-warning, and avoidance systems are to reduce the mistakes and operations made by the driver, and enhance effective and active measures for the vehicle safety[6,7]. To view the system's integrated safety as a whole, sensors can provide additional condition monitoring and situation awareness; then, based on the data obtained from sensors, security algorithms can conduct evaluation, estimation, calculation, perception, and judgment using various determined and random methods. Moreover, this algorithm can provide optimal and safe corrective actions, which are similar to the judgment of proficient drivers in emergencies. The judgment can serve as an intelligent co-driver, and finally transfer to particular control processes (i.e., driving, braking, steering, etc.). The actual functionality of interaction/interference/drive is also a part of system integrated design.

As there is not yet any applicable uniform performance standard, the further development and popularization of active and passive safety integrated design technologies has been hindered. Note that it is complicated to set appropriate criteria for active safety systems,

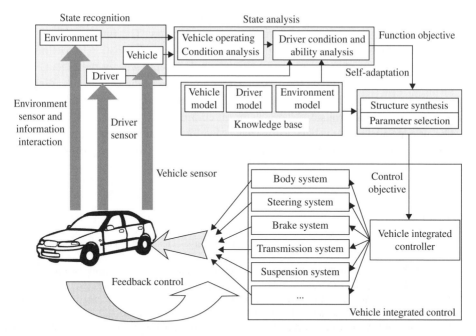

Figure 9.2   The adaptive intelligent vehicle.

due to great differencse between whole systems and the absence of uniform criteria between manufacturers; also, it is much harder to solve the problems of driver adaptability because the driver interactive system is quite complicated. Novel and effective criteria for the afore-mentioned technologies only can be set up by further studies and evaluations.

9.6   Assumption about the Vehicle of the Future

The assumption of vehicle development in the future is that the size of the vehicle body will vary according to different types of environment and available space, making passengers feel at ease and comfortable. The future vehicle will not have wheels, and will travel by suspending itself over the road, flying in a low level as a vertical take-off or landing aircraft. The vehicle velocity will reach up to 600km/h. The color of the vehicle body will be able to change to fit with different environments. Driving by wire will be used instead of current mechanical devices, and the drivers will not need to manipulate the car themselves, but will only have to send electronic signals and input destinations to the computer before departure. The vehicle will not only recognize people's voices, but also correctly determinate the relative position and driving situation of adjacent vehicles on the road, so that the parameters of the vehicle in control can be adjusted. Before a collision accident, the vehicle will be able to momentarily start the safety equipment by an electronic detection and sensor systems, ensuring the driving safety and rightness.

The future vehicle will be an adaptive intelligent vehicle as shown in Figure 9.2. This type of vehicle can be divided into two main modules: the action control module consisting of the steering, braking and driving subsystems, and the integrated control module which adapts to external environment and the driver's instructions. The future vehicle system utilizes a three-level structure: an adaptation layer for the driving situation, an optimization and coordination layer for the subsystems, and an adjustment layer for the execution subsystems. The driver module can form expected motion states of the vehicle according to all the information obtained by the drivers, and then conduct integrated control by sending instructions to each control subsystem via on-off control.

The ultimate goal of the vehicle is to be super intelligent and have zero emissions, which can greatly promote technological innovation and finally achieve the ideal goal—harmonious coexistence between man and nature.

References

[1] Jiang W, Yu Z P, Zhang L J. A review on integrated chassis control. Automotive Engineering, 2007, 29(5):420–425.

[2] Yang L Q, Chen W W, Wang H B. Optimal robust fault tolerant control for vehicle active suspension systems based on $H_2/H\infty$ approach. China Mechanical Engineering, 2012, 23(24): 3013–3019.

[3] Wang H B, Chen W W, Yang L Q, Xia G. Coordinated control of vehicle chassis system based on game theory and function distribution. Journal of Mechanical Engineering, 2012, 48(22): 105–112.

[4] Chen W W, Liu X Y, Huang H, Yu H J. Research on side slip angle dynamic boundary control for vehicle stability control considering the impact of road surface. Journal of Mechanical Engineering, 2012, 48(14): 112–118.

[5] Zhong J, et al. Coupling Design Theory and Methods of Complex Electromechanical Systems. Beijing: China Machine Press, 2007.

[6] Wang J E, Chen W W, et al. Vision guided intelligent vehicle lateral control based on desired yaw rate. Journal of Mechanical Engineering, 2012, 48(4): 108–115.

[7] Rajamani R. Vehicle Dynamics and Control. New York: Springer US, 2006.

Index

Integrated Vehicle Dynamics and Control, First Edition. Wuwei Chen, Hansong Xiao, Qidong Wang, Linfeng Zhao and Maofei Zhu.
© 2016 John Wiley & Sons Singapore Pte. Ltd. Published 2016 by John Wiley & Sons, Ltd.